Culture of Chemistry

Balazs Hargittai • István Hargittai
Editors

Culture of Chemistry

The Best Articles on the Human Side
of 20th-Century Chemistry from
the Archives of the Chemical Intelligencer

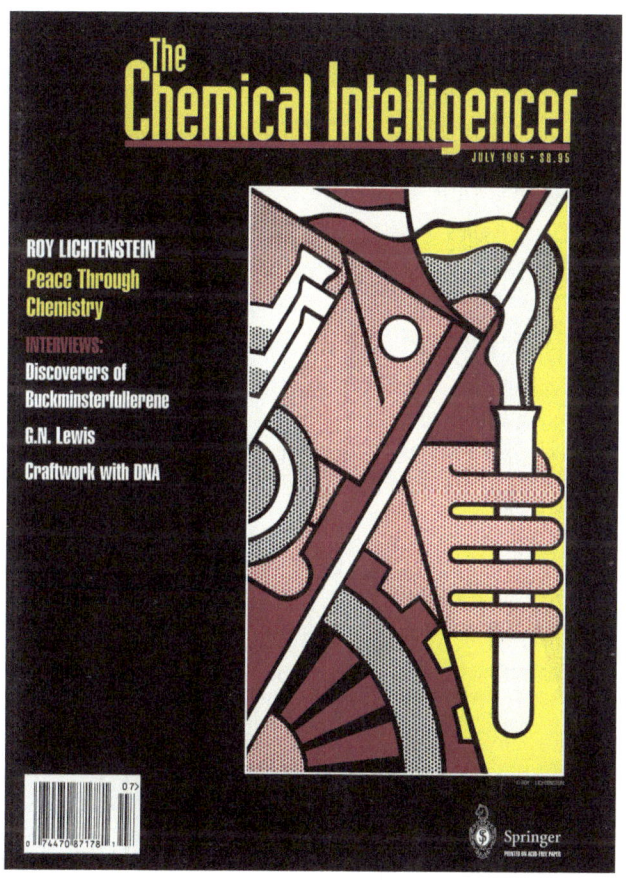

Editors
Balazs Hargittai
Department of Chemistry
Saint Francis University
Loretto, PA, USA

István Hargittai
Department of Inorganic and Analytical Chemistry
Budapest University of Technology
  and Economics
Budapest, Hungary

ISBN 978-1-4899-7564-5   ISBN 978-1-4899-7565-2   (eBook)
DOI 10.1007/978-1-4899-7565-2

Library of Congress Control Number: 2015936274

Springer New York Heidelberg Dordrecht London

On the cover and title page, the cover of the July 1995 issue of The Chemical Intelligencer with a detail of Roy
Lichtenstein's Peace through Chemistry 1970 (courtesy of the late Roy Lichtenstein (1923-1997), © Estate of Roy
Lichtenstein). The detail is the right panel of the full three-panel painting of oil and magna on canvas

Printed on acid-free paper

Springer Science+Business Media LLC New York is part of Springer Science+Business Media (www.springer.com)

*To those who create and
to those who benefit from
the Culture of Chemistry.*

Also by the Editors

István Hargittai and Magdolna Hargittai, *Budapest Scientific: A Guidebook* (Oxford University Press, 2015)

Balazs Hargittai, Magdolna Hargittai, and István Hargittai, *Great Minds: Reflections of 111 Top Scientists* (Oxford University Press, 2014)

István Hargittai, *Buried Glory: Portraits of Soviet Scientists* (Oxford University Press, 2013)

István Hargittai, *Drive and Curiosity: What Fuels the Passion for Science* (Prometheus, 2011)

István Hargittai, *Judging Edward Teller: A Closer Look at One of the Most Influential Scientists of the Twentieth Century* (Prometheus, 2010)

Magdolna Hargittai and István Hargittai, *Symmetry through the Eyes of a Chemist,* 3rd Edition (Springer, 2009; 2010)

Magdolna Hargittai and István Hargittai, *Visual Symmetry* (World Scientific, 2009)

István Hargittai, *The DNA Doctor: Candid Conversations with James D. Watson* (World Scientific, 2007)

István Hargittai, *The Martians of Science: Five Physicists Who Changed the Twentieth Century* (Oxford University Press, 2006; 2008)

István Hargittai, *Our Lives: Encounters of a Scientist* (Akadémiai Kiadó, 2004)

István Hargittai, *The Road to Stockholm: Nobel Prizes, Science, and Scientists* (Oxford University Press, 2002; 2003)

Balazs Hargittai, István Hargittai, and Magdolna Hargittai, *Candid Science I–VI: Conversations with Famous Scientists* (Imperial College Press, 2000–2006)

István Hargittai and Magdolna Hargittai, *In Our Own Image: Personal Symmetry in Discovery* (Plenum/Kluwer, 2000; Springer, 2012)

István Hargittai and Magdolna Hargittai, *Symmetry: A Unifying Concept* (Shelter Publications, 1994)

Ronald J. Gillespie and István Hargittai, *The VSEPR Model of Molecular Geometry* (Allyn & Bacon, 1991; Dover Publications, 2012)

Preface

The Chemical Intelligencer (often referred to as *CI*) lasted six exciting years from 1995 through 2000. The preparation for launching it started in 1993 and Gabriela Radulescu, our sponsoring editor at Springer New York, was an enthusiastic supporter of creating the magazine. The magazine had an intriguing name, which we liked, but which we did not coin; it copied *The Mathematical Intelligencer* to which one of us had been a contributor. The word "Intelligencer" is an archaic English word; it means newsletter or newspaper. It is not known commonly and it sounds as if it could be related to some secret activities.

This misinterpretation brought us some benefit. As we learned later, the title intrigued Arnold Kramish (1923–2010) who at one time worked as a liaison of the US Atomic Energy Commission to the Central Intelligence Agency and authored a book about a spy who passed information to the Allies about the German nuclear project during the war.[1] Kramish connected us with Clarence Larson and his wife Jane who had built up a video interview collection with famous scientists and technologists. Clarence Larson (1909–1999) was a former Commissioner of the US Atomic Energy Commission. After Clarence's death, Jane donated their original tapes to us, and excerpts of some of their interviews appeared in *The Chemical Intelligencer*.

Our own interviews project developed also due to the magazine. When prior to launching the publication we asked a dozen or so leading chemists about the desirability of starting such a project, the most enthusiastic support came from Linus Pauling. He only regretted that his busy schedule would not allow him to write for the magazine. We knew that Pauling was terminally ill by then. His response gave us the idea to send him a few questions to which he responded promptly. That interview, Pauling's last, as far as we know, became the lead entry of our charter issue. Many more followed.

We interviewed famous chemists for the magazine, including Nobel laureates, but the non-Nobel laureates represented a similar level. This is so much so that several among our inter viewees received the Nobel Prize following the publication of our interviews rather than before. Thus, we published interviews with the discoverers of buckminsterfullerene in 1995 and they received the award in 1996. In 1997, we communicated interviews with John Pople (award in 1998), Ahmed Zewail (award in 1999), and Dan Shechtman (award in 2011).

Our interviewing project continued for a few more years after the magazine had ceased publication. Most of our interviews, among them those that first appeared in *The Chemical Intelligencer*, were subsequently published in our six-volume *Candid Science* book series (see Appendix 1). This is why the present volume contains only two interviews (they did not appear in *Candid Science*), as a token of the interviews. Kurt Mislow's interview is, among others, about chirality in chemistry and about chemical topology for which he had published pivotal discoveries.

---

[1] Arnold Kramish, *The Griffin – The greatest untold espionage story of World War II* (Houghton Mifflin, 1986). Paul Rosbaud (1896–1963) was the Griffin. He was a metallurgist and a leading editor at Springer-Verlag. He rushed Hahn and Strassmann's manuscript on the discovery of nuclear fission for publication in the German magazine *Naturwissenschaften* to inform the world about the potential danger of a German atomic bomb. During the war, Rosbaud kept informing the British of German progress in war-related research. After the war, he co-founded Pergamon Press.

*István and Magdolna Hargittai (on the left) with Zipora and Dan Shechtman in December 2011
at the Royal Swedish Academy of Sciences in Stockholm during the 2011 Nobel
award celebrations (by unknown photographer).*

Speaking about personal aspects, Mislow mentioned his escape from Nazi Germany, spending a few years in Italy and England before arriving in the United States in 1940. He was grateful to Tulane University in New Orleans for generous help allowing him to get an education. It was also in New Orleans that Mislow got his first experience in racism in the United States. He was not invited to join the local chapter of Alpha Chi Sigma because he was Jewish. Not long before our conversation in 1997, Mislow checked whether this national chemical honor society had changed their rules and they had indeed: "Jewish chemists became eligible for membership in 1948. Black chemists became eligible in 1954. Women chemists became eligible in 1970. A chronology of progress, of sorts."

Most of our interviewees opened up more to a fellow scientist than they might have to a journalist. In particular, Mislow later wrote us that he told about things that he had never discussed before, not even with his wife. When we sent back the transcripts for checking and he could have deleted whatever he wanted, he left everything intact.

Eugene Garfield is an iconic pioneer of information science. He got his first degree in chemistry; then he went on to a unique career in a field that he created mostly himself. We talked with him in 1999, and we now asked him how he looked back to what he told us in the original interview reproduced in this Volume[2]:

> I just re-read the interview you did in 1999. That is now 15 years ago. I would not change anything in the interview. However, the influence of the journal impact factor is ever more pervasive. I wonder how often people even use databases like SCI or WebofScience for information retrieval. *Current Contents* has essentially been displaced by free contents page alerts from publishers. However, that is not the same and a lot of serendipitous connections are lost. I have to constantly remind people that the journal impact factor [JIF] should not be used to evaluate papers but at the same time, the JIF is justifiably used as a way of demonstrating the prestige of well-cited journals. Administrators and evaluators are always looking for new metrics even if they are complex and really do not understand them. So now, we have hundreds of bibliometricians churning out citation analyses and mappings.

Browsing the 24 issues of the magazine, quite a broad spectrum of topics and scientists emerge and we found the title "Culture of Chemistry" appropriate for this compilation. It does not cover everything this concept means, but all that there is is part of this culture. During the brief existence of the magazine, we constantly felt the interest and support of the community

---

[2] Eugene Garfield, e-mail message, August 28, 2014.

of chemists, and the sustained interest in various articles of the magazine during the ensuing years has been especially gratifying. This sustained interest was the main impetus for initiating this volume.

The arrangement of the material follows the structure of the magazine. The first half of each issue contained the so-called Departments and the second, the Articles. The Letters from the Editor-in-Chief and the Letters to the Editors will not be represented, although there were some interesting exchanges. For example, The chemistry Nobel laureate mathematician Herbert Hauptman published a paper "On the Packing of Spheres in the Regular Icosahedron" [*CI* 1(2), 26–30]. In a letter published in the next issue, the great Canadian geometer H.S.M. Coxeter pointed out a clever simplification for one of Hauptman's expressions [*CI* 1(3), 4]. In his turn, Hauptman readily admitted that Coxeter was right.

The Departments (with department editors if there were such) included Interviews, Notes, Beautiful Molecules (Balazs Hargittai), Chemical Tourist, Cooking Chemist (Nicholas Kurti and Hervé This-Benckhard), A Chemist's Photoalbum (Jack D. Roberts), Encounters with Chemistry (William B. Jensen), Book Reviews, and Stamp Corner (Edgar Heilbronner). The Articles follow the Departments. Within each section, the order of entries is chronological.

Of the various articles in the magazine, some had in them a certain time element. We have asked a few authors to comment on further developments although, sadly, a number of authors are no longer around; hence, we could collect such reflections on only a limited scale. The articles about the discoveries of buckminsterfullerene appeared prior to the 1996 Nobel Prize in Chemistry. The Nobel laureate discoverers have received ample exposure, but we note the premature death of Richard E. Smalley.

The watershed effect of the Nobel Prize has been amply demonstrated on the differences in the lives of scientists involved in the same award-winning field who had been honored with this award and those who were not. Thus, for example, Donald R. Huffman and Wolfgang Krätschmer could have also shared the buckminsterfullerene Nobel award, but they were not included, presumably only for the stipulation of the limited number of three awardees in any given category in any given year. On August 30, 1999, in Tucson, Arizona, Don and Wolfgang treated the two of us to a special privilege. They recreated their seminal experiment in which for the first time they had produced—rather than just observed—buckminsterfullerene.[3] Our pictorial account of the experiment appeared in the magazine and is reproduced in this volume.

Richard E. Smalley (1943–2005) in 2004 in New York (photo by I. Hargittai).

---

[3]W. Krätschmer, L. D. Lamb, K. Fostiropoulos, and D. R. Huffman, "Solid C_{60}: a new form of carbon." *Nature* **1990**, *347*, 354–358.

We asked Don and Wolfgang for their comments on the afterlife of fullerene research and their own activities. Appendix 2 contains Wolgang Krätschmer's and Donald R. Huffman's responses. We mention here two characteristic features of what Krätschmer had to say. One is the chemical nature of his and his colleagues' research at his nuclear physics research institute. The other is his magnanimity concerning the Nobel award for buckminsterfullerene discovery. He says that in our time most discoveries involve many scientists, and even though the Nobel Prize is given to individuals, it is a recognition of a scientific area. Huffman's activities since the fullerene discovery could be summarized, in his words, "as a reprise of earlier work if one simply replaces the phrase 'interstellar' with 'atmospheric'. Both Krätschmer and Huffman received significant awards, including the Hewlett Packard Europhysics Prize in 1994, which they shared with Harry W. Kroto and Richard E. Smalley.

We asked Kozo Kuchitsu ("Training of a Molecular Scientist, East and West") whether he would like to augment the article for the ensuing two decades. Here is what he had to say[4]:

> Communications using electronic webs, which started shortly after the publication of this article, may confuse my question in this Epilogue, because e-mail can be used for 'a serious but silent academic discussion' for training students. In addition, we have observed in these years a remarkable increase in the long- or short-term international tours of young scientists, either from Japan or from overseas, for their joint studies and/or presentations in academic meetings. These activities should significantly affect their mentality and ambition. As for the traditional Japanese religious and cultural heritages discussed in this article, I believe firmly that they will be handed down thoughtfully in our modern society.

The time element did not play a role in Alan Mackay's reflections about J. Desmond Bernal's activities in Mackay's article, "The Lab." Nonetheless, we asked him whether he would add anything to his account. In response, he sent us information about a paper by Bernal's daughter, Jane Bernal, in which she, a physician rather than science historian or political scientist, draws an illuminating picture of her father.[5]

We also asked Nadrian C. Seeman whether he wished to add anything to his paper "Molecular Craftwork with DNA." His article concerned a fast-developing field. His response was, "The article is largely still correct, although we have made a lot of advances in the last 20 years. Thus, we are no longer limited to topological characterization."[6] His more detailed response is reproduced in Appendix 3. As pioneer of DNA nanoscience, Seeman has been recognized with a host of prestigious awards and prizes of which we mention only one: "Nadrian Seeman is recognized with the Kavli Prize in Nanoscience, for inventing DNA nanotechnology, for pioneering the use of DNA as a nonbiological programmable material for a countless number of devices that self-assemble, walk, compute, and catalyze".[7]

The late Mordecai Rubin wrote a captivating article about the Wall of Fame in the Schulich chemistry department of the Technion—Israel Institute of Technology. We have asked Distinguished Professor and President Emeritus of the Technion Yitzhak Apeloig about further development since Rubin's 1997 account[8]:

> Since the death of David Ginsburg (1988), the founder of the Wall of Fame, only names of Nobel laureates in chemistry were added to the Wall. To accommodate the new names, the sidewalls of the auditorium were also used. Now these walls are also covered with nameplates and for the time being we have stopped adding new names. Until 2004, there were no names of Israeli chemists on the Wall (although several may have qualified), because according to the "Wall rules" Israelis can be included only if they win the Nobel Prize. Happily, this situation changed dramatically in the 21st century when we proudly added FOUR Israeli Nobel laureates; THREE of them from the Technion: Avram Hershko and Aaron Ciechanover (2004); Dan Shechtman (2011); and Ada Yonath (2009) from the Weizmann Institute of

---

[4]Kozo Kuchitsu, e-mail message, August 25, 2014.

[5]Jane Bernal, "J. D. Bernal," *LLULL: boletín de la Sociedad Española de Historia de las Ciencias* **2001**, *24*, 605–628 (the article is in English with an Abstract in Spanish and English).

[6]Nadrian C. Seeman, e-mail message, August 26, 2014

[7]http://www.kavliprize.org/prizes-and-laureates/prizes/2010-kavli-prize-laureates-nanoscience (accessed September 4, 2014). The Kavli Prize recognizes scientists internationally in astrophysics, nanoscience, and neuroscience.

[8]Yitzak Apeloig, e-mail message, September 6, 2014.

Science. The names of two more Israeli chemists who won the Nobel Prize, Arieh Warshel (a Technion graduate) and Michael Levitt (2013) will be added to the Wall soon. With this impressive achievement for a small country like Israel, it seems that David Ginsburg's dream to see names of Israeli Nobel laureates on the Wall of Fame has been fulfilled.

We would have thought that hardly anything could be added to Bart Kahr's article "Gibbs and Amistad," but this was not the case.[9] Recently at least three studies have appeared in the most diverse venues, such as *Modernism/Modernity*,[10] the *Journal of Narrative Theory*,[11] and the *Journal of Statistical Physics*.[12] They give us the impression that Kahr's paper in our magazine was quite pioneering.[13]

*****

Looking back to the time we spent with the magazine, we remember with gratitude the contributions of the Editorial Board members, especially Lennart Eberson[14] (1933–2000), Roald Hoffmann, William B. Jensen, George B. Kauffman, Harold W. Kroto, Nicholas Kurti (1908–1998), Torvard C. Laurent[15] (1930–2009), Jean-Marie Lehn, George A. Olah, Guy Ourisson (1926–2006), Lev V. Vilkov (1931–2010), and Ahmed H. Zewail. We express heartfelt thanks to Madeline R. Kramer, then at Springer, who was responsible for the production of the magazine to which she was very much dedicated. We also appreciate our current Publishing Editor Sonia Ojo's support for bringing out this volume and Production Editor Karin de Bie's dedicated and expert efforts in making it happen.

It is our pleasure to mention the multifaceted assistance in and the encouragement for the present project we have received from Magdolna Hargittai. Being not only her professional colleague, but also her son (BH) and her husband (IH), we strive to manifest activities worthy of her expectations of us.

Loretto, PA, USA   Balazs Hargittai
Budapest, Hungary   István Hargittai
Spring 2015

---

[9]Bart Kahr, e-mail message, September 4, 2014.

[10]Eben Wood, "The Private Lives of Systems: Rukeyser, Hayden, Middle Passage," *Modernism/Modernity* **2010**, *17*(1), 201–222.

[11]Stefania Heim, "'Another Form of Life': Muriel Rukeyser, Willard Gibbs, and Analogy," *Journal of Narrative Theory* **2013**, *43*(3), 357–383.

[12]Leo P. Kadanoff, "Reflections on Gibbs: From Statistical Physics to the Amistad V3.0," *Journal of Statistical Physics* **2014**, *156*, 1–9.

[13]We thank Bart Kahr for the quoted References.

[14]At the time, Eberson chaired the Nobel Prize Committee for Chemistry.

[15]At the time, Laurent chaired the Board of the Nobel Foundation.

Contents

Part III   Beautiful Molecules

Part IV   Chemical Tourist

Part V   Cooking Chemist

Part VI   A Chemist's Photo Album

Part VII   Encounters with Chemistry

Part VIII   Book Reviews

Appendix 1

Interviews First Published in *The Chemical Intelligencer* and Republished Subsequently Elsewhere

Many though not all of the interviews that first appeared in *The Chemical Intelligencer* were subsequently republished in the *Candid Science* book series.[16] Excerpts from some of these interviews then appeared in the *Great Minds* volume.[17] In addition to the name of the interviewee, there is *CI* standing for *Chemical Intelligencer*, volume(issue) numbers, and starting page number in the magazine. This is followed by the volume number of *Candid Science* with starting and ending page numbers and the starting page number in *Great Minds* in the relevant cases. The asterisks indicate that the entries contain excerpts from Clarence and Jane Larson's interviews (see Preface).

Altman, Sidney: *CI* 5(2), 12; *Candid Science II*, 338–349
Alvarez, Luis W.*: *CI* 5(1), 43; *Candid Science V*, 198–217
Anderson, Philip W.: *CI* 6(3), 26; *Candid Science IV*, 586–601; *Great Minds*, 6

Bader, Alfred: *CI* 4(3), 4; *Candid Science III*, 146–157
Bartlett, Neil: *CI* 6(2), 7; *Candid Science III*, 28–47
Barton, Derek H. R.: *CI* 3(1), 31; *Candid Science I*, 148–157
Bax, Ad: *CI* 5(1), 6; *Candid Science III*, 168–177
Berg, Paul: *CI* 6(3), 11; *Candid Science II*, 154–181; *Great Minds*, 247
Bodánszky, Miklós: *CI* 6(4), 42; *Candid Science V*, 366–377
Boyer, Paul D.: *CI* 6(2), 16; *Candid Science III*, 268–279
Brown, Herbert C.: *CI* 3(2), 4; *Candid Science I*, 250–269; *Great Minds*, 125

Calvin, Melvin*: *CI* 6(1), 52; *Candid Science V*, 378–389
Chargaff, Erwin: *CI* 1(1), 4; *Candid Science I*, 14–37; *Great Minds*, 129
Chulabhorn of Thailand, Princess: *CI* 6(1), 25; *Candid Science V*, 332–339
Cornforth, John W.: *CI* 4(3), 27; *Candid Science I*, 122–137; *Great Minds*, 135
Cotton, F. Albert: *CI* 3(2), 14; *Candid Science I*, 230–245
Cram, Donald J.: *CI* 2(1), 6; *Candid Science III*, 178–197; *Great Minds*, 138
Crutzen, Paul J.: *CI* 5(1), 12; *Candid Science III*, 460–465; *Great Minds*, 141

Dewar, Michael J. S.: *CI* 3(1), 34; *Candid Science I*, 164–177
Djerassi, Carl: *CI* 3(1), 4; *Candid Science I*, 72–91; *Great Minds*, 146

---

[16]I. Hargittai, M. Hargittai, and B. Hargittai, *Candid Science: Conversations with Famous Scientists*, Vols. I-VI (London: Imperial College Press, 2000–2006).

[17]B. Hargittai, M. Hargittai, and I. Hargittai, *Great Minds: Reflections of 111 Top Scientists* (New York: Oxford University Press, 2014).

Edelman, Gerald M.: *CI* 5(3), 18; *Candid Science II*, 196–219
Elion, Gertrude B.: *CI* 4(2); *Candid Science I*, 54–71; *Great Minds*, 149
Ernst, Richard R.: *CI* 2(3), 12; *Candid Science I*, 294–307
Eschenmoser, Albert: *CI* 6(3), 4; *Candid Science III*, 96–107; *Great Minds*, 152

Fukui, Kenichi: *CI* 1(2), 14; *Candid Science I*, 210–221; *Great Minds*, 155
Furka, Árpád: *CI* 6(2), 37; *Candid Science III*, 220–229

Galpern, Elena G.: *CI* 1(3), 11; *Candid Science I*, 323–331; *Great Minds*, 158
Gilbert, Walter: *CI* 5(2); *Candid Science II*, 98–113; *Great Minds*, 270
Gillespie, Ronald J.: *CI* 5(3), 6; *Candid Science III*, 48–57

Hauptman, Herbert A.: *CI* 4(1), 10; *Candid Science III*, 292–317
Herschbach, Dudley R.: *CI* 2(4), 12; *Candid Science III*, 392–399
Hoffmann, Roald: *CI* 1(2), 14; *Candid Science I*, 190–209; *Great Minds*, 165
Huffman, Donald R.; *CI* 6(3), 39; *Candid Science V*, 390–399

Klug, Aaron: *CI* 6(4), 4; *Candid Science II*, 306–329; *Great Minds*, 282
Krätschmer, Wolfgang: *CI* 2(1), 17; *Candid Science I*, 388–403
Kroto, Harold W.: *CI* 1(3), 14; *Candid Science I*, 432–357
Kuroda, Reiko: *CI* 6(2), 51; *Candid Science III*, 466–471; *Great Minds*, 175

Larson, Clarence E.: *CI* 5(1), 43; *Candid Science V*, 316–323
Laurent, Torvard C.: *CI* 4(1), 26; *Candid Science II*, 396–415
Lederberg, Joshua: *CI* 6(1), 4; *Candid Science II*, 32–49; *Great Minds*, 292
Lederman, Leon M.: *CI* 4(4), 20; *Candid Science IV*, 142–159
Lehn, Jean-Marie: *CI* 2(1), 6; *Candid Science III*, 198–205; *Great Minds*, 181
Lewis, Edward B.: *CI* 4(4), 12; *Candid Science II*, 350–363; *Great Minds*, 299
Lipscomb, William N.: *CI* 2(3), 6; *Candid Science III*, 18–27; *Great Minds*, 184

Mackay, Alan L.: *CI* 3(4), 25; *Candid Science V*, 56–75
Marcus, Rudolph A.: *CI* 3(4), 14; *Candid Science III*, 414–421
McCarty, Maclyn: *CI* 4(2), 20; *Candid Science II*, 16–31; *Great Minds*, 305
Merrifield, Bruce: *CI* 4(2), 12; *Candid Science III*, 206–219; *Great Minds*, 190
Milstein, César: *CI* 5(4), 6; *Candid Science II*, 221–237; *Great Minds*, 310
Mößbauer, Rudolf: *CI* 3(3), 6; *Candid Science IV*, 260–271; *Great Minds*, 62
Mullis, Kary B.: *CI* 5(3), 11; *Candid Science II*, 182–195
Müller-Hill, Benno: *CI* 6(1); *Candid Science II*, 114–129; *Great Minds*, 316

Nirenberg, Marshall W.: *CI* 6(4), 48; *Candid Science II*, 130–141

Oliphant, Marcus L. E.: *CI* 6(3), 50; *Candid Science IV*, 304–315; *Great Minds*, 69
Osawa, Eiji: *CI* 1(3), 308; *Candid Science I*, 308–321

Pauling, Linus: *CI* 1(1), 5; *Candid Science I*, 2–7; *Great Minds*, 196
Peierls, Rudolf E.*: *CI* 6(2), 54; *Candid Science V*, 282–289
Perutz, Max F.: *CI* 5(1), 16; *Candid Science II*, 296–305; *Great Minds*, 325
Polanyi, John C.: *CI* 2(4), 6; *Candid Science III*, 378–391; *Great Minds*, 199
Pople, John A.: *CI* 3(3), 14; *Candid Science I*, 178–189; *Great Minds*, 202
Porter, George: *CI* 4(4), 30; *Candid Science I*, 476–487; *Great Minds*, 205
Prelog, Vladimir: *CI* 2(2), 16; *Candid Science I*, 138–147; *Great Minds*, 207
Prigogine, Ilya: *CI* 5(4), 13; *Candid Science III*, 422–431

Appendix 2

2014 Update on Wolfgang Krätschmer's Activities [to "Rising to New Heights," *CI* 2000(3), 42–43.]

Research on Fullerenes

It may be remembered that the discovery of C_{60} was motivated by the search for carbon molecules in interstellar space. However, in the early 1990s, when the spectral features of fullerenes became known, astronomy provided no inkling of presence of C_{60} in space. This somehow puzzling situation lasted until C_{60} was finally detected by its IR emission features in circumstellar and interstellar environments in 2010. At that early time in the 1990s, we were hoping that at least a form of hydrogenated C_{60} might exist in space. After we had synthesized and characterized $C_{60}H_2$, we unfortunately found no spectral coincidences with prominent interstellar features. We realized however that $C_{60}H_2$ was extremely unstable and readily decaying into C_{60} and H_2. This in fact is an interesting feature of possible relevance to the formation of molecular hydrogen in space. It is believed that the abundant atomic hydrogen in space requires the surfaces of interstellar grains to convert into molecular H_2. C_{60} in this context may be looked upon as performing the role of such grains.

Almost by accident, we synthesized the dimer $C_{120}O$ from a mixture of $C_{60}O$ and C_{60}, a reaction which readily takes place at elevated temperatures in the solid state, i.e., in powders. We could completely characterize $C_{120}O$ as well as the related compound $C_{120}O_2$, which forms at still higher temperatures. Both compounds are dumbbell-like C_{60} dimers, in which two C_{60} units are bridged by oxygen (and direct C-C) bonds. Both show the characteristic low frequency cage-cage stretching modes in Raman and IR. We then aimed at synthesizing either the oxygen frcc C_{120}, the dimer of C_{60}, or possibly an elongated (i.e., zeppelin-shaped) C_{120} fullerene. What we obtained instead was the peanut-shaped "odd" fullerene C_{119}, a species, which was already known from MS. The complete characterization of C_{119} was a major effort and I still appreciate the help of William E. Hull from the DKFZ[18] Heidelberg in that work. C_{119} consists of two C_{58} (sp²-C) "fullerene baskets" twisted against each other and connected by an array of three (sp³-C) carbon atoms. Finally, I may remark that single-walled-carbon-nanotubes could not, so far, be produced mono-dispersed, i.e., in a unique structure. A somehow controlled "fusion" of mono-dispersed C_{60} units into a zeppelin-shaped fullerene may thus be challenging.

$C_{60}O$ was the first fullerene compound detected as an impurity in fullerene samples by MS. We tried without success to synthesize the related compound $C_{60}S$. For steric reasons $C_{60}S$ seems to be unstable. What we did succeed in preparing was $C_{120}OS$, a dimer similar to $C_{120}O_2$, in which one of the oxygen atoms is replaced by sulfur.

Fullerene formation is still a not well understood process. In order to shed some light on this problem, Yohji Achiba from Tokyo Metropolitan University designed an experiment based on a Smalley type laser furnace in which the region of the laser impact (onto the graphite rod) was monitored by a high-speed electronic camera. Optical filters also allowed observations with spectral resolution. In contrast to an arc discharge, fullerene formation in a laser furnace allows

---

[18] Deutsches Krebsforschungszentrum (German Cancer Research Center).

a controlled change of various parameters, such as buffer gas temperature and laser power, and one obtains reproducible results. I had the privilege to join this experiment (and other works performed by the group) and thus could watch the fullerenes forming "in their cradle." From the images of the laser plume obtained at different times after laser impact, the temperature of the laser-ablated material (nanoscopic carbon grains) could be monitored as a function of time. Thus, the cooling down history of the laser-produced carbon plasma could be recorded. Under conditions that were known to lead to measurable yields of fullerenes (when the buffer gas temperature was elevated above about 900 K), the cooling speed of the plasma was significantly reduced and a clear re-heating effect was observed. Apparently, the exothermic process of fullerene formation occurred. We named this effect "fullerene fire."

Work on Smaller Carbon Molecules

Some years before the dawn of fullerenes, Don Huffman and I did some spectroscopic work on carbon grains (which finally led to the synthesis of fullerenes in bulk quantities) and on carbon vapor which had been matrix-isolated in solid argon. These latter studies had been pioneered by William Weltner in Florida in the early 1960s, and I think we could add some new exciting data. The vapor molecules (mainly C, C_2 and C_3) partially polymerized in the matrix, yielding a variety of species. The most intense UV-VIS absorption bands that we observed formed a regular sequence, indicating the presence of a variety of linear chain species. Surprisingly, the pattern of bands looked quite reminiscent to the so-called "diffuse interstellar bands," the so far unexplained broad absorptions in the UV-VIS. These mysterious bands have been known for almost 100 years.

Using matrix-isolation of mass-selected molecular beams, the group of John Maier in Basel had found that each UV-VIS band belongs to a specific linear carbon chain molecule. The species with an odd number of carbon atoms (i.e., cumulenic structures) produce the most intense absorptions. In order to determine the IR bands for these carbon chains, we devised the method of selective molecular oxidation. For this purpose, we chose a matrix of solid oxygen, in which the linear carbon molecules were trapped, and in which their UV-VIS bands were excited by laser exposure. The selected species reacted photo-chemically with the surrounding oxygen, and upon laser exposure, the UV-VIS and the corresponding IR bands decreased in strength. In the range from linear C_9 to C_{21} we found an interesting regularity in the positions of the most intense IR absorptions as a function of size. Furthermore, we could characterize the IR spectra of various carbon chain oxides. We found no indication for the presence of cyclic species.

Bare carbon chains seem to be quite unstable when electronically excited and thus as free molecules should show broad UV-VIS absorptions, which in width are probably comparable to the diffuse interstellar bands. From our data, one may speculate about the carriers of these mysterious absorptions: If they are neutral species, either they may be bare carbon chains or, more likely, may possess a carbon chain backbone structure to which other cosmic abundant atoms like H, N, or O are attached.

Some Personal Remarks

With the decline of interest in fullerenes, my life reverted to normal. The hectic atmosphere of the 1990s became history. Naturally, from that time my memory has kept funny and bitter events. To the funny and in some sense bitter events I count that for two or three years in October, on the day when the winners of the Nobel Prizes are made public, journalists occupied my office, waiting in case I got the famous telephone call from Stockholm. Their waiting, and mine, was in vain. When the Nobel Prize was finally given to Curl, Kroto, and Smalley, I was not in my office; I was in Japan. Nevertheless, a clever journalist managed to reach me by phone. He asked me whether I intend to object to this decision of the Nobel committee. He was distinctly disappointed when I said, "No." When I explained him that in fact, these three colleagues represent the entire fullerene field and it was that field which was honored by the Nobel decision, he hung up. In fact, I knew this man: he was a science journalist from a German news magazine. He should have known how modern science works: Science is not a game with individual winners and losers, it is a collective effort of many researchers and the Nobel-Prize-winners are usually the main representatives of that particular field. Would a journalist ask a player of a winning soccer team, "Why didn't you score the decisive goal?"

The declining interest in fullerenes naturally leads to a decline in quoting our contributions. After almost 25 years this is understandable but makes me a little sad since the Nobel-Prize-winners' paper is still quoted. Perhaps in the meantime, we have entered the collective memory of science and do not need to be quoted anymore.

Otherwise, I am quite happy. I am retired since 2007, and have a guest status in my institute. I could keep my office, and can use the facilities here and take part in the scientific life. Occasionally I give lectures. My interest has turned to issues which I found interesting but did not have the time to study in my active life. This mainly concerns natural science, physics and in particular the rapidly developing field of cosmology. Because of the usual heavy mathematical overhead involved, this task, or better hobby, keeps me quite busy.

Max-Planck-Institut für Kernphysik   Wolfgang Krätschmer
Heidelberg, Germany

2014 Update on Donald R. Huffman's Activities [to "Rising to New Heights," *CI* 2000(3), 42–43.]

Following publication of our paper in *Nature* on 17 November, 1990, which revealed most of the basic properties and the production process for fullerenes, it seemed that almost every laboratory in the world involved with materials science, chemistry and physics became involved with research and applications-development on this new material. Since I have always enjoyed my research most when only a few colleagues were involved with me in non-pressure situations, I mostly laid aside fullerene research in favor of other applications of "Absorption and Scattering of Light by Small Particles" which is not only the title of my book with Craig Bohren, but a one-phrase summary of the last 45 years of my career in science. In recent years, I have involved myself in work on small particles in the earth's atmosphere. Surprisingly perhaps, carbon is just as important in modern atmospheric aerosol research as it was in the astrophysics of interstellar dust in the 1970s and 1980s, which led to our fullerene discovery.

In fact, just as our 1990 discovery broke, and I found myself travelling the world, I had just formed a company (DH Associates) to research, produce and market the only commercially available cloud condensation nucleus counter (CCNC), an important instrument for studying clouds by way of the CCNs that are necessary for the formation of any and all cloud droplets. Although the heyday of the fullerenes in the early 1990s caused, in part, my shutting down of DH Associates, I have continued to work on the physics of CCNs and their detection, since aerosols including clouds have been the largest unknown factor in greenhouse warming scenarios.

In the past three years I have collaborated with my son, Alex Huffman, in the Chemistry Department of the University of Denver, in studying biological particles in the atmospheric aerosol, including pollen, bacteria, and fungi, which have become more interesting lately because of both health and environmental concerns. We have just presented a paper on a new, inexpensive instrument for discriminating biological aerosols at the annual meeting of the American Association of Aerosol Research in Orlando, which includes a smart phone-based instrument that is small, portable and inexpensive. Our vision is to make it available at little or no cost to "Citizen Scientists", who would use their own smart phones for the measurements and relay the data to the central facility by way of "the cloud" thereby greatly increasing the global coverage of measurements.

An important event following the 2000 *Intelligencer* article was the granting of the U.S. Patents for the production of fullerenes, the first of which was granted in the year 2007, 28 years after its original filing in July of 1990. Wolfgang Krätschmer and I have continued to be close friends. When he received the award for European inventor of the year after the issuing of the patent, I took pleasure in joking with him about this honor, in view of the fact that, in those early days in 1990, he only grudgingly went along with being named a co-inventor on the patent.

Krätschmer and I have both used old vacuum bell jar systems that were instrumental in the extended work on carbon, which resulted in the production of C_{60} and the other fullerenes. Although Krätschmer's device has been enshrined in the Deutsches Museum in Bonn, I still have mine. I joke with Wolfgang that I wouldn't give mine to the Smithsonian Institution in Washington even if they should ask for it, as I still use it for teaching and demonstrations. Its latest task was to make nanoparticles of carbon that might mimic the properties of atmospheric carbon particles. This sounds very much like a reprise of earlier work if one simply replaces the phrase "interstellar" with "atmospheric".

Department of Physics, University of Arizona   Donald R. Huffman
Tucson, AZ, USA

Appendix 3

2014 Update on "Molecular Craftwork with DNA" [to N. C. Seeman, "Molecular Craftwork with DNA," *CI* 1995(3), 38–47.]

It has been nearly 20 years since my article appeared in *The Chemical Intelligencer*. At the time, mine was the only laboratory engaged in DNA nanotechnology. Today, there are about 100 laboratories working on DNA nanotechnology like our own, which is based on objects, tiles and lattices [1]. I cannot count the number of laboratories engaged in all forms of DNA nanotechnology, including that which focuses on colloids whose surfaces are derivatized by DNA [2, 3]. Consequently, there have been major advances in this time period. The first key advance was that a robust DNA motif was discovered [4], making it possible to build 2D lattices [5] and nanomechanical devices [6, 7]. Ultimately, a robust motif was discovered in three dimensions [8], that led to the self-assembly of DNA molecules into crystals [9]. However, we still haven't achieved the goal of scaffolding oriented macromolecules (as illustrated in Figure 4 of the original *CI* article); the crystals whose lattices can accommodate macromolecules do not yet diffract to adequate resolution.

The article emphasized the importance of minimizing the sequence symmetry of DNA strands that act as components of these structures. However, a major development has led to research that seems not to need tightly designed molecules. This is the relatively inexpensive nature of DNA. When the first four-arm junction (Figure 5 of the original *CI* article) was purchased, in large quantities, the cost was $312 per nucleotide. Today, the techniques we use (which certainly don't include ligation) have made it possible to buy adequate quantities of sequences for $0.08 per nucleotide. This cost drop has enabled investigators to try experiments with large numbers of sequences, which include the highly popular DNA origami [10] and the DNA brick methodology [11]. Thus, we do not optimize sequence so carefully any more, because if we make a mistake, it is cheap to correct it. Furthermore, origami has shown that any sequence can apparently be adapted to making a DNA object.

A major change in the methodology is the adoption of atomic force microscopy (AFM) [5] as a central tool for analysis. The complex catenanes and gel methodologies used to demonstrate the early constructs described in the article are gone, along with the ligations they required. More minor updates to the article include that 12-arm junctions have been built [12], that Z-DNA has been replaced with commercially available L-nucleotides [13], and simple branched molecules (but not complex ones) have been cloned [14]. Nevertheless, the key change is that so many laboratories are engaged in the effort. Consequently, the advancement of the field no longer depends on the imagination and talents of a single laboratory.

Department of Chemistry
New York University
New York, NY, USA

Nadrian C. Seeman

References to Appendix 3

1. N.C. Seeman, DNA in a Material World, *Nature* **421**, 427–431 (2003).
2. C.A. Mirkin, R.L. Letsinger, R.C. Mucic, J.J. Storhoff, A DNA-Based Method for Assembling Nanoparticles into Macroscopic Materials, *Nature* **382**, 607–609 (1996).
3. V.T. Milam, A.L. Hiddessen, J.C. Crocker, D.J. Graves, D.A. Hammer, DNA-Driven Assembly of Bidisperse Micron-Sized Colloids, *Langmuir* **19**, 10317–10323 (2003).
4. T.-J. Fu, N.C. Seeman, DNA Double Crossover Structures, *Biochemistry* **32**, 3211–3220 (1993).
5. E. Winfree, F. Liu, L. A. Wenzler, N.C. Seeman, Design and Self-Assembly of Two-Dimensional DNA Crystals, *Nature* **394**, 539–544 (1998).
6. C. Mao, W. Sun, Z. Shen, N.C. Seeman, A DNA Nanomechanical Device Based on the B-Z Transition, *Nature* **397**, 144–146 (1999).
7. H. Yan, X. Zhang, Z. Shen, N.C. Seeman, A Robust DNA Mechanical Device Controlled by Hybridization Topology, *Nature* **415**, 62–65 (2002).
8. D. Liu, M.S. Wang, Z.X. Deng, R. Walulu, C.D. Mao, Tensegrity: Construction of Rigid DNA Triangles with Flexible Four-Arm DNA Junctions, *J. Am. Chem. Soc.* **126**, 2324–2325 (2004).
9. J. Zheng, J.J. Birktoft, Y. Chen, T. Wang, R. Sha, P.E. Constantinou, S.L. Ginell, C. Mao, & N.C. Seeman, From Molecular to Macroscopic *via* the Rational Design of a Self-Assembled 3D DNA Crystal, *Nature* **461**, 74–77 (2009).
10. P.W.K. Rothemund, Folding DNA to Create Nanoscale Shapes and Patterns, *Nature* **440**, 297–302 (2006).
11. Y.G. Ke, L.L. Ong, W.M. Shih, P. Yin, Three-Dimensional Structures Self-Assembled from DNA Bricks, *Science* **338**, 1177–1183 (2012).
12. X. Wang, N.C. Seeman, The Assembly and Characterization of 8-Arm and 12-Arm DNA Branched Junctions, *J. Am. Chem. Soc.*, **129**, 8169–8176 (2007).
13. T. Ciengshin, R. Sha, N.C. Seeman, Automatic Molecular Weaving Prototyped Using Single-Stranded DNA, *Angew. Chemie Int. Ed.* **50**, 4419–4422 (2011).
14. C. Lin, S. Rinker, X. Wang, Y. Liu, N. C. Seeman, H. Yan, *In Vivo* Cloning of Artificial DNA Nanostructures, *Proc. Nat. Acad. Sci. (USA)* **105**, 17626–17631 (2008).

Part I

Interviews

Interview

Kurt Mislow[a]

István Hargittai[b]

Photos of Kurt Mislow during the interview are by I. Hargittai.

[a]*Chemical Intelligencer* 1998(3), 16–25.

[b]Department of Inorganic and Analytical Chemistry, Budapest University of Technology and Economics, Szt. Gellert ter 4, 1111 Budapest, Hungary
e-mail: istvan.hargittai@gmail.com

Kurt Mislow (b. 1923 in Berlin, Germany) is Hugh Stott Taylor Professor of Chemistry Emeritus in the Department of Chemistry of Princeton University. He attended Tulane University (B.S., 1944) and obtained his Ph.D. degree at the California Institute of Technology under Linus Pauling in 1947. The dominant theme of Dr. Mislow's research has been the development of stereochemical theory. His many distinctions include membership in the National Academy of Sciences and in the American Academy of Arts and Sciences, the Solvay Medal (Belgium), the Prelog Medal (Switzerland), the American Chemical Society's James Flack Norris Award in Physical Organic Chemistry and William H. Nichols Award, and the Chirality Medal. Our conversation was recorded in Professor Mislow's office at Princeton University on March 12, 1997, and was later augmented by figures and references.

ISTVÁN HARGITTAI (IH): A recent book stated, "Mislow introduced chirality into chemistry."

KURT MISLOW (KM): That's very kind, but it's putting it too strongly. It's true that my work has been focused on symmetry and chirality in chemistry for a long time. But credit as the originator of the concept of chirality in chemistry obviously belongs to Louis Pasteur. Pasteur connected chirality on the macroscopic scale to chirality on the molecular scale in his famous experiments with sodium ammonium tartrate [Pasteur, L. *Ann. Chim. Phys.* **1848** *24*, 442]. Eventually Pasteur became interested in other things and didn't pay attention to developments in structural theory, particularly the work of van 't Hoff and others. But there was really no need for that. In effect, what Pasteur said was that the chirality—he called it dissymmetry—of the atomic

B. Hargittai and I. Hargittai (eds.), *Culture of Chemistry: The Best Articles on the Human Side of 20th-Century Chemistry from the Archives of the Chemical Intelligencer*, DOI 10.1007/978-1-4899-7565-2_1, © Springer Science+Business Media New York 2015

arrangement is the necessary and sufficient condition for molecular enantiomorphism and optical activity. And that was enough. He didn't have to worry about structural theory because his conclusion was based purely on a symmetry argument. The chiral arrangement of atoms is all that matters, regardless of the detailed structure of the molecule.

So how did I get interested in this? Right after I got my Ph.D. with Linus Pauling at Caltech, in 1947, I got a teaching position at New York University. At that time, another new Ph.D. Henry Hellman, who came from Purdue University, joined the faculty. He brought with him the syllabus for *Advanced Organic Chemistry*, by George Wheland of the University of Chicago. These were mimeographed notes that were ultimately published in a second edition, as a book, in 1949. Wheland's uncluttered and logical way of thinking about stereochemistry opened my eyes to the power of symmetry arguments. It was a revelation to me that symmetry and chirality were at the heart of stereochemistry. So Wheland's book had a tremendous influence on my thinking and was a real inspiration. I became fascinated by stereochemistry, and it permanently changed the direction of my research. Wheland, incidentally, was also a student of Pauling's, though before my time.

Wheland's book inspired the first paper I published at NYU [Mislow, K. *Science* **1950**, *112*, 26]. Its title was "The Concept of Internal Compensation." In *meso*-tartaric acid, there are conformations with a plane of symmetry and conformations with a center of symmetry, and many others in between that are asymmetric. The question back then was, why is *meso*-tartaric acid optically inactive? There were people who said the reason was that one half of the molecule causes a rotation that is compensated by the rotation caused by the other half of the molecule. In other words, the two halves taken separately are mirror images of each other, so supposedly their rotations cancel and the molecule is optically inactive no matter what the conformation is. They called this "internal compensation." Obviously this can't be right, because if you have a chiral conformation, you expect it to be optically active. Pasteur taught us that. It shows the primitive state that stereochemistry was in at that time. Wheland's book discussed and debunked this notion, and my contribution was to propose an experiment to settle the issue once and for all.

I'll give you another example of Wheland's penetrating way of analyzing problems in stereochemistry. He asked the question: what is a diastereomer? Standard textbooks of stereochemistry, even in the early 1960s, defined diastereomers as stereoisomers some or all of which are chiral but that are not enantiomers. For example, mannose and glucose are

diastereomers, and they are both chiral. *meso*-Tartaric acid, which is achiral, is a diastereomer of D- and L-tartaric acids, both of which are chiral. They all conform to this definition. Now let's examine the case of some methyl-substituted cyclopropanes. If you take 1,2-dimethylcyclopropane, the *cis* isomer is achiral and the *trans* isomer is chiral. So this is like the case of the tartaric acids. But now take 1,2,3-trimethylcyclopropane: the all-*cis* isomer is achiral, of course, but the *trans* isomer is also achiral. There are no chiral isomers here. According to the conventional textbook definition, these two isomers are not diastereomers. And they are certainly not enantiomers. So then what are they? Wheland recognized that the issue of chirality is irrelevant. As far as he was concerned, maleic and fumaric acids, which certainly aren't chiral, are also diastereomers. In short, Wheland's point was that diastereomers are stereoisomers that are not enantiomers. Period. All of this seemed perfectly reasonable to me, and I argued for adoption of Wheland's definition at the first international stereochemistry congress in Bürgenstock, in 1965. I ran into heavy weather, though, because the stereochemistry establishment of the day just didn't like changing what had been the traditional definition. But in the end the argument was compelling, and ultimately everybody adopted Wheland's definition. Even the textbook writers finally saw the light. It's been officially sanctified by IUPAC. Today, nobody realizes that there was ever a problem to begin with. The irony of it all is that the definition of diastereomer that is accepted today was the original one. This appeared in the second (1907) edition of the textbook of organic chemistry by Victor Meyer (the father of the word stereochemistry) and Paul Jacobson, the junior author. We know all this thanks to some historical research by Günter Schiemenz, a professor at the University of Kiel.

In the fifties and early sixties, I kept pushing symmetry arguments at conferences, but I wasn't making many converts. For example, it seemed perfectly safe to me to predict that asymmetric molecules of the type Cabcd should have six different bond angles; or that carbenium ions in asymmetric molecules should be nonplanar; or that if nuclei are symmetry–nonequivalent, their NMR signals should be shifted relative to one another; and so forth and so on. Detailed structural information is completely unnecessary to arrive at these conclusions. Pasteur's legacy again, you see. When people found these things experimentally, they thought that it was a big deal, but it was all easily predictable from symmetry arguments without any recourse to detailed structural information.

All this seems obvious today, but it wasn't obvious at all to many of my colleagues at the time. Well, I found this intensely frustrating, and this gave me the impetus to write a textbook, *Introduction to Stereochemistry*, which, basically, was a radicalized version of Wheland's book. I introduced point groups for the first time; organic chemists had evidently never been exposed to them, at least not in stereochemistry textbooks. My message was that the major principles of stereochemistry rest on considerations of symmetry and group theory. I emphasized symmetry arguments throughout and, all in all, retextured stereochemistry. The book was published in 1965, the year after I moved to Princeton, the year of the first Bürgenstock conference. It did what I hoped it would do. That is, it changed the way people thought about stereochemistry. This way of thinking, via Pasteur and Wheland, has now been adopted by everyone, and you find it in all the modern textbooks. Today we are at the point where most people have accepted what I said before that symmetry and chirality are at the heart of stereochemistry. This is enormously gratifying to me, to have played a part in changing the way people think about my subject.

So it's amazing that there are still people who haven't gotten the message. I'll give you a particularly egregious example. A lot of excitement was caused by the announcement that some people had synthesized an enzyme, HIV-1 protease, from D-amino acids and had found that the D-enzyme was active only with D-configured substrates. This is in contrast to the naturally occurring L-enzyme, which reacts only with substrates of the L-configuration [Milton, R.C. deL; Milton, S.C.F.; Kent, S.B.H. *Science* **1992**, *256*, 1445]. This discovery made the cover of the issue of *Science* in which this article was published. Milton et al. claimed that "we can now state, based on experimental evidence, that protein enantiomers should display reciprocal chiral specificity in their biochemical interactions." That was an amazing thing to say. The outcome was totally dictated by symmetry, so where was the need for experimental evidence? What else did these people expect? If they had found anything else, they would have had to claim violation of parity at the chemical level, and nobody in their right mind would have believed that. Enough said.

As a follow-up to my book I published a paper on stereoisomeric relationships in molecules, with a graduate student, Morton Raban, who is now a professor at Wayne State University. This paper was published in the inaugural issue of the series *Topics in Stereochemistry* [Vol. 1, Chapter 1, p 1 (1967)]. In this paper we recognized local symmetry relationships and introduced the idea of enantiotopic and diastereotopic groups. This, again, had considerable repercussions in enzymology and NMR spectroscopy, and it's now in all the textbooks. Then, 17 years later, a further step was taken with the introduction of the concept of chirotopic

groups, and with local symmetry now the central issue. This work was done with Jay Siegel, then a graduate student and now a professor at the University of California, San Diego [Mislow, K.; Siegel, J. *J. Am. Chem. Soc.* **1984**, *106*, 3319]. The main point of that paper was to clarify a variety of stereochemical concepts. We critically reexamined some classical concepts in stereochemistry. For example, we pointed out that chirality elements, like chiral centers, chiral axes, and chiral planes, are very useful as nomenclatural devices, but they don't have anything to do with chirality. In order to determine whether something has a chiral center or a chiral axis or a chiral plane, you have to decide where the bonds are in the molecule. And, of course, that is not what geometrical chirality is all about. Incidentally, this work was the direct outgrowth of some earlier work with Siegel, in collaboration with Frank Anet [Anet, F.A. L.; Miura, S.S.; Siegel, J.; Mislow, K. *J. Am. Chem. Soc.* **1983**, *105*, 1419], on the so-called "coupe du roi." That's a parlor trick in which an apple is divided into homochiral halves. Figure 1 shows reversible bisection and reconstitution of an apple (the two homochiral segments on the left are the enantiomorphs of the two homochiral segments on the right). It's a wonderful example of how a playful activity can lead to significant science [Mislow, R. *Bull. Soc. Chim. Fr.* **1994**, *131*, 534].

IH: You never revised the book.

KM: No. I have been thinking of revising it off and on, but I can't seem to get around to it.

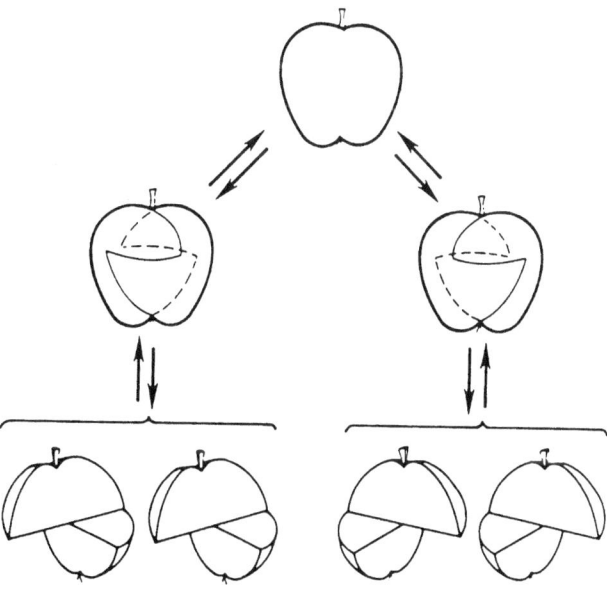

Fig. 1 Bisection of an apple into homochiral segments.

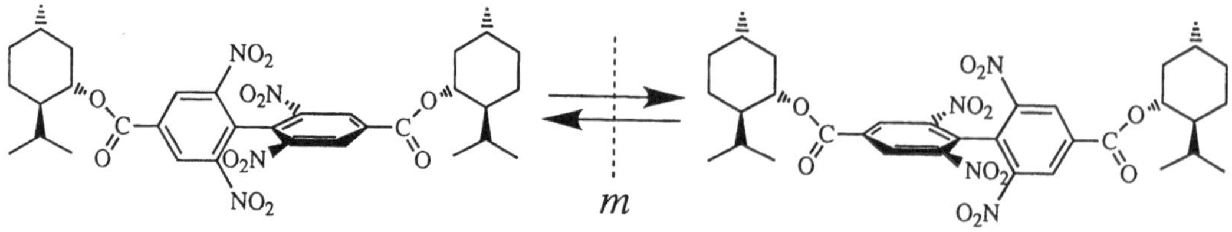

Fig. 2 Interconversion of enantiomorphs by a chiral pathway.

Fig. 3 Model of a bridged biphenylketone viewed down the twofold axis.

IH: Looking back, which of your contributions would you single out as the most important?

KM: Well, I guess I can summarize the gist of my work by saying that it has provided some of the theoretical underpinning of modern stereochemistry. The role of symmetry and chirality in stereochemistry has been the principal motif. This opus, if I may call it that, is made up of lots of individual contributions that fit together, sort of like a jigsaw puzzle. There was the introduction of the topicity concept, which is now taught to every undergraduate in organic chemistry. Another example was an offshoot of the internal compensation problem that I mentioned before. In *meso*-tartaric acid, there are a couple of achiral conformations. But then, in 1954, I came up with the idea of a chemically achiral molecule like *meso*-tartaric acid but one in which all the conformations are asymmetric. The mirror-image conformations interconvert, but they can't enantiomerize through an achiral state [Mislow, K. *Science* **1954**, *120*, 232]. Subsequently, we actually made such a molecule, (1*R*)-menthyl (1*S*)-menthyl 2,2′,6,6′-tetranitro-4,4′-diphenate [Mislow K, Bolstad R. *J. Am. Chem. Soc.* **1955**, 77, 6712]. Figure 2 shows how the

two mirror-image-related (*m*) biphenyls are interconverted by a 90° twist of the biphenyl moiety about the two single bonds to the menthyl groups. This caused quite a stir, because it raised a question: if you never go through an achiral conformation, at which point do you cross the dividing line between left and right? This was discussed in a number of textbooks, including in Wheland's third edition of *Advanced Organic Chemistry*. Since that time, we have found other molecules of this type among asymmetric molecular propellers and gears and, most recently, in a catenane, a chemical link.

Thinking about biphenyls inspired the design of the first configurational correlation between a classically asymmetric molecule of the type Cabcd and one that owes its chirality purely to restricted rotation. This work was done with my student Paul Newman [Mislow, K.; Newman, P. *J. Am. Chem. Soc.* **1957**, *79*, 1769]. The idea was to partially reduce a racemic biphenyl symmetrically bridged in the 2,2′-positions with CH_2-CO-CH_2, and held conformationally rigid by bulky substituents in the 6,6′– positions, using an alcohol of known absolute configuration, like (+)-(*S*)-methyl-*tert*-butyl-carbinol (pinacolyl alcohol). The ketone has twofold symmetry, so it doesn't matter which face of the carbonyl group gets attacked. Figure 3 shows the enantiomer with the S configuration, viewed down the C_2 axis. The two carbonyl faces are homotopic. But the faces of the carbonyl groups in the enantiomers are enantiotopic, so the reduction gives rise to diastereomeric transition states. From a knowledge of the transition structure and the shape of the model, we knew that the R enantiomer of the ketone was reduced more rapidly by (S)-pinacolyl alcohol. By interrupting the reaction before completion of the reduction, we got a mixture of the more slowly reduced (+)-S ketone and the alcohol derived from the more rapidly reduced (−)-R ketone, both in optically active form. This correlates sign of rotation with absolute configuration. So the idea worked just fine and inaugurated a whole new era of biphenyl stereochemistry. Absolute configurations were established for the first time for a wide variety of optically active biphenyls [Mislow, K. *Angew. Chem.* **1958**, *70*, 683]. This was all later confirmed by X-ray crystallography. Among other spin-offs were the first example of a photoracemization, which involved racemization of a bridged biphenyl ketone by irradiation, a generalization of the octant rule, was done in collaboration with Al Moscowitz and Carl Djerassi, and the first experimental demonstration that steric isotope effects play a role in conformational interconversions. Specifically, what we found was that 9,10–dihydro–4,5–dimethylphenanthrene with CD_3 in the hindered positions racemizes more rapidly than the CH_3 analogue. That's evidence that CD_3 has a smaller van der Waals radius than CH_3.

Raban and I were also the first to show how NMR could be used to determine enantiomeric composition [Raban, M.; Mislow, K. *Tetrahedron Lett.* **1965**, 4249; ibid. **1966**, 3961]. The idea was to react an alcohol of unknown enantiomeric composition with an enantiopure acid chloride. The ratio of diastereomeric esters can be estimated by integration of appropriate NMR signals, and, if the reaction goes to completion, that's of course the same as the ratio of enantiomers in the starting alcohol. Our acids were 2-phenylpropionic acid and α-methoxyphenylacetic acid. Well, Harry Mosher picked up this idea and developed an improved derivatizing agent, α-methoxy-α-(trifluoromethyl) phenylacetic acid (MTPA), based on our method. That's widely known as Mosher's reagent. I was happy to see that our idea had fallen on fertile ground and has found wide application in chemistry. In our 1965 paper, we also predicted that a chiral solvent would be effective in the same sense and serve the same purpose. This was a perfect example of a symmetry argument. Bill Pirkle shortly after found just such a solvent, and the subsequently developed chiral shift reagents also fall into this class.

In later years I concentrated mainly on developing the concept of correlated rotation in internally mobile molecules, like the molecular propellers and gears that I mentioned

before. Working with molecular propellers, we provided the first examples of molecules whose conformers are collected in sets that are distinguished by the phase relation of all correlated rotors. This imposes a constraint that results in a novel kind of stereoisomerism [Mislow, K. *Acc. Chem. Res.* **1976**, *9*, 26]. Stereoisomerism of this kind does not result from restricted rotation, and so it differs fundamentally from stereoisomerism due to hindered rotation as is found in biphenyls, for example. I'll try to explain this with an example from work that we did with molecular gears [Iwamura H.; Mislow, K. *Acc. Chem. Res.* **1988**, *21*, 175; Mislow, K. *Chemtracts-Org. Chem.* **1989**, *2*, 151]. It's a little bit complicated. Let's look at what I'll arbitrarily call the D-isomer of bis(2,3-dimethyl-9-triptycyl) methane (TpCH$_2$). The structural formula is shown in the middle of Fig. 4 (a=b=H). Each benzene ring in one triptycyl (Tp) group acts like a cog that fits snugly into the V-shaped notches formed by two benzene rings in the other Tp group. The two Tp rotors behave essentially like frictionless and securely meshed bevel gears. The system undergoes dynamic gearing, or correlated disrotation, with a very low barrier and no gear slippage, at least under normal conditions of observation. Figure 4 shows the gearing circuit of this molecule. [The view is along the bisector of the C(9)-CH$_2$-C(9) angle, the horizontal line

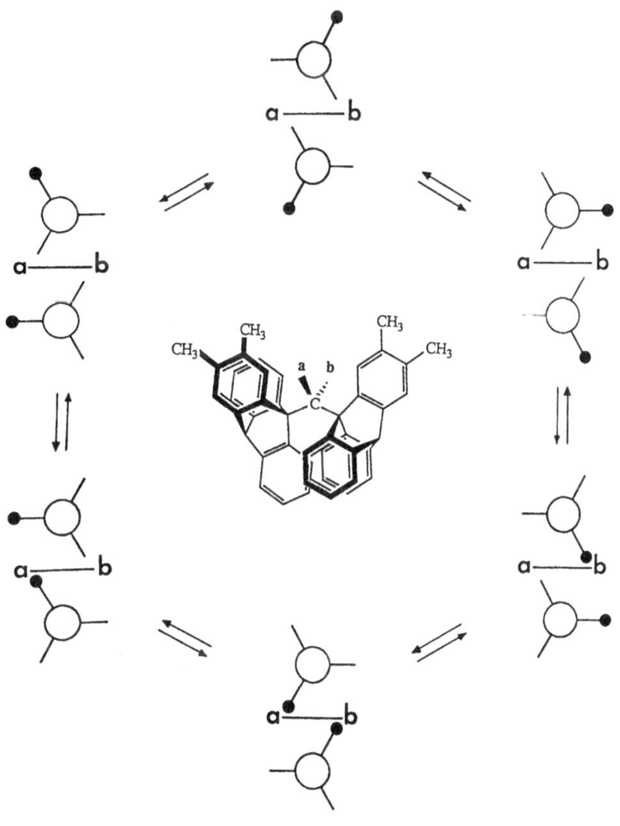

Fig. 4 Gearing circuit for a chiral bitriptycylmethane.

symbolizes H-C-H, the open circles and the three lines radiating from them represent the bridgehead carbons of the two Tp's and the three benzene blades, and the small filled circle identifies the 2,3-dimethyl-substituted ring.] The point here is that the D-isomer does not change into the L-isomer, no matter how fast the rotation is or how many complete revolutions the system undergoes. The reason is that the phase relationship between the labeled cogs remains invariant, just as it does in real mechanical gears. The gearing circuit for the L-isomer is the mirror image of Fig. 4. There is also a third gearing circuit for the third isomer, which is chemically achiral.

IH: What about the quantification of chirality?

KM: It's perfectly natural to attach a quantitative meaning to chirality [Mislow, K.; Bickart P. *Israel J. Chem.* **1976/77**, *15*, 1]. Consider an acetic acid molecule in which two of the methyl hydrogens are replaced by deuterium and tritium, CHDT-COOH. Its chirality properties, like optical activity, are surely going to be less pronounced than the chirality properties of an acetic acid derivative in which two of the methyl hydrogens are replaced by a methyl and a *tert-butyl* group. One can put this in geometrical terms. Take an isosceles triangle and distort it a tiny bit. It becomes a barely scalene triangle, which is chiral in two dimensions. Next, you create a chiral triangle with very different angles, say 30, 60, and 90°. The new triangle will be "more chiral" than the one that is barely different from an isosceles triangle. All this is intuitively obvious, and the degree of chirality can be formulated in a precise way. We looked into this quite a bit [Buda, A.B.; Auf der Heyde T.; Mislow K. *Angew. Chem. Int. Ed. Engl.* **1992**, *31*, 989; Weinberg N.; Mislow K. *J. Math. Chem.* **1995**, *17*, 35]. The problem is that there are many different ways of quantifying chirality; there is no unique function for doing this. This can lead to difficulties. Let's say you have two different functions $f(x)$ and $f(y)$, that are zero only in case of achirality. It can then happen that when you measure the chirality of two objects A and B, $f(x)$ finds that A is more chiral than B while $f(y)$ finds the opposite. In quantifying symmetry, you face basically the same problem on a broader scale, and with similar limitations. Going back in history, just a few years after van 't Hoff came out with his tetrahedral model of the carbon atom, a French physicist, Guye, developed a function that he used to quantify the chirality of an asymmetric tetrahedron [Guye, P.-A. *C.R. Hebd. Seances Acad. Sci.* **1890**, *110*, 714]. This was the first example of a chirality function in chemistry. The idea was to find a relationship between the shape of a tetrahedron and the optical activity of a compound. So the concept of quantifying chirality and relating this degree of chirality to observable properties is not new. It's just a question of working out suitable functions. I really don't know whether this is going to turn out to be very useful in chemistry.

Professor George W. Wheland, ca. 1949.
Photo by Betty C. Wheland.

IH: In the meantime, the business of chirality has become huge.

KM: Huge and profitable. That growth has been fueled by the pharmaceutical companies, which are under pressure to produce enantiopure drugs. There are now three journals specifically devoted to chirality: *Chirality, Tetrahedron Asymmetry,* and *Enantiomer.* How did interest in chirality originally get started? It goes back a long way. Immanuel Kant was the first person who meditated extensively on the paradoxical nature of chirality. The paradox is this. Idealized left and right hands are isometric. They have exactly the same metric properties. So in that sense they are congruent. Yet they are also not congruent because they can't be superposed, they can't be made to coincide in space. This problem puzzled Kant and has given rise to a lot of philosophical speculation. Kant coined the phrase "incongruente Gegenstücke" [incongruent counterparts] to describe nonsuperposable mirror images [Kant, I. Von dem ersten Grunde des Unterschiedes der Gegenden im Raume (1768). In *Kant's gesammelte Schriften;* Königl. Preuss. Akad. Wissensch., Vol. 2; Verlag Georg Reimer: Berlin, 1905; pp 375–383].

IH: Is there any basic research going on nowadays in stereochemistry?

KM: Much of the really novel stuff is in chemical topology. The first person to look into the problem of topological chirality was Johann Listing at the University of Göttingen. Carl Friedrich Gauss, who was also at the University of Göttingen, was himself passingly interested in knots. Listing introduced topology into mathematics, and in a paper in 1848 he recognized the mirror-image relationship between enantiomorphous knots. This was, of course, the year of Pasteur's paper on the optically active tartrates. About 30 years later, Peter Guthrie Tait started the first systematic investigation of the mathematics of knots and links. Tait was a professor at the University of Edinburgh, and he had been inspired to get into the knot business by the work of Lord Kelvin, who was then professor at the neighboring University of Glasgow. Kelvin, by the way, was the one who first introduced and defined the words chiral and chirality. Chemical knots and links are fascinating structures, and the problem of chirality gets a lot more complicated because molecular models get treated as though they were infinitely deformable. Topology has been called rubber-sheet geometry. We've done a fair amount of work in this area, and I've recently written a commentary on the subject [Mislow, K. *Croat. Chem. Acta* **1996**, *69*, 485].

IH: What is the difference between structural chemistry and stereochemistry?

KM: Chemical structure is an invention of chemists. Structure in the chemical sense is an undefined concept in quantum mechanics. You need the Born-Oppenheimer approximation to get the atomic nuclei in the molecule to behave like classical particles, like little balls with more or less fixed positions in space sitting inside a cloud of electrons. This classical model is an extremely powerful concept, though. Stereochemistry is based on it. Historically, stereochemistry was considered as secondary to constitution. Constitution is the bonding connectedness among atoms. Two molecules with the same constitution can still have different stereochemical arrangements. These are called stereoisomers. Stereoisomers can then be further classified into enantiomers and diastereomers. To me, a different classification makes a whole lot more sense [Mislow, K. *Bull. Soc. Chim. Belg.* **1977**, *86*, 595]. In my scheme, constitution takes second place to symmetry. I start with the question, are two isomers isometric or not? If the answer is yes, you can ask whether they are superposable. If not, they are enantiomers; if yes, they are homomers. The point here is that homomers and enantiomers have pairwise exactly the same scalar properties because they are isometric. Enantiomers differ only in pseudoscalar properties, like sign of optical rotation. So enantiomers have more in common with homomers than with diastereomers. Let me try to say this in other words. The relationship between enantiomers doesn't require any knowledge of molecular constitution, any more than the

relationship between homomers. It's established on the basis of symmetry alone. On the other hand, the relationship between diastereomers requires specifying molecular connectivity. If you don't do this, you can't tell the difference between diastereomers and constitutional isomers. All right, back to the classification scheme. If the answer to the isometry question is no, then we could have either diastereomers or constitutional isomers. If the isomers have the same constitution, then they are diastereomers; if not, then they're constitutional isomers. That's the end of it. If it comes to physical properties, there is not really much to choose between pairs of constitutional isomers and pairs of diastereomers. They differ in all scalar properties, in just about every physical measure you take, and sometimes even in chemical properties. Look at maleic and fumaric acid, for example. Only maleic acid can form an anhydride. So this is my classification scheme. You'll notice that there is no mention of stereoisomers in this scheme. The concept becomes totally superfluous. That's the beauty of giving primacy to symmetry. Still, in chemical reaction schemes, it's normally convenient to lump enantiomers and diastereomers into a common class. So the class of stereoisomers will endure, as a sort of auxiliary concept, for purely practical reasons.

István Hargittai and Kurt Mislow at the symmetry meeting in Darmstadt, Germany, in 1986 in front of a display of Javanese batik patterns.

IH: How about different conformers, *gauche* and *anti,* for example?

KM: They are different diastereomers, as far as I am concerned. They have distinctly different properties.

IH: Let's go back now to your own history.

KM: I officially retired in 1988. My retirement brought about a big difference in my lifestyle. It has been a sort of liberation. My time isn't structured by external demands anymore. No more committee meetings and numerous other departmental services. My time is my own, and I devote it fully to my work. My external funding continues. The money comes from the National Science Foundation, which has given me unstinting support throughout my career. It's my only support, actually. I don't take any more graduate students, but I usually have one or two postdocs.

IH: You were born in Berlin. When did you leave Germany?

KM: I was born in Berlin in 1923. My family left Germany in 1936. My father was a businessman so we moved around a lot. I spent my formative years in Düsseldorf. Even before Hitler came to power, in 1933, there were lots of street fights between the Communists and the Nazis. I witnessed all this. It was a violent environment. After 1933 there were no more Communists on the street, but the violence continued in a different way. For Jews, it was mostly a matter of humiliation and intimidation; the physical violence against Jews didn't really get started until *Kristallnacht,* in 1938. So this was merely the prelude to the Holocaust. I remember banners on buildings, in addition to the ubiquitous "Deutschland, erwache!", proclaiming things like "Die Juden sind unser Unglück" [The Jews are our misfortune] and "Juda ver-recke!" [Jews, croak!]. One of the Nazi marching songs contained the words "Wenn's Judenblut vom Messer spritzt dann geht's nochmal so gut" [When Jewish blood spurts from the knife, it's going even better]. Pretty scary to hear this yelled by the "braunen Battalionen," especially for a kid. Restaurants had signs on the outside that said that Jews and dogs were "unerwünscht" [unwelcome]. Jews were the target of implacable, irrational hatred. Fortunately, my father had the sense to see what was coming and got us out in time. I still have my old brown passport, with the "Hoheitszeichen," the spread eagle holding a swastika in its claws on the cover, and a big red "J" printed over it. I always assumed that this was the idea of some German bureaucrat. It turns out, though, that this was actually the idea of the head of the Swiss Federal Police, a guy by the name of Heinrich Rothmund. He didn't want Jews to escape into Switzerland. So he asked the Germans to put this mark of Cain on our passports, and they obliged him.

In 1936 we left for Italy, where my father had business connections, thanks to my grandfather. My grandfather stayed behind, I don't know why, and was killed by the

Gestapo, in Frankfurt. My father was in sales of electric equipment. Not a glamorous business but he was a good provider, a good father. In Italy, in Milan, I was sent to a so-called *liceo classico;* it was called Istituto Edmondo de Amicis. I had to learn Greek and Latin taught in Italian, so I had to learn Italian fast. I became Italian in spirit at that time. Italian Fascism was a joke compared to the demented racism of National Socialism. The Fascists were Italians in black shirts. They were relaxed about everything, they had a very urbane outlook on life, the outlook of a people who have had a long, distinguished culture. Afterwards, we learned that Jews were treated relatively well by the Italians even during the war. I never felt any sort of problem of any kind there and have nothing but good memories of my years in Italy.

In 1938 we went to England. I went to a grammar school there, in Shoreham-by-Sea, and found out what boarding school life was like in a boy's school. The less said, the better. In that school I was exposed for the first time to serious physics and mathematics. By accident, in my senior year, I also picked up a chemistry book. It was Partington's *Textbook of Inorganic Chemistry*. I loved everything I read in that book and decided that chemistry was my thing. Then the war started, and my father and I were interned on the Isle of Man as "enemy aliens." Then we were released, got affidavits from the States, and crossed the Atlantic in 1940 under adventurous conditions. My European experience was over. I went to live with relatives in New York and earned my living as a page boy in the Columbia Main Library. Then I got a generous scholarship from Tulane University, in New Orleans. I got my bachelor's degree there, in 1944, and then went on to graduate school at Caltech.

Without the generous help from Tulane University, I couldn't have gotten my education. But in New Orleans I also got my first experience of racism in the U.S. That was in 1941. I took a streetcar on St. Charles Street. The streetcars were segregated. There were little movable shields on the backs of the seats saying "for whites" or words to that effect. I got on this streetcar, and all the seats for whites were taken but there were some empty seats in the black section. So the conductor came over and moved a shield to give me a seat. It was a shock to me because I had just got away from this cauldron of racism in Europe. In those days there was quite a bit of anti-Semitism too. There was a local chapter of Alpha Chi Sigma on campus, the national chemical honor society for undergraduates. If you were an upperclassman majoring in chemistry, you could become a member. I was a major in chemistry all right, but I wasn't invited to join because I was Jewish.

IH: Have they changed their rules since?

KM: I recently checked, and they have. Originally, to become a member of Alpha Chi Sigma, you had to be white, Christian, and male. Jewish chemists became eligible for membership in 1948. Black chemists became eligible in 1954. Women chemists became eligible in 1970. A chronology of progress, of sorts.

IH: How about your Caltech years?

KM: I was there from 1944 to 1947. I mentioned before that I was one of Linus Pauling's students. He was a remarkable man in many ways. I took his course in quantum mechanics, using the text by Pauling and Wilson, and he would give his lectures in a totally impromptu way. It was beautiful. For example, he would disappear for a couple of weeks on one of his lecture tours. When he came back into class, we had to remind him where he had left off, and he picked up the thread, just like that, and went on. Obviously, without any preparation whatever. I learned a lot from him. When I graduated, I wanted to postdoc with Melvin Calvin at Berkeley, but Pauling got me the job at NYU instead.

IH: You obviously thrive on ideas of stereochemistry and symmetry. Did you manage to bring some of this home to your family?

KM: Of course I've tried to share my enthusiasms with my wife, Jacqueline. As an internist, she has the requisite scientific background. This means that I can tell her about chemistry that relates to cutting apples, tying knots, flipping propellers, turning gears, distorting triangles, and so forth. More important, though, she is an attentive listener and takes nothing on faith. She is a live wire and great fun, and she keeps me on my toes.

IH: It is now 50 years since you started your first job at NYU and got an inspiration that set you on a course for your entire career. Imagine a young beginning faculty member arriving here today; what could be an inspiration for her or his career?

KM: That's really hard to say. My own experience was abnormal. It was pure serendipity that I was exposed to Wheland's book at the beginning of my career. And there was no way to predict that I would be turned on so powerfully by his way of thinking about chemistry. So let's assume the more normal thing, that the new faculty member is excited about a scientific problem. What should that problem be? Well, in my opinion, chemistry is now pretty well understood. There are still plenty of problems to be solved, of course. How do proteins fold? Nobody knows. Nobody can predict, with any degree of confidence, how molecules will pack in crystals. And so forth. So people are going to be kept busy for a very long time. But problems like these are merely complicated; there are no deep mysteries here, at least that I know of. There are deep mysteries in cosmology and maybe in particle physics, and of course in biology. Like the molecular basis of heredity before Watson and Crick. I am thinking particularly of the neurosciences. How does memory work? How do we think in words and pictures? How does consciousness work, the subjective feeling of the "I" inside us? Nobody has a clue. We don't have the tools yet to answer these questions, but there are plenty of theories around. I don't know how much chemistry can contribute to this, but talking about inspiration—that's extremely exciting stuff. Our younger son, John, who is in medical school now, is planning to go in that direction.

Interview

Deeds and Dreams of Eugene Garfield[a]

István Hargittai[b]

Eugene Garfield during the conversation (Photo by I. Hargittai).

Dr. Eugene Garfield (b. 1925) is President and Editor-in-Chief of The Scientist, *a biweekly professional newspaper, which he founded, and Chairman Emeritus of the Institute for Scientific Information (ISI), Philadelphia. He is probably best known for* Current Contents (CC) *and* Science Citation Index (SCI). *Dr. Garfield instituted an information revolution in scientific research. He received a B.S. in chemistry in 1949 and an M.S. in library science in 1954, both from Columbia University, and a Ph.D. in structural linguistics from the University of Pennsylvania in 1961. He was President of Eugene Garfield Associates from 1954 to 1960 and President and CEO of the Institute for Scientific Information from 1960 to 1992. He has published numerous books and articles on scientific information retrieval and related topics. We recorded our conversation in Dr. Garfield's home in Bryn Mawr, Pennsylvania, on March 7, 1999.*

ISTVÁN HARGITTAI (IH): You introduced *Science Citation Index* and changed the way scientists are employed, professors at universities are given tenure, and research journals are judged for their quality. This is heavy stuff. The Sputnik in 1957 had a tremendous impact on American science. Is there anything comparable to your impact worldwide?

EUGENE GARFIELD (EG): Thanks for the pleasant hyperbole but if there is any truth to the statement, I'm not acutely conscious of it. Of course, *CC* and *SCI* are widely used, but I don't hear people say much about it. If you use *SCI* especially for evaluative purposes, you don't advertise it. If the *SCI* is used in tenure evaluations, hopefully it is done intelligently. I described this in an essay on faculty evaluation [1], one of my most popular. Undoubtedly, this use of citation analysis is due to the paucity of objective data for

[a]*Chemical Intelligencer* 1999(4), 26–31.

[b]Department of Inorganic and Analytical Chemistry, Budapest University of Technology and Economics, Szt. Gellert ter 4, 1111 Budapest, Hungary
e-mail: istvan.hargittai@gmail.com

such evaluations. I can't imagine how you would evaluate the impact of my work. How would you measure it? The Internet is having an impact but how would you measure it? When we talk about intellectual impact, it is very subjective—economic impact is another thing.

Nevertheless, I do find it hard to keep up with the large literature involving journal impact factors. I am especially frustrated that I can't respond to the portion containing misstatements or misuses. There is much controversy about the validity of impact factors, which are used for many purposes. As you have implied, *SCI* and *Journal Citation Reports* (*JCR*) data have become institutionalized. People often criticize the impact factor because it is so pervasive. Editors, especially of new journals, are using *JCR* to demonstrate how quickly their journals are accepted or whether they measure up to the best-known journals. Some of the most respected journals do not hesitate to use impact factors in their advertising.

IH: In *The Chemical Intelligencer*, there were a couple of papers comparing the impact factors of the *Journal of the American Chemical Society* and *Angewandte Chemie* [2, 3]. It was alleged that the impact factor of *Angewandte Chemie* was overinflated because it is published in the original German and in an English translation edition. The people at *Angewandte Chemie* were rather unhappy about these papers because they thought that *Angewandte Chemie* should have a higher impact factor for the very reason that it carries reviews in addition to research papers.

EG: I think these allegations are overstated. The analysis is not as simple as it is made to seem. There is some inflation in the impact factor due to dual citation of both editions. But the journal's self-citation only represents about 10–15 % of the citations that it receives. Undoubtedly, these disputes indicate that there is more citation consciousness among editors and publishers today.

In the studies that I did in the past, citation analysis "exposed" the political nature of East European science academies—many academicians were administrators, not world-class scientists. That was true also in some other European countries. In Italy, the *SCI* was like salvation to some scientists even though it did not immediately correct the unfair allocation of credit and resources. It called attention to the disparities in funding and publication. There are still many politically based science decisions—who gets tenure, who gets research funding, money, and so forth. The Italians started using the *SCI* data over 20 years ago, not only to measure citation impact but simply to determine if particular grantees had published any papers in peer-reviewed journals. The younger scientists resent the power of the old guard, who continue to get the money. The younger ones

publish in reputable journals and do significant research. So there is no doubt that *SCI* had some effect, in particular in Europe. In those days, I don't think we had that much of a problem in the United States. There may not have been enough money available, but, in general, our peer-review system is not nearly as politicized. The use of citation data in Italy led to the publication of an interesting monograph [4].I'm curious as to of what effect *SCI* had in Israel. Gideon Czapski, a Professor of Chemistry at Hebrew University, has made an extensive citation analysis of Israeli science, especially in chemistry [5]. Nevertheless, he likes to point out that one of his papers [6] is rarely cited because it disproved a theory that was investigated heavily. There is no need to continue citing the proof that a theory is wrong. Falsification in science is also important. However, I don't think he disagrees with the idea that citation frequency is associated with creativity, but it is always important to note that there are exceptions. Some important discoveries are not matched with high citation. And false ones, like cold fusion, may be cited heavily but they are the exception. In general, Nobelclass work is accompanied by significantly higher citation, as we demonstrated over 30 years ago. And every Nobelist has published one or more *Citation Classics*.

IH: You have brought many of your ideas to fruition. Have there been any that did not happen?

EG: Sure. When I sold ISI, new management almost immediately emasculated certain projects. We had started *The Atlas of Science* and later changed its name to *Research Reviews*. We used the results of our global co-citation analyses to identify the newly emerging research fronts that needed to be reviewed. We published several volumes. JPT, the new partners, killed it because it still was not profitable, and it might have taken five years for it to break even. It would have been an encyclopedic treatment of current science. There are, of course, plenty of review articles published. The *Current Opinion* series, published by Vitek Tracz, came out later. He is a brilliant Polish-Russian-Israeli scientist who now lives in London. His company, Current Science, also is located in Philadelphia. He understood the mapping concept very well but, to my knowledge, neither he nor anyone else has used co-citation mapping to produce an international encyclopedia of science. Systematic examination of the literature is necessary to identify what is *not* being reviewed.

I'm on the Board of Directors of the nonprofit *Annual Reviews*, which produces about 30 annual review volumes in print and online each year. They have not used citation data as yet. Their methods for choosing topics are purely subjective—not that that is bad, just different. Their editorial boards are top-notch. Derek Price used to say that for every 50 new papers in each field, you need a review, which then becomes the paper that people cite as a surrogate for those references. One could do an interesting historical mapping based on the network of review papers.

So, returning to your original question, there are a lot of things I wanted to do that have not happened. I wanted to publish a constantly current dictionary of science. What could be a better source of new terms than the ISI database? There is constant input of new terminology. The nomenclature from indexing services is not systematically being exploited to compile dictionaries. Libraries spend a lot of energy compiling thesauri. Most of those terms eventually do get into dictionaries, but it could take many years.

Catheryn and Eugene Garfield in their home (Photo by I. Hargittai).

I would also like to see the algorithm finished for creating historiographs. The *Citation Index* is a gold mine for the history of science. Mapping all the key references for a given topic, you should be able to graphically portray the development of a field. My brother, Ralph Garner, wrote a graph theoretical description of such networks [7]. And there has been some recent work done on visualizing citation networks [8].

I find it very frustrating that so many scientists are ignorant of what they could do with information retrieval systems. I think it is important not only to be literature-minded, but to develop citation consciousness. I'm not sure how you teach this. It requires indoctrination by informed mentors.

I also wanted to use ISI Press to launch a systematic series of scientific biographies. It would have been an extension of our *Citation Classics* series. We published 4000 of those in *Current Contents*, and 2000 were reprinted in a series of books called *Contemporary Classics in Science* [ISI Press: Philadelphia, 1986].

We could have easily published many more thousands of *Citation Classics*. And there are always more recent ones to

be covered. A systematic series of biographies could include not only most-cited authors and members of the academies. Josh Lederberg was a strong supporter of this idea. The National Academy of Sciences publishes their *Memoirs*, but they appear only after members die. In addition to monographic autobiographies and biographies, I thought a journal of scientific biography would also be an interesting project.

IH: Early on, you had a meeting with J. D. Bernal. He was very much concerned with the ways of science publishing. He considered the unit of scientific publication the article, not the journal. How much impact did Bernal have on you?

EG: On the Internet we now have a preprint depository in physics and other topics. That was essentially what Bernal had in mind. He gave a paper in 1958 at the Information Conference on Scientific Information in Washington. That's where I met him for the first time.

My uncles were Marxists. One of them gave me Bernal's book *The Social Function of Science* in 1939 when I was 14 years old. It may have had some influence on me. But it was not until 1951 that I realized that he was involved in the "science of science" movement, the predecessor of scientometrics and science policy studies. He was involved in the 1947 Royal Society Scientific Information Conference. The *Proceedings* volume was my bible when I worked at Johns Hopkins from 1951 to 1953. My interest, however, was in information retrieval, not in research evaluation. Bernal was a Nobel-class scientist who might have received more recognition for his science if he had not been so openly leftist. His politics undoubtedly affected his influence. In 1962, when we had the first experimental printouts of *SCI* from the *Genetics Citation Index*, I sent samples to him, Robert R. Merton, and Derek de Solla Price. He responded very positively as did Bob and Derek.

IH: Looking back, was there anything in your family background that steered you in the direction of your future career?

EG: There were political discussions with my uncles but not much science. Only one of my uncles finished college. At first, I attended a science high school, Stuyvesant, but I left for a variety of reasons. I had no real mentors there and throughout high school. We lived in the Bronx, and Stuyvesant High School was a long subway ride to lower downtown Manhattan. And I wanted to study more foreign languages. So I transferred to DeWitt Clinton High School in the Bronx. Except for math, I was not a good student in high school. My grades were not exceptional. I still was interested in science and I wonder what might have happened if I had stayed at Stuyvesant. My regret is that I didn't encounter a scientist or teacher there who could have steered me in the right direction. And Stuyvesant was very competitive.

Eugene Garfield relaxing after the Interview (Photo by M. Hargittai).

In my last year at Clinton, I met an English teacher and former journalist, Wilmer T. Stone, who gave me some direction. Almost 10 years after I graduated, I visited him in Maryland, where he had retired, but he really didn't want to be bothered. Of course, I was not his child, just one of many students to whom he had described his experiences as a free-lance journalist interviewing Jack London, among others. He taught in high school because of the depression. When I was an undergraduate in college, none of my professors had a significant impact. At 17, I started out in chemical engineering at the University of Colorado, but it was wartime, so I left soon for San Francisco, worked in a shipyard, and eventually was drafted even though I had been accepted for the merchant marine. After the war, I returned to Berkeley. Classes were huge but I did encounter famous chemists like Joel Hildebrand and Melvin Calvin. But I was a premed student at that time and switched to chemistry later.

IH: What did your parents do?

EG: My mother was a housewife. My father became a successful newspaper–magazine distributor, but I never lived with my father. My parents separated before I was born and shortly afterward were divorced, when my sister Sylvia was 2 years old. I was 5 years old when I saw my father for the first time. And then, again, four or five years later. Our relationship is a long and sad story. My mother's oldest brother became my surrogate but absentee father. My uncle helped support us, but he never was there in person. The only time we would see him was at my grandmother's house on Friday night. He was a successful ladies' coat and suit manufacturer.

On my mother's side, they were Lithuanian Jews. I'm not sure about my father's parents. I once heard that they came from Galicia. Garfield is not my original name. It was the name of my father's firm, the Garfield News Company. My father opposed my changing my name but my uncle forced the issue since they had had a long, bitter rivalry.

My stepfather was a butcher and later drove his own taxicab. He was an Italian immigrant, so we were a nondenominational family. I was never bar mitzvahed. My half-brother Ralph was born when I was 12.

IH: There is a lot of change going on in journal publishing: the American Chemical Society is bringing out new journals and the European chemical societies are consolidating their national journals.

EG: Science is still growing so there's more capacity for journal growth. Inevitably, there is twigging of journals to accommodate new fields.

IH: You have written about the connection between publishing, impact, and the Nobel Prize.

EG: It became an interesting game. But I never tried to predict who would win the prize. It was more relevant to suggest the fields that might be recognized. We might have predicted a prize for nitric oxide. Certainly, among those names would be Salvador Moncada. Moncada was certainly among the most-cited authors. Nobel prizewinners are almost invariably well cited. The Nobel Committee didn't include him, and it has created a lot of controversy. I'm not suggesting that the committees should select on the basis of citation analysis, but they should be aware of the most cited scientists for each field considered. The same thing happened to Moncada for the Lasker Award. Something odd is going on there. I find it very strange that members of many lesser prize committees prefer to choose Nobel laureates. Why not pick someone who hasn't been so visibly recognized? I have fought this battle many times. Most award committees like to play it safe. I think awards should go to people for whom the award would make a significant career difference. Why give a lesser prize to Nobel laureates? They've already had the highest recognition. But there will always be exceptions even to that generalization.

IH: Speaking about publishing, sometimes people complain that they cannot get their message through.

EG: Hans Selye said that to get his general adaptation syndrome accepted, he published everywhere and over and over. He didn't care if he repeated his message. But take an opposite example, Eiji Osawa in Japan, who had the basic idea of what later became known as buckminsterfullerene. Did he do what he should've done to get across his message? The question is, to what extent does a scientist sell his ideas? The word "sell" is not usually used, but that's part of it. Scientists all have to get their ideas across to fellow scientists. Consider the *SCI*, for instance. It didn't happen just by itself. Long after I published my 1955 paper in *Science*, I had to publish dozens of articles and give hundreds of talks. I became a propaganda machine. Merton described this very well in his Foreword to my book *Citation Indexing* [9]. It is the same with scientific ideas. I'm awfully curious to know what was missing in the Moncada affair.

In the case of the Japanese, their problem often is that they don't learn to speak English well. So they are at a disadvantage at conferences. At least in the past, the Japanese authorities didn't insist that scientists learn to speak and write English, from an early age. If Japan wants to have its fair share of recognized scientists, they have to emphasize good linguistic skills. A lot of good work in Japan is probably underappreciated because they are so timid about promoting their ideas, especially to authoritative figures.

IH: Would you care to single out what you consider to be the most important thing you have done?

EG: To many people, *Current Contents* had the most pervasive influence. *Current Contents* is a ridiculously simple idea. Curiously, there has never been a scholarly article written about *Current Contents*. But it is still the model that has been adopted and copied. Its simplicity is what made it so successful. You say that I have a strong influence on science. Well, for a 25-year period I had a captive audience worldwide. The readership was larger than that of *Science* or *Nature*. The number of printed copies was as high as 40,000, but the average readership was tenfold that number. Some copies were read by hundreds of people in Eastern Europe and China, where they also copied it. When I went to Eastern Europe and elsewhere, people respected me because I did not attempt to criticize their political systems. I used citation data to demonstrate the relative strengths of their science. I didn't have to tell them what they knew. Rather, I provided them a window on the rest of the world. Since *Current Contents* was just a bibliographical tool, the Russian censors did not touch my essays. They allowed my essays to go through. Of course, I was proselytizing about citation indexing and not capitalism. Many people still think that I'm writing those weekly essays. Recently, I met a senior scientist who said that he loved my essays and read them every week! I wrote the last one six years ago. Maybe he's thinking of my occasional editorials in *The Scientist*. In fact, many people don't know that *The Scientist* has not been an ISI publication for 10 years.

IH: Do you have any children?

EG: I have a 52-year-old son, Stefan, who is a crane operator. My second son, Joshua, 40 years old, graduated in marine biology but is now a computer scientist. Both live in Florida. I had two daughters, Laura and Thea. Laura is 41, but I don't hear from her. My younger daughter committed suicide 20 years ago. I have a stepdaughter, Cornelia, who lives in Philadelphia and we visit regularly. From my third marriage, I have a 14-year-old son, Alexander Merton, who is a violinist and a good student in math and science. My wife, Catheryn, originally taught biology. Then she got an information sci-

ence degree and worked at ISI as a lecturer. Eventually, she became Vice President but left ISI after we sold the company to Thomson. We have been married for 16 years.

IH: How would you formulate the lessons to be learned from your career?

EG: Too often, people are afraid of failure. They worry that they cannot manage financially. Money never drove me; it came to me. Nevertheless, if I had worried about money, I might never have achieved financial success. I don't know what accounts for this quality of persistence. My mother never stopped until she finished the task at hand. You learn a certain doggedness. I grew up working. When I was 9, I was delivering orders in a grocery store and worked in a laundry for hours just to earn a quarter. Later, I went to work for my uncles. I delivered orders in my Uncle Lou's liquor store. Then I worked in the garment district after school and summers. Maybe that was another reason that I didn't do that well in high school. I certainly enjoyed the work but realized I didn't want to remain in the garment business, much as my uncle Sam would have liked me to.

IH: How did your Ph.D. happen to be in structural linguistics?

EG: I got my B.S. in chemistry from Columbia. I had a good friend who was working on mechanical translation of Russian at Georgetown University. I was supposed to join him there. However, I was broke and had to support my son. I got sidetracked by some people from Smith Kline and French. I met them when I was at Johns Hopkins. They offered me a consulting job to set up a punch card system on thorazine. That's why I came to Philadelphia in 1954. My friend Casimir Borkowski later came to Philadelphia to get his Ph.D. under Zelig Harris, the chairman of structural linguistics at Penn. Noam Chomsky was graduating that year. I introduced Zelig Harris to the field of information retrieval. Within a few months, he had a half a million dollar grant from the National Science Foundation.

I had started my doctoral program at Columbia but couldn't get my interdisciplinary committee to meet. So I left and subsequently made a deal with Professor Harris. He agreed that I could transfer my credits and do my dissertation in chemical linguistics. My task was to create an algorithm for translating chemical nomenclature into molecular formulas using a computer. Today it seems ridiculously simple but in those days it seemed impossible. That's how I got my degree in the linguistics department. It was as much a chemistry topic as structural linguistics. Allan Day, Chair of Chemistry, was a great help to me.

Later on, I taught at Penn in the electrical engineering school. I gave a course in information retrieval for computer and information science graduate students. Many of them worked on Department of Defense contracts involving information retrieval.

IH: Do you keep track of citations of your own work?

EG: I have received a weekly printout for over 30 years, which lists every paper that cites my work. Because of my essays in *Current Contents*, I am probably the most self-cited person in the world. There are still quite a few papers published that cite my papers and books, but lately impact factors are very popular. I've posted almost all my work full-text on my web site [http://165.123.33.33/eugene_garfield] and that's a good place to end.

References

1. (a) Garfield, E. "How to Use Citation Analysis for Faculty Evaluations, and When Is It Relevant? Part 1," *Current Contents* **1983**, No. 44, p. 5–13. Reprinted in *Essays of an Information Scientist*, Vol. 6; ISI Press: Philadelphia; 1984; pp. 354–362. (http://www.the-scientist.library.upenn.edu/eugene_garfield/essays/v6p354y1983.pdf); (b) Garfield, E. "How to Use Citation Analysis for Faculty Evaluations, and When Is It Relevant? Part 2," *Current Contents* **1983**, No. 45, 5–14. Reprinted in *Essays of an Information Scientist*, Vol. 6; ISI Press: Philadelphia, 1984; pp 363–372. http://www.the-scientist.library.upenn.edu/eugene_garfield/essays/v6p363y1983.pdf

2. Braun, T.; Glänzel, W. "The Sweet and Sour of Citation Rates," The Chemical Intelligencer 1995, 1(1), 31–32.

3. Van Leeuwen, T. N.; Moed, H. F.; Reedijk, J. "JACS Still Topping Angewandte Chemie: Beware of Erroneous Impact Factors," The Chemical Intelligencer 1997, 3(3), 32–36.

4. Spiridione, G.; Calza, L. *Il Peso Della Qualita Accademica [The Weight of Academic Quality]*; Cooperativa Libraria Editrice Universita di Padova: Padua, 1995.

5. Czapski, G. "The Use of Deciles of the Citation Impact to Evaluate Different Fields of Research in Israel," Scientometrics 1997, 40(3), 437–443.

6. Czapski, G. "Radical Yield as a Function of pH and Scavenger Concentration," *Adv. Chem.* **1968**, *81,* 106–130.

7. Garner, R. "A Computer Oriented, Graph Theoretic Analysis of Citation Index Structures." In Flood, B., Ed.; *Three Drexel Information Science Research Studies* Garner, R.; Lunin, L.; Baker, L. Drexel Press: Philadelphia, 1967; pp 1–43.(http://165.123.33.33/eugene_garfield/rgarner.pdf)

8. Mackinlay, J. D.; Rao, R.; Card, S. K. "An Organic User Interface for Searching Citation Links," presented at CHI Conference on Human Factors and Computing Systems, Denver, May 7–11, 1995. (http://turing.acm.org/sigchi/chi95/proceedings/papers/jdm_bdy.htm#colorplate1)

9. Merton, R. K. Foreword to: Garfield, E. *Citation Indexing: Its Theory and Application in Science, Technology, and Humanities*; ISI Press: Philadelphia, 1979; pp v–ix. (http://165.123.33.33/eugene_garfield/cifwd.html)

Part II

Notes

Notes

The Naming of Buckminsterfullerene[a]

E.J. Applewhite[b]

Systematic chemical nomenclature has always been corrupted—or enhanced, depending on your point of view—by the prevalence of eponyms. The fact that C_{60} was named buckminster-fullerene could be construed as (a) an erratic departure from the etiquette of attributing discoveries to individuals, (b) trivial, or (c) the validation of an intuitive vision of a designer of geodesic domes. H.W. Kroto said that the newly discovered carbon cage molecule was named buckminsterfullerene "because the geodesic ideas associated with the constructs of Buckminster Fuller had been instrumental in arriving at a plausible structure" [1]. It is becoming, in Fuller's case, that he made no claim; the honor was bestowed by others.

The Israeli poet Yehuda Amichai once described naming as "the primary cultural activity," the crucial first step anyone must take before embarking on thought. John Stuart Mill declared that "The tendency has always been strong to believe that whatever received a name must be an entity or being, having an independent existence of its own."

When Harry Kroto and Richard Smalley, the experimental chemists who discovered C_{60}, named it buckminsterfullerene, they accorded to Richard Buckminster Fuller (1895–1983), the maverick American engineering and architectural genius, a kind of immortality that only a name can confer—particularly when it links a single historical person to a hitherto unrecognized universal design in the material world of nature: the symmetrical molecule C_{60}. Smalley's laboratory equipment could only tell them how many atoms there were in the molecule, not how they were arranged or bonded together. From Fuller's model they intuited that the atoms were arrayed in the shape of a truncated icosahedron—a geodesic dome. Only after a novel phenomenon or concept is named can it be translated into the common currency of thought and speech.

This newly discovered molecule, a third allotrope of carbon—ancient and ubiquitous—transcends the historical or geographical significance of most named phenomena such as mountains of the moon or Antarctic peaks and ridges. Cartographers named two continents for Amerigo Vespucci, because he asserted (as Columbus did not) that the coasts of Brazil and the islands of the Caribbean were a landmass of their own and not just obstacles on the route to Asia. C_{60} is a far more elemental discovery; it is more ancient; and it pervades interstellar space. Fuller has no reason to envy Vespucci.

Buckminsterfullerene was discovered by chemists who were not looking for what they found. Kroto was looking for an interstellar molecule. Smalley said he hadn't been very interested in soot, but they agreed to collaborate. Smalley's laboratory at Rice University had the exquisite laser-vaporization and mass spectrometry equipment to describe the atoms of newly created molecules. Scientific experimenters investigate nature at a level where revelation is often unpredictable and sometimes capricious. This is a phenomenon that Fuller (who was not a scientist, but a staunch defender of the scientific method) generalized into the dogmatic statement that all true discovery is precessional. For Fuller, the escape from accepted paradigms is precessional. (Vespucci precessed; Columbus did not.) Fuller had a life-long preoccupation with the counter-intuitive, gyroscopic phenomenon of precession. He defined precession, quite broadly, as the effect of bodies in motion on other bodies in motion. Every time you take a step, he said to me many times, you precess the universe.

[a]Chemical Intelligencer 1995(3), 52–54.

[b]1517 30th Street NW, Washington, DC 20007, USA
e-mail: edapple@aol.com (deceased)

For that matter, one may say that Kroto and Smalley, in recognizing the shape of the C_{60} molecule made a precessional discovery. Earlier, Osawa, in a paper published in Japanese in 1970, had described the C_{60} molecule with the truncated icosahedral shape; so had Bochvar and Gal'pern in 1973 when they published a paper in Russian on the basis of their calculations. They all recognized the novelty of the molecule and conjectured that its structure should afford great stability and strength. However, neither Osawa nor Bochvar and Gal'pern had experimental evidence, nor did they consider their result important enough to follow up their finding with further work or to convince others to do so. Curiously, in 1984 a group of Exxon researchers made an experimental observation of C_{60} along with many other species. They failed, however, to discern the shape of this species and did not recognize its special importance. These precursors to Kroto and Smalley apparently lacked the requisite—precessional—insight to appreciate the significance of what they had found. Kroto and Smalley's precessional insight was best manifested by their decision to give a name to the C_{60} molecule of the truncated icosahedral shape.

Buckminster Fuller and E.J. Applewhite at a midnight supper at the Waldorf-Astoria Hotel celebrating their completion of the final galleys of Synergetics. *Fuller in his own hand has inscribed the mat of this photograph as follows: "Entering the home stretch of the ½-century long,* Synergetics *galley race. B.F. and E.J.A. Jr. at the Waldorf Jan 9, 1974."*

As a longtime close friend of the Fuller family, as his collaborator on his two volumes of *Synergetics* (1975, 1979), and as a trustee of the Buckminster Fuller Institute (BFI), I rejoiced vicariously in the molecular celebration of his name. I preserved the copy of its first publication in *Nature* (November 1985), with the C_{60} molecule on its cover, and, with the compulsion of an archivist, I documented the proliferation of reports on this molecule in the professional literature for some while thereafter. While I sensed that Professors

Kroto and Smalley had granted the name for perhaps trivial reasons, I felt that there was a greater resonance between C_{60} and Fuller's writings and design philosophy than the mere congruence of the topology of that molecule and Fuller's geodesic domes. Fuller did not develop his peculiar geometry in order to build a dome. Of course, he delighted in building domes and built a great many of them (though all were replicable, no two of his prototypes were the same), and he succeeded admirably in containing a greater volume of space in an enclosed stable structure than any architect or engineer before him had ever done. (He had a dozen or so patents relating to his domes.) But I knew that Fuller was one of the most celebrated but least understood original thinkers of his day. Fuller did not develop his original great-circle coordinate geometry in order to build domes; he built domes because otherwise people would not understand the geometry—which rejected the *XYZ* coordinate system of standard mensuration. He advanced synergetics as nothing less than a new way of measuring experience and as a new strategy of design science which started with wholes rather than parts.

Although I felt that it was presumptuous for me, as a nonscientist, to address Kroto and Smalley on Fuller's behalf, I nevertheless offered them copies of Fuller's *Synergetics* books and drew their attention to collateral aspects of Fuller's work that might be relevant to their major discovery. I was careful to disavow any claim for priority of discovery on Fuller's behalf. He did not anticipate C_{60}, but its discovery did validate his intuitions that geodesic design plays a more significant role in nature's arrangements than had hitherto been recognized. Fuller would have been less surprised than any of us to learn that the 60-atom array possessed an extraordinary property of stability. Although he regarded the hydrogen atom as the simplest—and hence the most beautiful—design in nature, Fuller had a lifelong interest in the carbon atom, and, in many of his writings and lectures, he celebrated J.H. van 't Hoff's 1874 concept of the tetrahedral configuration of carbon bonds.

Some years later, on March 21,1991, on a visit to Houston, I had the opportunity to call on Professor Smalley in his laboratory at Rice University and pay him homage, specifically on behalf of the Fuller family and the BFI—expressing our gratification in the luster that he and Professor Kroto had added to Fuller's name. He greeted me with a hospitality, a sympathy, and an enthusiasm matching the cordiality of the correspondence I had initiated with Professor Kroto at the University of Sussex in Brighton. A sense of destiny permeates his large, comfortable office; he told me I was sitting on the very couch where he and Kroto christened the new molecule on September 9, 1985. He told me about how he and his colleagues had sat up all night making models out of Gummy Bear jelly beans and paper cutouts of pentagons and hexagons. I recalled that Fuller as a child had made models out of toothpicks and dried peas, and he had always felt that geometry should be taught as a hands-on laboratory discipline. Smalley said that he had overcome any initial reservations he might have had to Kroto's proposal to name C_{60} buckminsterfullerene. For one thing, the standard IUPAC name for the molecule was impossibly awkward and difficult to read, much less speak. When I asked him why he found the name so appropriate, he said that it was because it conveys in a single word so much information about the shape of the molecule, and he found a happy congruence in the fact that its 20 letters match the 20 faces of the icosahedron—a letter for each facet. All even-number carbon cluster-cage molecules are now termed fullerenes. The root name Fuller lent itself to generic applications with the various other conventional suffixes, producing not just fullerenes, but fulleranes, fullerenium, fullerides, fullerites, fulleroids, fulleronium, metallofullerenes, and so forth. Colloquially—even affectionately—they are subsumed as buckyballs.

As Smalley escorted me out of the laboratory complex on that steaming hot March afternoon (Houston is like that), I was exhilarated by his conviction that C_{60} is one of the most stable and photoresistant molecules known to chemistry, and also probably the most proliferating, and possibly the oldest. A new branch of organic chemistry indeed—and countless textbooks had instantly been rendered out of date.

After a few letters objecting to the name buckminster-fullerene had appeared in the columns of *Nature*, Harry Kroto gallantly defended its choice on the grounds that no other name—none of the forms of the classic Greek geometers—described the essential three properties of lightness, strength, and the internal cavity that the geodesic dome affords. To the protest that nobody had ever heard of Fuller, he submitted that the name would have educational value. A fine exercise of onomastic prerogative.

Fuller was not a chemist. He was not even a scientist, and made no pretension of adhering rigidly to an experimental and deductive methodology, and he did not follow the rules of submitting published papers to peer review. But he had an extraordinary facility for intuitive conceptioning. Jim Baggott, in his superb account *Perfect Symmetry: The Accidental Discovery of Buckminsterfullerene* [2] quotes Fuller in an epigraph: "Are there in nature behaviors of whole systems unpredicted by the parts? This is exactly what the chemist has discovered to be true." Baggott goes on to describe how Fuller had derived his vector equilibrium (cuboctahedron, in conventional geometry) from the closest packing of spheres of

energy. What he had was a principle that led to the design of geodesic structures capable of a strength-to-weight ratio impossible in more conventional structures. Fuller had a highly generalized definition of the function of architecture that put him outside the scope of the academicians' view of their discipline. Bucky said "architecture is the making of macrostructures out of microstructures."

Baggott concludes: "Fuller's thoughts about the patterns of forces in structures formed from energy spheres had led him to the geodesic domes....That his geodesic domes should serve as a basis for rediscovering these principles in the context of a new form of carbon microstructure has a certain symmetry that Fuller would have found pleasing, if not very surprising."

References

1. Kroto, H. W. *Nature (London)* **1987**, *329,* 529.
2. Baggott, J. *Perfect Symmetry: The Accidental Discovery of Buckminsterfullerene;* Oxford University Press: Oxford, 1994.

E.J. Applewhite grew up in Newport News, Virginia, except for two years spent in Tahiti. He graduated from Yale University in 1941 and later went to the Harvard Business School. After Navy service on an aircraft carrier, he worked with Fuller in a housing project in Wichita, Kansas. He joined the CIA in 1947 and served in Bonn and Beirut. In 1979 he completed nine years of collaboration with Fuller on his *Synergetics* books. He describes himself as a layman.

Notes

Peace Through Chemistry[a]

István Hargittai[b]

Last November during a lecture tour in the Washington, DC area, I visited the National Gallery. One of the temporary exhibitions was a showing of Roy Lichtenstein's works. I had known some of his pictures either as reproductions or from visits to modern art museums. His unique style was strongly imprinted in my mind, the large-scale comic-strip-like paintings with bright colors in which benday dots were visible in an exaggerated way. There were some striking examples of such paintings in this exhibition. I noticed, however, a large picture, consisting of three adjacent panels, of a different style. It differed from the rest in being rigorously geometrical, although the Lichtenstein trademark benday dots and the bright colors were there, too.

I liked the picture as soon as I noticed it from a distance. I liked its geometry and colors, and then, as I got closer, I was completely taken by its title, *Peace through Chemistry.* It was painted in 1970.

Chemistry has such a bad popular image, due most of all to ignorance about it among nonscientists, that this was a pleasant surprise. Here was a major artist and such a title and a truly beautiful piece of art.

Following this visit to the National Gallery, I wrote a letter to Mr. Lichtenstein, and eventually got a color transparency for reproducing the painting in our magazine. I also got some information about the painting through the artist's kind assistant, Ms. Shelley Lee. I sent the questions and received the answers. This is why there are no follow-up questions where the reader might feel there should have been some.

ISTVÁN HARGITTAI (IH): How did Mr. Lichtenstein come to the idea of making this picture?

SHELLEY LEE (SL): He was thinking about the WPA [Works Project Administration] murals done in the United States in the 1930s on public buildings. *Peace through Chemistry* is similar in style to the painting *Preparedness*, 1968, which is owned by the Guggenheim Museum. Both paintings were part of his Art Deco series.

IH: Did he have any previous encounter with chemistry?

[a]*Chemical Intelligencer* 1995(3), 55–56.

[b]Department of Inorganic and Analytical Chemistry, Budapest University of Technology and Economics, Szt. Gellert ter 4, 1111 Budapest, Hungary
e-mail: istvan.hargittai@gmail.com

B. Hargittai and I. Hargittai (eds.), *Culture of Chemistry: The Best Articles on the Human Side of 20th-Century Chemistry from the Archives of the Chemical Intelligencer,* DOI 10.1007/978-1-4899-7565-2_4, © Springer Science+Business Media New York 2015

ROY LICHTENSTEIN *Peace through Chemistry, 1970; oil & magna on canvas; 3 panels; 100" × 60", overall: 100" × 180"; 0440. Used with permission © Estate of Roy Lichtenstein.*

SL: He studied chemistry at de Paul University in Chicago while in the Army, as part of an engineering course in 1942.

IH: Where did the title originate from?

SL: Preparedness was a theme of government during World War II and the Cold War. Of course, in spite of his serious interest in science, the title is meant to be ironic.

IH: How did Mr. Lichtenstein create the picture? Did he have some tools as models for the picture?

SL: He was influenced by Art Deco, but the painting was all thought up. There was no original source.

IH: Is there any meaning in the change of style stressing geometry?

SL: The style is Art Deco, and the geometry is connected with Art Deco style. He thought it was fitting for the government-sponsored mural style.

IH: Has he created any other picture related to chemistry?

Roy Lichtenstein (Used with permission © Estate of Roy Lichtenstein).

SL: He created several versions of *Peace through Chemistry* in prints done with Gemini G.E.L. in Los Angeles.

IH: Has he created any other picture related to any other fields of science?

SL: No.

Some biographical data on Roy Lichtenstein: He was born in 1923 in New York City. Attended summer art classes in 1939 in New York, and studied in the School of Fine Arts, Ohio State University, 1940 to 1943. Served in the U.S. Army in Europe from 1943 to 1946. Obtained his BFA (1946) and MFA (1949) at the Ohio State University. He taught there in 1949 to 1951. He lived in Cleveland working as a graphic and engineering draftsman among other jobs in 1951 to 1957. Worked as Assistant Professor first at the State University of New York at Oswego, and then at Douglass College of Rutgers University until he moved to New York City in 1963.

He was inducted as Fellow into the American Academy of Arts and Sciences, Boston, in 1971, and since then received numerous recognitions, including a Doctorate of Fine Arts from the California Institute of the Arts (1977), membership in the American Academy and Institute of Arts and Letters, New York (1979), and honorary doctorates from Southampton College, New York, Ohio State University, Columbus, and Royal College of Art, London. He has had numerous one-man exhibitions since 1951.

The ironic overtone in the title *Peace through Chemistry* might almost suggest that I scored a goal against my own team in "discovering" this painting. I don't think so, though. The painting is beautiful and it is about chemistry. Also, in the context of the rest of his works, it could not be vintage Lichtenstein without some humor, without some irony.

Notes

Food for Thought[a]

Alan L. Mackay[b]

Eating together is the most basic and important social activity, and the enjoyment of eating, and even the mere process of ingestion and digestion, is connected with one's mental disposition. Only now is it beginning to be understood, through delicate magnetoelectric recording techniques, "how the brain and the gut talk to each other." Very recent work shows that the gastric–intestinal tract even seems to be differently lateralized in the brains of different individuals. To eat comfortably goes with social ease. The opposite, social dis-ease or disquiet, shows in the phrase "I cannot stomach his opinions." Eating together is therefore one of the basic social institutions of science. If eating is mixed with talking, then nobody can talk all the time. A practical everyday version of this is the department tea club, of which Aaron Klug remarked "visiting Americans often think that the tea and coffee breaks [at the MRC Laboratory for Molecular Biology] are a waste of time, but some of them learn better" [1].

Commensality implies equality and democracy and the practice of the dialectic—the emergence, if not of truth, then of shared opinions and the recognition of differences, from the expression of argument and counterargument. Even Jesus ate with publicans and sinners, no doubt because the conversation as well as the drink was better.

In Plato's Symposium, which was literally drinking together, the duty of the chairman was to regulate the talk by adjusting the ratio of alcohol to water. If the talk was too slow, then more alcohol, and conversely. The Ciba Foundation is a master of the technology of the modern symposium, arranging more parameters. There are not too many participants; they have just the right amount of food and drink; the temperature is right; the seating is right—everyone can see and hear and be seen and heard; the projection is perfect. Few other organizations take such care, but the results justify it.

It is not a coincidence that the most developed institution of social dining is associated with the great scientific distinction of Cambridge University. For example, the Fellows of Trinity College may mention that Trinity College has produced as many Nobel prizewinners as France. In discussions with Chinese colleagues, for example, about how science is organized in China and how top rank scientists may be encouraged, the question "When do your professors eat together?" almost uniformly led to the answer "almost never." (For their part, the Chinese scientists were anxious to understand the role of the Trinity in Trinity Science Park—consulting the dictionary had only increased their mystification.)

In 1912 P.P. Ewald, A. Sommerfeld, and Max von Laue were drinking in the Café Lütz in Munich and discussing physics when von Laue proposed an experiment, based on Ewald's recent thesis, to see whether crystals diffracted X-rays. No doubt they penciled diagrams on the marble-topped table—every café for scientists needs something to write on. This experiment was quickly done by von Laue's students Friedrich and Knipping, and von Laue duly received the Nobel Prize for it. Moreover, a string of other Nobel Prizes followed directly from this discovery.

Perhaps the most important dining table ever was that of Baron d'Holbach in Paris, talk around which shaped the intellectual climate of Western Europe in the eighteenth century and nourished the Enlightenment [2]. The Baron entertained about 20 people to dinner at his house in the Rue Royale every Thursday and Sunday for 35 years (from 1750 to 1785). This was a subsidy to learning and culture of immense value. There was a constant nucleus with a fluctuating stream of visitors. The British visitors included John Wilkes, David Hume, Lawrence Sterne, Horace Walpole,

[a]*Chemical Intelligencer* 1996(1), 52–54.

[b]Department of Crystallography, Birkbeck College, University of London, Malet Street, London WC1E 7HX, UK

Samuel Romilly, David Garrick, Allan Ramsay, Edward Gibbon, Dr. Burney, and Joseph Priestley. There was also Benjamin Franklin from America. Besides Holbach, the key nucleus comprised Denis Diderot, the most celebrated figure, Friedrich-Melchior Grimm, the Abbé Raynal, and the Abbé Galiani. Although this "Coterie Holbachique" was reputed to be a nest of atheists, and atheist views were freely expressed, atheism was not pressed to the disadvantage of other views. There was, to use a modern phrase, no hidden agenda. Many of those attending contributed to Diderot's gigantic project of the "encyclopedia of the arts and technologies" [*Encyclopédie ou Dictionnaire Raisonné des Sciences des Arts et des Métiers,* 1751–1772], but this was not central to the group. The arts, the sciences, and politics were all discussed. People could speak and speculate without fear in the company of exciting minds in the "inviolable sanctity of the haven where they assembled."

H.G. Wells has reported on the influential dining club "The Coefficients," founded by Mrs. Beatrice Webb, where long-term geopolitical discussion about the Empire took place monthly from 1902 to 1908. The formidable Beatrice Webb had written: "You do not, as a matter of fact, get to know any man thoroughly except as his beloved and his lover—if you could have been the beloved of the dozen ablest men you have known it would have greatly extended your knowledge of human nature and human affairs" [3], However, this ambitious project was beyond her, and her dining club provided in substitution a deep familiarity with political affairs. J.D. Bernal has described the "Tots and Quots" (from "Tot homines, quot sententiae"), a club mostly of scientists who met once a month in wartime London around 1939/1940. From its discussions, ideas about how to manage the war emerged. Allen Lane, the publisher of Penguin Books, said that if Bernal and Solly Zuckerman would produce a text, then he would publish it in a month, and they did. The book was *Science in War* (1940), and a copy was at each place for the next month's meeting.

Earlier, Bernal was a member of The Kapitsa Club, which met in Cambridge in the 1930s, an informal "interdisciplinary" group at whose meetings participants heard about every scientific discovery as soon as it happened, or sometimes before. This was a typical "invisible college" at the growing front of science. It is an important character of such groups that everyone is interested in everything, like the ancient Athenians in the *agora* (Greek for the Latin *forum*)—"For all the Athenians and strangers which were there spent their time in nothing else, but either to tell, or to hear some new thing" [Acts 17:21]. This is the way to get new ideas for solving old problems.

What a pity then that, in the present period of cost-effectiveness and ranking of university departments by the citation index, leisured regular dining has gone by the board. Indeed, the commensal board has been replaced by the coffee machine. In the old days, when everyone in the laboratory knew what everyone else was doing, one could see beakers on the shelf with notices: "Brucine crystallization—do not use for coffee!" Everyone knew the significance. But today there might be people around who do not even know what brucine does. Staff meet only for committees and judge each other by bureaucratic statistics of numbers of papers and weight of money brought in. Sydney Brenner says that "Before we develop a pseudo-science of citation analysis, we should remind ourselves that what matters absolutely is the scientific content of a paper and that nothing will substitute for either knowing or reading it" [4].

The global market economy affects science, and an important analysis of the way in which the work of Third World scientists is cut off from the body of world science by a selfish scientific Mafia, using the science citation index from a distance, has just appeared [5]. In this way, we avoid meeting a large part of the world face-to-face at a communal table. If we do not know of their work, then we do not need to acknowledge their existence. "Science is public knowledge"—this is Merton's aphorism [6]. As in the Matthew Principle, also enunciated by Robert Merton: "To him that hath, shall be given. From him that hath not. shall be taken away even that which he hath" [Matt. 13:12].

Proper estimates of our colleagues develop only after long discussion of actual problems based on reading their work. There is nothing more interesting than professional shop talk, but with how many people can we productively interact? Nowadays, through the Internet, we may communicate with hundreds and realize many aspects of H.G. Wells' *World Encyclopedia* [7], but how many can we really get to know as well as if we had dined with them twice a week for a decade? The administrators have carried out the classic military tactic of cutting their opponents into small units and demolishing each in isolation. The captains do not now eat with the crew, and they find out what the crew are doing only from "management statistics" and not by discussing actual research face to face over dinner.

What hurried, mechanized eating in "fast food" cafeterias does for the digestion is another question, but it is a matter for reflection that half the profits of Britain's largest company come from the sale of copies of one molecule, ranitidine hydrochloride [8], for stomach ulcers.

References and Notes

1. *Sunday Times* (London), October 24, 1982, p 17.
2. Kors. A. C. *D'Holbach's Coterie:* Princeton University Press: Princeton, New Jersey, 1976.
3. Dickson, Lovat. *H. G. Wells;* Penguin, 1969; p 115.
4. Brenner, S. *Curr. Biol.* **1995,** *5,* 568.
5. Gibbs. W. Wayt. *Sci. Am.* **1995** (August), 76–83.
6. Merton, R. *Science, Technology, and Society in Seventeenth Century England;* Festig: New York, 1938; p 219.
7. There was no such book. *The World Brain* and *The World Encyclopedia* were ideas of H. G. Wells which he assiduously propagated in the 1930s. In many ways he anticipated the Internet.
8. Ishida. T.; In, Y.; Inoue, M. *Acta Crystallogr., Sect. C* **1990,** *46,* 1893–1896.

Notes

Hans Hellmann of the *Hellmann–Feynman Theorem*[a]

Michail A. Kovner[b]

Hans Georg Gustav Adolf Hellmann was born on October 14, 1903, in Wilhelmshaven, Germany. He got his doctorate in engineering as a student of V.E. Regener in 1929 in Stuttgart. He worked in Stuttgart, Kiel, and Hannover, last teaching physics in Germany at the Hannover Veterinarian College. From 1934 he was Professor of Theoretical Physics at the Karpov Institute of Physical Chemistry in Moscow. He became a Soviet citizen in 1936 and received various Soviet awards and decorations.

Hellmann was the author of the first quantum chemistry book, *Quantenchemie,* which was translated into Russian by myself and colleagues. By doing so, we had to create the Russian terminology for quantum chemistry [1].

Hellmann [2] formulated the famous Hellmann-Feynman theorem [3], to which the American physicist Richard Feynman came independently [4] and which has retained its fundamental importance in molecular dynamics.

Hellmann organized a quantum-chemical seminar in Moscow which became popular, especially among young chemists. I prepared my dissertation, "Quantum Theory of the Ammonia Molecule," under Hellmann's supervision. I was often a house guest of the Hellmanns, and I knew his wife and his little son well. Hellmann was an antifascist, and he hated Hitler and anti-Semitism. His wife was Jewish, and that is why they left Germany and came to Moscow as immigrants.

At the beginning of 1938, Hellmann was charged by the Soviet authorities as a German spy, and soon after these accusations he perished. After Hellmann's arrest his son was placed in an orphanage and his wife was sent into exile in Kazakhstan where she remarried. Because the little Hellmann did not have any papers, he got a new name, Minchin. He grew up in Kharkov and graduated from the Mining Institute. Eventually he also found his mother.

One day in 1988 a middle-aged person appeared at the door of my home. He was Hellmann's son and had come to visit. He was visiting all his father's former pupils in Moscow. Then one day, on another visit, he told me that he had come to say goodbye. He was leaving for Germany. I never heard from him again.

References

1. Gel'man, G. *Kvantovaya Khimiya;* Glavnaya Redaktsiya Tekhniko-Teoreticheskoi Literaturi: Moscow, Leningrad, 1937. Hellmann's name was transliterated into Russian, and its reverse transliteration now produces G. Gel'man. Hellmann makes reference in this book to his course in 1935–36, which served as partial basis for the book. The original text was prepared in German, and it was translated into Russian. Hellmann also mentions in the Foreword that he was planning to publish a somewhat abridged and revised version of this book in German in 1937.
2. Hellmann. H. *Einführung in die Quantenchemie;* Deuticke: Leipzig, 1937.
3. According to the Hellmann–Feynman theorem, for the exact wave function the energy gradient is equal to the expectation value of the derivative of the Hamiltonian, $E^a = \langle \Psi | H^a \Psi \rangle$; see, e.g., Pulay, P. In *Ab Initio Methods in Quantum Chemistry—II;* Lawley, K. P., Ed.; *Advances in Chemical Physics,* Vol. 69; Prigogine, I., Rice, S. A., Series Eds.; John Wiley & Sons: New York, 1987.
4. Feynman, R. P. *Phys. Rev.* **1939,** *56,* 340.

[a] *Chemical Intelligencer* 1996(1), 54–55.

[b] Institute of the History of Science, Russian Academy of Sciences, Staropansky pereulok, 1/5, 103012 Moscow, Russia

Notes

The Difference Between Art and Science[a]

Roald Hoffmann[b]

1

It's the land of the Slinky, warped mirrors,
the seemingly misfit gears of eccentric
motion. It's the modern science museum.
You would have to drop a crowbar on a gong
to hear it above decibels of ten-year old
visitors. The masterpieces—Planck's quantum
hypothesis, the quinine synthesis—are missing.
Only the photos of the makers, the tangible,
billboards explaining the mystery of common sense.
In the hushed temple of high art one is moved
from the discreet space carved out by a Simone
Martini to the Master of the Urbino Annunciation.
It's the untouchable preserve of patrimony,
cautiously labelled for the farsighted, all
masterpieces, at least until deaccessioned.
But there it hangs, my Crivelli with a fly,
in the palace of unique resolutions, once done
waiting patiently to be done again, differently.

2

He thinks of the unique
molecule friends in Moscow made,
tin in the middle,
linked to two niobiums, two chlorines.
Around tin, like carbon, there should be

a rough tetrahedron,
but that ancient figure opens an angle wide
vs. the opposing one.
So he puzzles with a student
who tweaks the supple molecule in the computer,
gauging its resistance until
from the electrons' chanced clouds, inner space,
the reason snaps clear.
So that one could kick oneself
for not having seen
how unexceptional
it really
is.

She takes the common,
here young eucalyptus,
and with neat saw-cuts sketches
the aura of its absent leaves and trunk.
She hard-wires its give
into a limber lattice-work of chambers
partially open, the pliant mystery
of shaped emptiness passing
through emptiness,
tough for simple space to bear.
A burl of the giving mind, out
of the ordinary, no one
like any other one.

[a]*Chemical Intelligencer* 1996(1), 55.

[b]Department of Chemistry, Cornell University, Ithaca, NY
14853-1301, USA

B. Hargittai and I. Hargittai (eds.), *Culture of Chemistry: The Best Articles
on the Human Side of 20th-Century Chemistry from the Archives of the Chemical Intelligencer*,
DOI 10.1007/978-1-4899-7565-2_7, © Springer Science+Business Media New York 2015

3

for Jorge Calado

From this Munch painting
of someone pained on a bridge,

hands held to ears, the observer
could scrape an orange

micron speck. He could
mount it on a slide, fine-tune

the fast beams that circle
under parking lots and football

fields, prodded on by magnets'
handless shove, focus, for that

is his craft, the probe particles
(fancy calibrated stones)

to jarring graphed impact
in the paint. The search

is for the force of the scream.
But the particles' pry is

too strong—they shock loose
the paint molecules, in sound

demonstration of the uncertainty
principle. The painting hangs;

Norwegian sky and harbor
pick up the scream, beam

it into the observer's skull.
There, echoing, effect change.

*This poem has a publication history.
Part 1 appears here for the first time.
Part 2 was published in the literary journal*
Webster Review *1989,* **14(2),** *155. Part 3 was
published in my first book* **Gaps and Verges,**
*and has been republished in French,
Portuguese, and Spanish translations in*
Nouv. J. Chim., CTS *(Lisbon), and*
La Republica *(Montevideo), and in English
in* **Chemisch Weekblad.**

Notes

A Memory of Dorothy Hodgkin[a]

Sir John Cornforth[b]

My wife Rita and I were good friends with Dorothy Hodgkin for 50 years. Dorothy was, technically, Rita's tutor at Somerville College; but we first got to know her better in 1943 during the chemical work on the structure of penicillin. We did a chemical synthesis of penicillamine, a degradation product of penicillin, and Dorothy did the X-ray work that proved conclusively the identity of the synthetic and the natural material. After penicillin, Dorothy took up work on the structure of calciferol, the D vitamin. She wanted a crystalline derivative containing a heavy atom, and this was proving difficult to obtain. She enlisted our help (via Robert Robinson, I think), and after failing to get a crystalline iodobenzoate, we tried the 3-nitro-4-iodobenzoate, then a new acyl group. Rita did one experiment with 0.1 g of calciferol (it was precious at the time) and got a crystalline derivative with which Dorothy solved the structure. There was an amusing sequel, some years later. A Frenchman, L. Velluz, showed that in the preparation of calciferol from ergosterol by ultraviolet irradiation the final stage, from precalciferol to calciferol, is a reversible thermal process. So in theory, since Rita heated her reaction mixture when making the derivative, she could have obtained a precalciferol derivative and the X-ray work could have given a wrong structure for calciferol. Dorothy wrote to us about this dilemma (we were at Mill Hill by then), and I said I'd make some more of the derivative and hydrolyze it, without heat, back to calciferol. I made some more of the iodonitrobenzoyl chloride and started work, and I had 10 g of purest calciferol to play with, and I couldn't get any crystals at all! Finally, in despair, I wrote to Dorothy and asked if she had any of the original specimen left. She said no, but she sent me the tiny tube that had contained Rita's specimen. I touched with a glass rod first the inside of the tube and then my specimens; they all crystallized rapidly. And the derivative was the same as Rita's, and it did give calciferol when saponified at 0 °C. So that was a happy ending, but it emphasized that some crystallizations are games of chance, and in this case we must have been lucky to get crystals the first time. We wrote an appendix to her full paper [1] on calciferol, describing the derivative and the method for making the reagent.

Dorothy Hodgkin, O.M., F.R.S., by Bryan Organ (From Ref. 2, reproduced with permission from The Royal Society, London).

[a]*Chemical Intelligencer* 1998(4), 57–58.

[b]School of Chemistry, Physics & Environmental Science, University of Sussex, Falmer, Brighton, East Sussex BN1 9QJ, UK (deceased)

We used to meet at the Royal Society from time to time, and we corresponded at intervals. And I was instrumental in raising money from Fellows and others to have her portrait painted for the Royal Society. It hangs in the restaurant there; she didn't like it much but everyone else did. And the project had other consequences, which I have described [2]. She was genuinely great and remained the same unassuming, compassionate, dedicated person for her whole life.

References

1. Hodgkin, D. C.; Rimmer, B. M.; Dunitz, J. D.; Trueblood, K. N. *J. Chem. Soc.* **1963,** 4945–4956; Cornforth, R. H.; Cornforth, J. W. *J. Chem. Soc.* **1963,** 4955–4956.
2. Cornforth, J. W. *Notes and Records of the Royal Society of London* **1982,** 37(1).

Notes

What Did Carl Wilhelm Scheele Look Like?[a]

Torvard C. Laurent[b]

Carl Wilhelm Scheele was born on December 9, 1742, in Stralsund in Pomerania, a province of present Germany. However, at that time Stralsund belonged to Sweden, and Carl Wilhelm was born as a Swedish citizen, but his native language was German. He was the 7th of 11 children of Joachim Christian and Margareta Eleonora Scheele. His older brother, who has been sent to Göteborg to be trained as a pharmacist, died of typhoid, and Carl Wilhelm, at the age of 14, succeeded him as an apprentice in the "Unicorn" pharmacy. From the beginning, Scheele showed a great interest in chemistry and, despite not having any formal training in the subject, he devoted all his free time to his personal experiments. In this, he was encouraged by his superiors.

Medal minted for the Royal Swedish Academy of Sciences in 1789 commemorating Carl Wilhelm Scheele. Artist: Carl Johan Wikman. Inscription on the back side: Ingenio stat sine morte decus (His genius makes his fame immortal).

[a]*Chemical Intelligencer* 1999(3), 28–30.

[b]Department of Medical Biochemistry and Microbiology, University of Uppsala, BMC, Box 575, S-751 23 Uppsala, Sweden (deceased)

Scheele stayed in Göteborg for eight years until 1765 and then worked in pharmacies in Malmö, Stockholm, and Uppsala before he got his own pharmacy in the small town of Köping in 1775. The previous owner of the pharmacy had died and, as part of the deal, Scheele committed himself to supporting his widow, 23-year-old Sara. She became his housekeeper. Scheele died in 1786, when he was only 43½ years old, probably due to heart failure caused by rheumatic fever. He married Sara a few days before he died.

Scheele never received an academic education; he was self-taught. He lived a secluded life with limited resources for scientific work, but he became one of the foremost chemists of all times. As a sign of respect from his contemporaries, he was elected a member of the Royal Swedish Academy of Sciences already at the age of 32. However, he only attended one meeting, the only time he left Köping and went to Stockholm. On the same visit, he took the formal examination required for managing a pharmacy.

The discoveries that Scheele made are numerous. He discovered oxygen, chlorine, manganese, and barium and did the basic work leading to the discoveries of other elements. He produced hydrogen at the same time as Cavendish. He described nitrous acid and nitrogen oxides, many inorganic and organic acids (citric acid, oxalic acid, tartaric acid, etc.), glycerol, hydrogen cyanide, and hydrogen sulfide. He was involved in the discovery of calcium phosphate in bone together with Johan Gottlieb Gahn. He studied the effect of light on chemicals and prepared papers with silver chloride which were blackened by light and showed that the effect was wavelength dependent. Although he communicated many of his discoveries to the scientific community during his lifetime, the true depth of his work did not become apparent until a century after his death, when Adolf Erik Nordenskiöld, the famous polar explorer who discovered the Northeast Passage, went through Scheele's notes and published them.

B. Hargittai and I. Hargittai (eds.), *Culture of Chemistry: The Best Articles on the Human Side of 20th-Century Chemistry from the Archives of the Chemical Intelligencer*, DOI 10.1007/978-1-4899-7565-2_9, © Springer Science+Business Media New York 2015

Scheele's most famous discovery was that of oxygen, at the same time as Priestley and Lavoisier described the gas. The book by Scheele *Chemische Abhandlung von der Luft und dem Feuer* was written in the fall of 1775 but was not published until 1777 due to a delay in the printing. Both Priestley and Lavoisier published their work in 1775. However, the story behind the publication date is less well known. For some years, Scheele had produced oxygen in several ways. In reply to Lavoisier, who had sent him his *Opuscles*, Scheele wrote on September 6, 1774, and asked Lavoisier to repeat an experiment in an apparatus which had been described in the *Opuscles*. Scheele wrote that when silver carbonate was heated and carbon dioxide removed by adsorption onto sodium hydroxide, a gas was produced which could sustain a burning candle and keep animals alive. Priestley made his discovery in 1774 and traveled to Paris and described it to Lavoisier. When Lavoisier, at Eastertime in 1775 at the French Academy, spoke about a new type of air, until then unknown, he mentioned neither Scheele nor Priestley.

Statue of Scheele on "Flora's Mound" in Humlegården, Stockholm. Artist: John Börjesson. The statue was inaugurated on December 9, 1892, on the 150th anniversary of Scheele's birth. The son of John Börjesson was the model. (Photo by I. Hargittai, 1998).

Although Scheele has been depicted both on stamps and in a famous statue, there seems to be no one who really knows what Scheele looked like. We have no picture of him from his lifetime. The closest to the original is probably his face on a coin minted by the Royal Academy of Sciences in 1789, four years after Scheele's death. At that time, people probably still remembered him. When a stamp was issued in 1942, 200 years after Scheele's birth, a miniature portrait owned by a distant relative was used. However, the portrait cannot be of Scheele because the person is wearing a necktie, which did not become fashionable until around 1800.

On December 9, 1892, the 150th anniversary of Scheele's birth, a statue of him was inaugurated on "Flora's Mound" in the Humlegården park in Stockholm. Present at the time were the King, the Crown Prince, two other princes, and representatives of the government, the diplomatic corps, and various universities and learned societies. There was military music, songs, and speeches. Count Carl Snoilsky, Sweden's foremost poet at the time, had written a poem about Scheele. The secretary of the Academy of Music had written music to the poem, and it was sung by a well-known choir, Par Bricole.

The money for the statue had come from a nationwide collection started by Nordenskiöld, and the artist contracted for the work was Professor John Börjesson. The statue weighted 900 kg and was 2.26 m high. Smaller replicas of this statue can be found elsewhere. One is kept by the Faculty of Pharmacy in Uppsala. But who is the man on the statue? It is generally thought that the son of John Börjesson was the model.

The poem written by Carl Snoilsky for the inauguration of the Scheele statue
(English translation by Roald Hoffmann and Torvard Laurent)

Quiet, modest, hidden from the world
You sat in your obscure corner
But from your labor at the hearth
A light spread, where there was only night

The square's noise is not for you,
A woody mount is best
And Linnaeus in Humlegården
Will have a worthy guest in you.

In the solid blend of metals
Sit with the crucible in your grove
Great seeker of the truth,
Searcher for matter's base

Bathe yourself in waves of air
Under the sighs of your tree
And when evening turns on its flames
Live on, in fire and light

One hundred years later, on the evening of February 25, 1992, an explosion was heard from the Humlegården. Scheele's statue had been blown into pieces. It later turned out that the wicked deed had been performed by a couple of schoolboys, who also had sabotaged other statues in Stockholm. Fortunately, the damage was not too extensive, and there was still some money left from the collection made in the nineteenth century. This was used to repair the statue, which was reinaugurated on December 6, 1992, three days before the 250th anniversary of Scheele's birth. The actual anniversary coincided with the Nobel festivities.

A Scheele Symposium on "Oxygen—Perspectives on the Element of Life" was held at the Royal Academy of Sciences on December 5–6. Oxidation is the main source of energy for mankind. Another source of energy celebrated its 50th anniversary at the same time. On December 2, 1942, the first nuclear chain reaction took place in Fermi's laboratory in Chicago.

In the original discussions about the location of the statue of Scheele, the town of Köping argued that it should be placed where his home and pharmacy had been. To protest the selection of a location in Stockholm, representatives from Köping did not attend the original inauguration of the monument. Later, in 1912, Köping got its own memorial, made by Carl Milles, but it has never achieved the fame of the one in Humlegården.

Stamp issued in 1942 to commemorate the birth of Carl Wilhelm Scheele in 1742. The stamp shows a painting previously owned by a distant relative of Carl Wilhelm. It cannot be a genuine picture of Scheele because the man is wearing a necktie, which did not become fashionable until 1800. © Posten Frimärken.

Retorts

TOP: The retort at Scheele's feet in John Börjesson's sculpture. (Photo by I. Hargittai.).
BOTTOM: Insignia of the U.S. Army Chemical Corps (A Guide to U.S. Army Insignia. Whitman Publishing Co., 1941).

At Scheele's feet in John Börjesson's sculpture there is an oven, a mortar, and a retort. The retort has been known from ancient times. It often appears in engravings depicting chemistry. If there is any single apparatus that symbolizes chemistry, it is the retort. A curious appearance of the retort is in the insignia of the Chemical Corps of the United States Army, originally called Chemical Warfare. The colors of the insignia are cobalt blue and golden yellow, and the insignia displays crossed retorts behind what may be taken as a benzene ring. Postage stamps, too, often display the retort to symbolize chemistry. Four examples are shown here from (**1**) Switzerland (1982), (**2**) Mexico (1972), (**3**) Kuwait (1969), and (**4**) the Soviet Union (1965).

I thank Dr. Louis Adcock (University of North Carolina at Wilmington) for calling my attention to the U.S. Army Chemical Corps insignia.

ISTVÁN HARGITTAI

Stamps 1, 2, 3, 4

Notes

Insider Trading[a]

Michael Lederer[b]

In recent years, stories in the daily papers about court cases dealing with "insider trading" have been common. A typical case concerned a columnist for the *Wall Street Journal* who analyzed the position of a firm and wrote an article showing that the prices of its shares should rise. He tipped off his friends that this article was to appear, thus permitting trades to be made based on this information. This case was brought to trial and resulted in jail sentences and fines.

The analogous situation with not so obvious financial implications exists, unfortunately, in the chemical literature. Sir Robert Robinson in his *Memoirs of a Minor Prophet* [Elsevier: Amsterdam, 1976] mentions under the heading "Piracy in Chemical Research" (p 84):

> As a rule this kind of invasion of territory is carried out with fresh forces using new methods. It is not a question of copying so much as of unnecessarily occupying ground, which somebody else is clearly tilling. Sometimes, others have to complain of leakage. I am quite unable to confirm that the following has any real basis in fact, but the statement was made, so that one supposes there must have been some reason for it. Samuel Smiles had established the tetrahedral environment of quadrivalent sulphur (trialkylsulphonium salt) and lectured on the subject after his publication. He then referred to the fact that simultaneously with his own paper another had appeared, the authors of which appeared to be Messrs. Peep and Poachey (Spoonerism for Pope and Peachey).

The late Dr. F.H.P. told me 30 years ago that when he submitted a paper to a journal, which we shall call "A," the referees often took more than six months to review it. In the meantime, his "peers" would publish a paper based on the same idea as his in another journal. I think that there are few chemists who have not experienced something similar. I have.

The "Ethical Guidelines to Publication of Chemical Research" [*Anal. Chem.* **1986,** **58,** 265] state that:

> A reviewer should treat a manuscript sent for review as a confidential document. It should neither be shown to nor discussed with others except, in special cases, to persons from whom specific advice may be sought; in that event, the identities of those consulted should be disclosed to the editor.

However, the referees consulted do not swear an equivalent of the Hippocratic oath. They may even carelessly mention something about the paper they have refereed to other colleagues without evil intentions. In our present state of science, where the development of a new instrument (or principle thereof) or a new observation in gene technology or biotechnology in general may translate into large sums of money, the reviewing process is far from satisfactory.

In the last 30 years, I have given a lot of thought to how journal editors can keep the situation under control. Here are some suggestions:

1. Manuscripts should be sent only to referees who return them quickly and, if accepted for publication, should be published as fast as possible. Since it usually does not take a referee much more than three hours to review a paper, manuscripts should be returned at the latest within a few days, and the whole review process should not take longer than two weeks. The time between submission and publication should not ever exceed six months. This makes it practically impossible for somebody else to develop an idea he has picked up and get it published earlier than the original author.

2. Most journals pride themselves on sending manuscripts to lots of referees and feel that this practice guarantees a thorough review and a maximum of assistance to the author. Before sending a paper to a reviewer, the editor should ask himself or herself the following questions:

2.1. Can the reviewer tell me something that I cannot establish by going to the library? If not, instead of asking the reviewer, check the literature yourself.

[a]*Chemical Intelligencer* 2000 (2), 47.

[b]Institut de Chimie Minérale el Analytique, BCH, Université de Lausanne, CH-1015 Lausanne, Switzerland (deceased)

2.2. Do I know for sure that the selected reviewer is not an enemy of the author? If you are not certain, then don't sent the paper to this reviewer.

2.3. Do I know the reviewer well enough personally that I can trust him or her with an original work entrusted to me? If not, don't send it.

I think that if these principles were followed, there would be very few cases of "insider trading." Of course, this would result in much work for the editor, and that may not be to every editor's liking.

Notes

The Pursuit of Happiness[a]

Liberty Medal Address, City of Philadelphia, 4 July 2000

James D. Watson[b]

We have assembled here this Independence Day to reaffirm that freedom is at the heart of human existence. When in control of our individual destinies, we thrive and look forward to the future. In contrast, when our aims and actions are determined by others, we feel stifled and unable to live up to our potentials as human beings. Without life, liberty and the ability to pursue happiness, human beings have no chance to realize the great talents that let Galileo see the moons of Jupiter, or Rembrandt catch the essence of humanity in his portrayal of our faces.

In preparing our country for the war that would soon envelop us, Franklin Roosevelt spoke of four essential freedoms—freedom of speech, freedom of religion, freedom of want, and freedom from fear. That bleak January 1941 day, Roosevelt emphasized freedom from fear, knowing well the mortal potential that attainment of Hitler's ugly aspirations held for human life. But if Thomas Jefferson were then mobilizing our nation, he would have added a fifth freedom—freedom from ignorance. The uneducated man can never be in control of his destiny. Had newspapers, the BBC, and cinema newsreels not informed the general public that Hitler was evil incarnate, the fateful Battle of Britain might very well had a different outcome.

As a product of the eighteenth century intellectual enlightenment, Jefferson saw truth arising from observations and experiments. So he wanted his state of Virginia to select, for special educational enrichment, youths of inherent genius who were sprinkled as liberally among the poor as the rich. He saw the knowledge so learned as the ultimate safeguard

of liberty. Correspondingly tyrannies thrive when education is prevented. Cromwell's victorious march across Ireland was soon followed by abolition of education for its Catholic denizens.

Essentially a deist who saw a role for God only in the creation of the universe and its life forms, but not in events afterward, Jefferson did not see organized religions as the basis for moral virtue. Instead he accepted the idea going back to the Greeks of natural rights which arose out of the essence of the human being as created by God. To Jefferson it was self evident that all humans were created equal with inalienable rights that transcended where or in what period of history one was living.

Today, 224 years after Jefferson so eloquently expressed these ideas in the Declaration of Independence, biology is witnessing the completion of an intellectual renaissance that Charles Darwin began in the nineteenth century. Through his Theory of Evolution through Natural Selection, Darwin forever changed our view of human life. He saw ourselves as the products not of creation by a God as revealed in Genesis but as arising through a series of evolutionary events going back to a common ancestor of many eons ago.

Much more recently we have learned that the variation upon which natural selection acts reflects mutational changes in DNA, the molecule of heredity. Differences between different forms of life reflect differences in the sequences of the four letters—A, T, G, and C—of the DNA alphabet. When the double helix was first revealed in 1953, neither Francis Crick nor I ever thought that, within our lifetimes, the three billion letters that compose the human genetic message would ever be close to being deciphered. But just a week ago, elegant technology and innovative science, combined with much human perseverance, allowed the world of

[a]*Chemical Intelligencer* 2000(4), 47–48.

[b]Cold Spring Harbor Laboratory,
Cold Spring Harbor, NY 11724, USA

science to give to humanity this true book of human life. Already we can see the outline of some 40,000 genes, the discrete packets of DNA information that are used to determine the structure of proteins, the actors in cellular life.

The inborn equality of all humans that Jefferson so forcefully believed in we now see arising from our common ancestor that existed in Southern Africa only some 100,000 years ago. Most likely the hunter-gatherer Bushmen of the Kalahari Desert are a direct and largely unchanged representation of human life as it then existed. On a recent trip to Botswana, I was struck by the Bushmen's quickness to learn despite speaking a language that only counts "one, two, three, and many." Underlying the close resemblance of all humans to each other is the close similarity of our books of DNA instructions. Individual variations in our DNA sequences amount to less than one letter in 1,000 with most of these differences arising long before modern humans spread across Africa into the Middle East.

Modern biological thought, however, is much less compatible with Jefferson's concept of undeniable rights. Evolution has not endowed ourselves or the fox or the chicken for that matter, with the right to live or to be treated well. Instead, every successful animal form has evolved with its own individual needs, say for specific foods. In turn, we all have evolved capabilities that largely satisfy such needs. High among the needs for virtually all vertebrates is liberty; for being free to move and act unimpeded by others is an indispensable condition for evolutionary survival. At the same time, our various brains have been programmed by our genes to initiate actions that keep us alive. Animals that do not seek out food or evade fast moving objects will not likely give rise to offspring.

Jefferson's most unique insight with regard to freedom was his identification of the pursuit of happiness as a fundamental prerequisite for human advancement. Under normal circumstances, most individuals are only fleetingly happy, say, after we have solved a problem, either intellectual or personal, that then lets our brain rest for a bit. Equally important, happy moods also reward higher animals after they make behavioral decisions that increase their survivability. Successfully replenishing fat cells not only turns off appetites but leads to the appearance of pleasure-bringing natural opiates—the endorphins. A desire for more endorphin enriched moments may well be the primary motivation for ourselves to seek out food or to bask in Vitamin D producing sunshine. Likewise, the happiness we feel upon strenuous exercise should be seen as a Pavlovian reward for the physical exertion needed for food gathering and sexual satisfaction.

These moments of pleasure best be short-lived. Too much contentment necessarily leads to indolence. As Shakespeare has Julius Caesar say, "Let me have men about me that are fat and sleek-headed, and such as sleep o' nights. Cassius has a lean and hungry look; he thinks too much; such men are dangerous." But it is discontent with the present that leads clever minds to extend the frontiers of human imagination. During a low moment in World War II, Joseph Stalin wanted to eliminate one of Russia's most brilliant individuals, the theoretical physicist Lev Landau. Fortunately, one of his colleagues, Peter Kapitsa, saved his life by arguing successfully that Landau was not subversive, only unpleasant.

Every successful society must possess citizens gnawing at its innards and threatening conventional wisdom—individuals like Thomas Jefferson, Tom Paine and Benjamin Franklin. Without the changes that radical ideas and actions like theirs bring about, established orders go stale and crumble before brasher peoples accepting the new. Now, more than ever, successful nations must be free societies where diversity of thought is not only tolerated but seen as the intelligent response to a constantly changing world. As long as we can see happiness ahead, the worries and faults of today are bearable. So in the perfect world we want some day to exist, humans will be born free and die almost happy.

Part III

Beautiful Molecules

Beautiful Molecules

Spherand[a]

Donald J. Cram[b]

Very often, molecules appeal to our sense of aesthetics as well as to our chemical interest. We would like to enhance such considerations by devoting a small column in The Chemical Intelligencer *to the beauty of molecules that have been recently isolated, synthesized, or computed, or whatever way they appear in the literature.*

We would like to have a brief description of the molecule and its properties and an illustration of its structure. We would also like to know why that particular molecule is considered to be beautiful, fully aware of the fact, of course, that molecular beauty may exist in the eyes of the beholder only.

We shall be looking for such entries by authors of new molecules or by anybody else who has an eye for them in reading the literature.

A beautiful molecule to me is one whose structure elicits a pleasureful aesthetic reaction that symbolizes other positive experiences. Compound **1** is shaped not quite like a snowflake. As a person happily raised in Vermont whose winters abounded in snowflakes, I designed **1**, a potentially specific complexing agent for Li⁺ and Na⁺ ions. In space-filling models, the 24 unshared electrons of the 6 octahedrally arranged oxygen atoms of spherand **1** line an enforced cavity, whose diameter is slightly larger than the diameter of six-coordinate Li⁺, and slightly smaller than that of Na⁺. The six O-methyl and six phenyl groups isolate these electrons from stabilizing interactions with solvent.

Subsequently my research group synthesized **1**, **1**⊙Li⁺ and **1**⊙Na⁺, whose crystal structures were essentially those predicted by model examination. These three molecular entities all possess one three-fold axis of symmetry, three twofold axes, and three mirror planes. As hoped, **1** bound these two cations more strongly than any other

---

[a]*Chemical Intelligencer* 1995(2), 58.

[b]Department of Chemistry, University of California, Los Angeles, Los Angeles, CA 90024, USA (deceased)

host, but would not touch other ions (e.g., K^+, NH_4^+, or Ca^{2+}). The noncyclic counterpart of **1**, in which two terminating hydrogen atoms replace one aryl-to-aryl bond, did not complex detectably any cations. The contrast in behavior between the rigidly-structured **1** and its highly mobile (>1000 conformations) analogue gave rise to the principle of preorganization. *The more highly hosts and guests are organized for binding and low solvation prior to their complexation, the more stable will be their complexes* [1].

The structure of **1** exhibits unusual symmetry, calls to mind the pleasures of childhood, and symbolizes the threefold satisfactions of prediction, creation and generalization. Spherand **1** is my entry in the beauty contest for organic compounds.

References

1. Cram, D. J.; Cram, J. M. *Container Molecules and their Guests. Monographs in Supramolecular Chemistry.* Stoddart, J. F., Series Ed.; The Royal Society of Chemistry, Thomas Graham House, Science Park: Cambridge, 1994; pp 20–41.

Part IV

Chemical Tourist

System of Elements in Anagni[a]

István Hargittai[b] and Aldo Domenicano[c]

Anagni is an ancient little town, beautifully situated on top of a hill about 60 km southeast of Rome, off the Rome–Naples motorway. Originally a Hernic settlement, it was conquered by the Romans in 306 B.C. Anagni became wealthy and important in the thirteenth century, during which it gave four popes to the Roman Catholic church.

Anagni Cathedral (Fig. 1) was built between 1072 and 1104, originally in the Romanesque style. Gothic elements were added later in the thirteenth century. A famous feature of the cathedral is its mosaic floors, created by the Cosma family in the first half of the thirteenth century.

To the chemical tourist though, the most interesting feature may be some of the frescoes covering the walls and ceiling of the crypt, built in the same period as the upper church. These twelfth- and thirteenth-century frescoes are due to Benedictine painters of the Roman-Byzantine school. They blend religious topics and representations of the physical world, namely, medicine, astrology, and alchemy [1, 2]. In one of the 21 vaults, a human figure symbolizes the allegory of life in relation to the astronomical cycles. The four ages of man are presented in relation to the four seasons and *the four elements*. The fresco is thought to have been inspired by Platonic cosmology (Plato's teachings were spread in southern Italy by the Salerno medical school). Another fresco displays two physicians, Hippocrates (fourth century B.C.) and

Fig. 1 Anagni Cathedral (the transept and two apses). (Photograph taken by I. Hargittai, June 1995).

Galenus (second century A.D.), sitting together as Teacher and Disciple.

Next to the two physicians, there is a diagram of the four elements (Fig. 2), Earth, Water, Air, and Fire, and six properties, *immobile, corpulent, obtuse, mobile, subtle,* and *acute.* The straight connecting lines indicate correspondence (e.g., fire is mobile, subtle, and acute) whereas the curved lines connect opposite qualities. There are Roman numerals beneath the names of the elements: for Earth, $8 = 2^3$; for Fire, $27 = 3^3$; for Water, $12 = 3 \times 2^2$; for Air, $18 = 2 \times 3^2$. The equality containing these numbers, i.e., $8/12 = 18/27$, unifies the whole universe in its perfection according to Platonic philosophy [3]. This relationship may be generalized as $x^3/[(x + 1)x^2] = x(x + 1)^2/(x + 1)^3$.

A detailed description of the six properties and their relationship to the four elements, corresponding closely to the

[a]*Chemical Intelligencer* 1996(1), 56.

[b]Department of Inorganic and Analytical Chemistry, Budapest University of Technology and Economics, Szt. Gellert ter, 1111 Budapest, Hungary
e-mail: istvan.hargittai@gmail.com

[c]University of L'Aquila, I-67100 L'Aquila, Italy

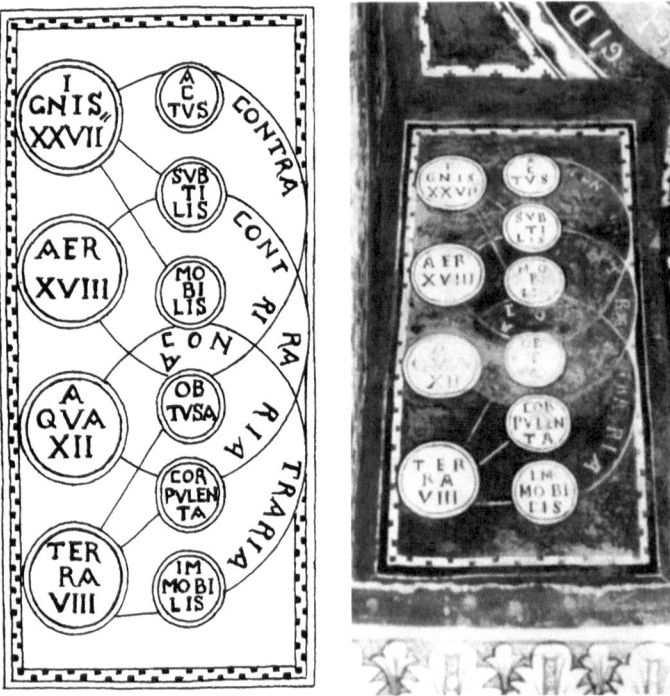

Fig. 2 System of four elements in the crypt of Anagni Cathedral (twelfth- or thirteenth-century fresco).
(a) Scheme after Ref. 1; (b) photograph (taken by I. Hargittai, June, 1995).

Anagni diagram, was already given by Chalcidius (ca. fourth century A.D.), a Latin philosopher who translated and commentated Plato's *Timaeus* [3].

Acknowledgments Alan L. Mackay (Birkbeck College, University of London) suggested that we visit Anagni Cathedral to see its mosaic floors. The Cathedral authorities graciously gave us permission to take one photograph in the crypt.

References

1. Ribaudo, C.; Scascitelli, S. *Anagni: Historical and Artistic Guide to the City:* ITER: Subiaco, 1990.
2. Toesca, P. *Gli Affreschi della Cattedrale di Anagni;* Roma, 1902. Reprinted, with color photographs, by ITER: Subiaco, 1990.
3. Plato, *Timaeus.* XXXI–XXXII.

Chemical Tourist

The Snub Cube in the Glanville Courtyard of the Beckman Institute[a]

at the California Institute of Technology

William P. Schaefer[b]

In the central (Glanville) courtyard of the recently constructed Beckman Institute building at the California Institute of Technology is a fountain, placed there by the architect, Mr. Tim Vreeland, to create some "white noise" and thus separate acoustically four areas of the courtyard designed for conversational groups. The architect asked for help from the future occupants of the building in designing the fountain itself; several suggestions were made and rejected by the Caltech administration as not having any relationship to the purpose of the building. Arnold O. Beckman, the donor of the building, had specified that he wanted this Institute to develop new methods and instruments that would advance research in the fields of biology and chemistry, including their interface. After our latest suggestion had been rejected, Harry B. Gray, then the Director-designate of the Beckman Institute (now Director), recalled a paper [1] describing the tertiary structure of the iron-containing protein ferritin; the molecule of ferritin was found to have 432 (read as four, three, two) symmetry; i.e., it has fourfold axes, threefold axes, and twofold axes relating the 24 subunits of the protein.

Now, the ferritin protein seemed to Harry Gray to be an excellent symbol for the work that would be done in the new building. Ferritin is found in plants and animals alike; it is an iron-storage protein containing up to 4500 iron atoms in a hydroxyphosphate complex form in the core, surrounded by the organic protein shell. Thus, the molecule can be claimed by biology, organic chemistry, and inorganic chemistry, all three fields that were to be emphasized in the Beckman Institute. Harry asked me to design something for the courtyard fountain that would capture the essence of the ferritin structure.

[a]*Chemical Intelligencer* 1996(4), 48–50.

[b]California Institute of Technology, Pasadena, California 91125, USA

B. Hargittai and I. Hargittai (eds.), *Culture of Chemistry: The Best Articles on the Human Side of 20th-Century Chemistry from the Archives of the Chemical Intelligencer*, DOI 10.1007/978-1-4899-7565-2_14, © Springer Science+Business Media New York 2015

changing the position of the arbitrary "atom" I started with. The solid had 6 square faces and 32 triangular ones, with 24 corners. The corners, then, would represent conceptually the subunits of the ferritin molecule. Some of the triangular faces could be either acute or obtuse, and I made paper models of both kinds to see which was more pleasing to the eye. I favored the solid with somewhat acute triangular faces, but my colleague Verner Schomaker pointed out that the solid with all equilateral triangles was special: it is called the snub cube, and Verner said that it was one of Linus Pauling's favorite solids. (The other was the icosahedron.) It is in fact an Archimedean semiregular solid, derived from a cube and having only two kinds of faces, squares and equilateral triangles, with all its edges of equal length.

With its pool and its tiling. (Photo by William P. Schaefer).

Three views of the snub cube sculpture (1996) in this article. Close-up. (Photo by I. Hargittai).

The essence of any structure is its symmetry [2]; this was the obvious starting point for the design. And because I am a crystallographer, symmetry was a handy tool for me to use. I looked in the *International Tables for X-Ray Crystallography* [3] and found the simplest space group that had 432 symmetry; that turns out to be space group #207, a cubic space group with symmetry $P432$ and 24 general equivalent positions, just the same as the number of subunits in the ferritin molecule. In order to visualize this structure, I used the computer program ORTEP, written by Carroll Johnson [4], and placed an arbitrary atom in the unit cell. The program used the 432 symmetry of the space group to generate the other 23 equivalent atoms and then drew a picture of the result. I discovered that by joining the "atoms" I had generated by "bonds," I had the outline of a solid; I could vary the shape of the solid by

(There are two other facts about the snub cube that may be of interest. First, despite its apparently high symmetry, with all sorts of rotational axes running through it, it has no planes of reflection; it exists in two forms, one left-handed and the other right-handed. Second, as with any semiregular solid, the snub cube can be inscribed in a sphere. In this case, the 24 points on the sphere represent the distribution for which the smallest distance between any two is as great as possible [5].)

The model I made of the snub cube pleased the architect as well as the Administration, and we decided to use a snub cube as the decorative element in the fountain of the Beckman Institute. The contractor who was to build this, though, insisted on making a half-sized model first to see if water could be made to flow evenly over the surface of such a solid. He was used to building much more symmetric fountains and was skeptical about this. A wooden model, though, showed that with a sufficiently strong flow, the entire surface of the solid could be wet; we were given the go-ahead to install a five-foot-tall, granite snub cube in the fountain. The granite chosen was a green variety from Africa. It was quarried there and shipped to Italy for cutting into slabs, and the slabs were shipped to California. The subcontractor charged with fabricating the actual fountain claimed not to be able to build such a complicated form, so I used the ORTEP program again to calculate all of the inter-facial angles that he needed to know, and I gave him precise measurements to work from. With these measurements and angles, the man went ahead with fabrication, first flaming the outer surface of the granite to roughen it and produce something that would be as hydrophilic as possible, and then attaching the cut slabs of granite to a stainless-steel armature he had built to my specifications. The plumbers would later run a pipe up through the snub cube to discharge water over the top, so it would flow down the sides and into the pond at the bottom, to create the white noise the architect wanted. The final granite construction is five feet across, from square face to square face, and, because of its cubic symmetry, also five feet tall. It rests on a cylindrical pedestal of green granite about 18 inches high, so the top of the snub cube is visible only to quite tall people, or from the upper floors of the building.

With the Beckman Institute in the background. (Photo by William P. Schaefer).

The fountain, with its impressive granite snub cube, has been functioning for nearly six years. The Beckman Institute building won an award given by Pasadena Beautiful for the most beautiful noncommercial building built in 1991, and the snub cube fountain itself was recognized by the City of Pasadena in 1992 as one of the ten best examples of public art in the city. The citation recognized as "artists" of the sculpture Harry B. Gray and William P. Schaefer, the first time either of us had won such a distinction. We continue to be pleased with our work.

Jay A. Labinger, William P. Schaefer, and Verner Schomaker on February 19. 1996. (Photo by I. Hargittai).

References

1. Smith, J. M. A.; Ford, G. C.; Harrison, P. M. *Biochem. Soc. Trans.* **1988**, *16*, 836.
2. Holden, A. *Shapes, Space and Symmetry*; Dover Publications: New York, 1991; 45.
3. *International Tables for X-Ray Crystallography*, Volume A; D. Reidel: Dordrecht, Holland, 1983; 624.
4. Johnson, C. K. ORTEP II, Report ORNL-5138, Oak Ridge National Laboratory, U.S.A. 1976.
5. Coxeter, H. S. M. *Introduction to Geometry*, 2nd ed.; John Wiley & Sons: New York, 1969; 276.

Chemical Tourist

Rising to New Heights[a]

Balázs Hargittai[b,c,d] and István Hargittai[e]

On the campus of the University of Arizona (UA), in front of the Chemistry-Biology Teaching Building, there is an intriguing sculpture in the shape of an archway, called "25 Scientists," which was inaugurated in 1993. It is made of welded steel and painted in bright colors. It resembles a "Glockenspiel," popular in Germany and elsewhere in Europe, in which little figures come out of the clock in a clock tower and go around at certain times. The figures on the archway at UA, which do not move, represent some important branches of science and some discoveries.

One of the central units of this sculpture at UA shows a buckyball, C_{60}, somewhat compressed like a filled pancake, with the carbon atoms and the bonds between them painted on a blue background. Two figures are holding this buckyball. One of them resembles Donald Huffman of UA. The other could have been Wolfgang Krätschmer but doesn't look like him. The similarity to Donald Huffman is no accident. The creator of the archway, New York artist George Greenamyer, painted Huffman as one of the two people flanking the buckyball, with emphasis on his gray beard.

Huffman was pleased to serve as the model for the figure on the four-meter-high archway. Referring to this experience, he said, "I've risen to new heights."

The students walking underneath may not be paying too much attention to the topics depicted on the sculpture. Since Dr. Huffman is rather taciturn, even his own students are not aware of his being the model for one of the figures flanking the buckyball.

We noticed the archway on our very first day on the campus, August 30, 1999, upon our arrival in Tucson. By lucky coincidence, Wolfgang Krätschmer just happened to be visiting Donald Huffman in the Department of Physics for a couple of days. Donald and Wolfgang graciously posed for us beneath the archway. Then, the next day, they generously recreated for us the experiment that produced the C_{60}-rich soot. The pictorial report is from these encounters.

The artistic entranceway to the Chemistry-Biology Teaching Building of the University of Arizona, Tucson, with Wolfgang Krätschmer and Donald Huffman under the archway. (All photos by I. Hargittai).

[a]*Chemical Intelligencer* 2000(3), 42–43.

[b]Department of Chemistry, University of Arizona, Tucson 85721, Arizona

[c]Department of Biochemistry, Molecular Biology, and Biophysics, University of Minnesota, Minneapolis, MN 55455, USA

[d]Currently at Saint Francis University, Loretto, PA

[e]Department of Inorganic and Analytical Chemistry, Budapest University of Technology and Economics, Szt. Gellert ter, 1111 Budapest, Hungary

B. Hargittai and I. Hargittai (eds.), *Culture of Chemistry: The Best Articles on the Human Side of 20th-Century Chemistry from the Archives of the Chemical Intelligencer*, DOI 10.1007/978-1-4899-7565-2_15, © Springer Science+Business Media New York 2015

1. Closeup of the buckyball section of the archway with Huffman and Krätschmer.

2. Huffman and Krätschmer before the experiment. The clean bell jar is visible between them. This is a historic piece of equipment since Huffman has been using this bell jar ever since these experiments were started at UA. Krätschmer's original bell jar is no longer in his Heidelberg lab. It is now on display in the Deutsches Museum in Munich.

3. The graphite rods (in somewhat displaced position) in the apparatus.

4. While Krätschmer is adjusting the helium pressure, Huffman is getting ready to start the experiment.

5. The resistive heating experiment producing the fullerene-rich soot is on. The bell jar is rapidly logged by soot.

6. Krätschmer is collecting the soot with Huffman looking on.

7. Huffman is dissolving the soot in carbon disulfide.

8. Donald Huffman is transferring the sample of C_{60} solution.

Part V

Cooking Chemist

Soufflés, Choux Pastry Puffs, Quenelles, and Popovers[a]

Nicholas Kurti[b] and Hervé This-Benckhard[c]

"The advantage that would result from an application of the late brilliant discoveries in philosophical chemistry and other branches of natural philosophy and mechanics to the improvement of the art of cookery are so evident that I cannot help flattering myself that we shall soon see some enlightened and liberal-minded person of the profession to take up the matter in earnest and give it a thoroughly scientific investigation. In what art or science could improvements be made that would more powerfully contribute to increase the comforts and enjoyments of mankind?" (From the 400-page essay "On the Construction of Kitchen Fireplaces and Kitchen Utensils together with Remarks and Observations relating to the various Processes of Cookery and Proposals for improving that most useful Art," by Sir Benjamin Thompson, Count Rumford, 1794.)

What could be better for introducing the Cooking Chemist column to readers and contributors alike than the above quotation from the writings of Count Rumford, the English soldier, statesman, natural philosopher, inventor, and social reformer?[1]

It is true that Rumford's wish of a thorough application of science to the art of cookery has been fulfilled to some extent: good basic science and engineering have greatly helped the development of the food industry in the last 50–100 years. However, it still seems to be very rare to see the professional scientist-cum-amateur cook using his or her physics, chemistry, or mathematics to explain, to explore, or to improve the everyday processes in the domestic kitchen and, in doing so, perhaps even to create new dishes. It is in this spirit that we invite readers to contribute to this column. There are no rules about the lengths of the contributions (but see Instructions to Authors). On the other hand, brevity is a virtue, and often a paragraph of a few lines suffices to make a point, correct an error, or illuminate a fact. Nor are there any rules about the nature of the contents. Contributions may provide scientific explanations of culinary processes, suggest improvements, comment on traditional cooking practices, or, if justified, debunk entrenched myths.

If potential contributors feel apprehensive about being accused by their peers of debasing science by using it for ultimately hedonistic purposes, they may answer their critics with the following quotation from Rumford's essay: "These minute investigations may perhaps be tiresome to some readers; but those who feel the importance of the subject and per-

[a]*Chemical Intelligencer* 1995(1), 54–57.

[b]Department of Engineering Science, University of Oxford, Parks Road, Oxford OX1 3PJ, UK (deceased)

[c]6 allée Georges, 78530 Buc Versailles, France

[1]Benjamin Thompson was born in 1753 in the Commonwealth of Massachusetts. During the American War of Independence, he was a loyalist and a spy working for the Governor of Massachusetts and later raised and commanded the "King's American Rifles." a regiment noted for its atrocities on Long Island. He spent most of his working life in England, where among other things he founded the Royal Institution of Great Britain, and in Munich. Bavaria, in the service of the Elector of Bavaria, who made him a Count of the Holy Roman Empire. It was in Munich that, with his famous canon-boring experiments, he demolished the caloric theory of heat. He also designed the famous Englischer Garden and put the beggars of Munich into workhouses, feeding them on "Rumfordsche Suppe," providing a daily intake of 1000 calories!

B. Hargittai and I. Hargittai (eds.), *Culture of Chemistry: The Best Articles on the Human Side of 20th-Century Chemistry from the Archives of the Chemical Intelligencer*, DOI 10.1007/978-1-4899-7565-2_16, © Springer Science+Business Media New York 2015

ceive the infinite advantages to the human species that might be derived from a more intimate knowledge of the science of preparing food, will be disposed to engage with cheerfulness in these truly interesting and entertaining researches."

And if that fails, let them have Brillat-Savarin's eternal aphorism: "La découverte d'un mets nouveau fait plus pour le bon-heur du genre humain que la découverte d'une étoile" (the discovery of a new dish does more for the happiness of mankind than the discovery of a star).

In this first article to appear in the Cooking Chemist column, we discuss four dishes, namely, soufflés, choux pastry puffs, quenelles, and popovers. We could have added meringues, sponge cakes, and some others. They have in common that they are, wholly or partly, solid—or at least firm—foams, which means that they contain air bubbles encased in rigid or semirigid walls. The size and the number of these bubbles are different for the different dishes, and so is the appearance and the consistency of the starting material, that is, the dish before it is cooked. Thus, whisked egg whites mixed with sugar, which become meringues when cooked, look like a heap of bubbles; a soufflé mixture is definitely foamy; but it is hard to discern any bubbles in the traditional choux pastry or pancake batter. However, they all expand when they are cooked and finish up as rigid or spongy foams.

The expansion is clue to the increase in size of the air bubbles brought about by two distinct mechanisms. There is first the expansion of air on heating, which, however, could account for only a 25% increase between 20 °C and 100 °C, whereas soufflés are known to increase twofold or threefold. This is made possible by a second mechanism, namely, the rapid rise of the vapor pressure of water with temperature. At 60 °C, the vapor pressure is one-fifth of an atmosphere, and, because the water evaporates from the walls of the bubbles, the pressure inside them is at least one-fifth of an atmosphere above that of the surroundings. The size of the air bubbles will thus depend on the elasticity of the bubble walls and, of course, on temperature.

We shall now describe experiments carried out recently on these dishes. It should be noted that where the personal pronoun "we" is used in this and any further contributions by the two of us, it means that the conception and the design of the experiment was joint but not necessarily that it was carried out by us together.

Soufflés

The first quantitative experiments on the variation of temperature in a soufflé were done some 25 years ago

(see Nicholas Kurti, The physicist in the kitchen, *Proceedings of the Royal Institution of Great Britain*, Vol. 42, No. 189, pp 451–467, 1969). A standard vanilla soufflé mixture was used: béchamel sauce (butter, flour, milk, sugar, vanilla) to which egg yolks and finally whisked egg whites are added. The temperature was measured with a thermocouple anchored to the soufflé dish (20 cm diameter and 10 cm deep) with its tip 2 cm from the surface of the mixture at the start. When the dish was placed in the preheated oven, the temperature rose for the first 10–15 minutes, then leveled out at between 45 °C and 50 °C (in some experiments it even dropped by a few degrees), and then, after about 25–30 minutes, rose again steadily, and the soufflé was found to be perfectly cooked when removed from the oven a few minutes after the temperature rose to about 65–70 °C.

Since experiments with a baked custard (egg and milk mixture baked in a bain marie in the oven) showed no similar effect, the temperature leveling could not be ascribed to the endothermicity of the protein denaturation and coagulation. It was probably caused by the rise of some cold layers to the tip of the thermocouple.

We recently carried out further soufflé experiments to find out to what extent the quality of a soufflé (characterized by its rise in height) depended on the history of the mixture between folding the whisked egg whites into the béchamel and placing the dish in the oven. These experiments were done in small 10-cm-diameter, 4-cm-deep ramekins filled to a depth of 3 cm whereas the previously described large soufflés started with a 6-cm-deep mixture. No plateau was observed, but only a point of inflection at around 50 °C.

Five experiments were carried out, some of them repeatedly: (1) soufflé ramekin put into the oven immediately after folding in the whisked egg whites; (2) ramekin kept for half an hour in a 40 °C bain marie; (3) ramekin kept at kitchen temperature for about two hours; (4) ramekin kept for six hours in the refrigerator; and (5) ramekin deep frozen for two days and placed in the oven after rewarming to room temperature.

Not surprisingly, the soufflé cooked immediately after the whisked egg whites had been folded in (experiment 1) was the best (a 2.5-fold rise), but the next three experiments produced only marginally worse soufflés (rises between 1.5-and 2-fold), and even the deep-frozen soufflé (experiment 5) produced a respectable rise of 1.8-fold. These results will be greeted with pleasure by hosts who are also the cooks. They will be able to ladle their soufflé mixture into the dish or the ramekins well before the arrival of their guests.

All recipes emphasize that, although it is important to mix the whisked egg white intimately with the béchamel without retaining any blobs of egg white, the folding in must be done gently so as to prevent the air bubbles from bursting or coalescing. However, recipes rarely give any indication about the firmness of the whisked egg whites. Pierre Hermé, Head Pâtissier of Fauchon, Paris, noticed that the use of firm egg foams improves the quality of the soufflés. Experiments were carried out with him in Paris to test this effect: a chocolate soufflé base (melted chocolate plus milk plus sugar plus egg yolk) was divided into two equal parts, and differently prepared egg-white foams were added to them. For the first one, the whisking was stopped immediately after the first soft peaks appeared. For the second, the whisking was continued until the foam was so firm that it could support an egg in its shell. The two mixtures were placed in identical soufflé dishes, 15 cm in diameter and 10 cm deep, filled up to 8 cm. They were cooked side by side, and after 20 minutes they were both taken out of the oven. The difference was remarkable: the "firm" soufflé rose to 15 cm, that is, nearly doubled in size, while the "soft" soufflé rose only to 11 cm. Furthermore, the firm soufflé was well cooked and light brown whereas the other one was dark brown and still liquid in the center. This result seems to indicate that the smaller the air bubbles are, the higher the heat transfer in the soufflé and the speed of migration of the water vapor.

In a second experiment carried out with smaller soufflé dishes, both soufflés were cooked until they were done, and while the rise of the firm soufflé was nearly threefold, that of the soft soufflé was barely twofold.

Choux pastry puffs

Choux pastry puffs have practically the same ingredients as soufflés but they are made differently: first, water (or milk) and butter are boiled, then flour is added in one batch, slightly dried, and finally eggs are mixed in to form a rather firm batter.

We became interested in choux pastry puffs because many cookery books (Escoffier, Pellaprat, Saint Ange, *Larousse gastronomique,* Flammarion's *L'art de la cuisine française,* Mathiot) insist that the eggs must be added one by one to the batter, some books even prescribing that the mixing time after each addition should be the same. This seemed strange because it was thought that, as long as the resulting paste was homogeneous, the mode of adding a given quantity of eggs should have no influence.

A simple microscopic examination provided the answer because it showed that in mixing the eggs with the batter, one also introduces many tiny air bubbles. So, as in the case of

soufflés, the expansion seems to be due to the increase of existing air bubbles and accumulation of vapor in these bubbles.

To test this assumption, the following experiments were carried out. The water-butter-flour mixture, the "panada," was divided into two equal parts, and the eggs (4) were incorporated using a wire whisk in different ways in the two batters. For the first batter, the eggs were added one by one, the number of whisk turns being counted; for the other half of the panada, the four eggs were added together. Since it was assumed that mixing was more important than adding the eggs one by one, the panada with eggs introduced together was given twice the number of whisks. The puffs were baked in batches, each having an equal number of the two types of puffs arranged symmetrically on the baking sheet. A blind tasting by 50 people gave the unanimous judgment that the thoroughly mixed puffs were better.

Following this experiment, it was reasoned that if air bubbles were indeed the key to the success in making choux pastry puffs, a better expansion could be obtained if the egg whites were whisked separately and then folded into the panada-yolk mixture. To test this, conventional choux pastry puffs were compared with choux pastry puffs of the same composition made by mixing thoroughly the panada and egg yolks and then folding in whipped egg whites. It was found that the diameter of the new kind of choux pastry puffs after cooking was 35 % higher than that of the conventional pastry puffs. From a gastronomic point of view, however, the more foamy choux pastry puffs were less appreciated than the traditional ones because their surface was not as smooth.

Quenelles

Quenelles are popular in France, especially in Alsace and in the Lyons region. They are generally made by mixing finely ground meat or fish with panada, cream, and eggs. They expand when they are poached. The principle of the quenelles' swelling is obviously the same as in choux pastry puffs or in soufflés: vapor accumulates in the air bubbles introduced by mixing. If introduction of egg foam separately into the choux pastry improves its quality, the same should be true for quenelles. As the aim is to obtain a solid foam, the various means of introducing air bubbles into the mixture before cooking should be analyzed both from the technical point of view (degree of expansion) and then from the gastronomic point of view: the two are complementary, and neither must be forgotten.

Durham popovers

Durham popovers are light bread rolls made from a pancake mixture, consisting of 50 g of plain flour, 75 ml of milk, one egg, 20 g of melted butter, and salt to taste, to give a batter of the consistency of single cream. Truncated conical metal cups, of about 2.5-cm base diameter, 6 cm high, buttered and floured, can be used. In an experiment, half of the batter was given 15 seconds with a low-speed electric whisk and poured into two of the cups. The other half was given 150 seconds whisking. The four cups were then put into a 200 °C oven and were taken out some 30–40 minutes later, by which time the crust was brown and hard and the volume increase was guessed to be 6–8-fold for the first two and somewhat larger for the second two (experiments done earlier without any whisking gave a smaller increase, but since the mixture was not the same, this result cannot be used to speculate about the effectiveness of the whisking). It should also be emphasized that the resulting popovers were not fluffy or spongy. Most of the bubbles coalesced and joined forces to push the dough up and out. This happens, though to a lesser extent, with éclairs and profiteroles.

Epilogue

The experiments in culinary physicochemistry we have discussed had as the central theme to make "bigger and better" or lighter and more uniform foams. When scientists develop new methods or techniques, they often display an almost missionary zeal to get them adopted by the professions. There is no harm in this since technical feasibility and economic advantages act both as brakes and as accelerators.

However, when it comes to the applications of science in the arts, in our case the culinary art, we scientists should be careful in making propaganda for new techniques and for purportedly improved results. Are we sure that a soufflé that has increased threefold is better, more enjoyable, than one that has merely doubled in size?

Thus, meringues increase only little while being dried out in the oven, but we enjoy biting into them and chewing their light, crisp insides. It is possible to make lighter meringues by drying them out at subatmospheric pressure, so that they expand to six or eight times their original volume. The sensation of biting into a "vacuum meringue" is the same, but there is hardly anything there to chew. The meringues are certainly bigger, but are they better?

We conclude as already mentioned in the discussion of the choux pastry puffs: technical improvements do not necessarily go hand in hand with gastronomical improvements.

Nicholas Kurti is Emeritus Professor of Physics in the University of Oxford and Emeritus Fellow of Brasenose College. Born in Budapest, Hungary, in 1908, he studied at the University of Paris (Licence és sciences physiques) and at the University of Berlin (Dr. Phil.). He settled in Oxford, England, in 1933, and worked until retirement in 1975 in the Clarendon Laboratory. He first applied his professional skills to cookery while preparing a "Friday Evening Discourse" on The Physicist in the Kitchen, at the Royal Institution, in 1969. This was televised by the BBC, and there followed further TV programs with the Bayerischer Rundfunk, the BBC, and the Australian television. With his wife, he edited an Anthology on Food and Drink by Fellows and Foreign Members of the Royal Society (*But the Crackling is Superb*, Institute of Physics Publishing, Bristol, 1988). He is a Fellow of the Royal Society and is an honorary member of several academies and learned or professional societies, including the Fachverband Deutscher Köche (Guild of German Chefs).

Hervé This-Benckhard is Joint Editor-in-Chief of *Pour la Science*, the French edition of *Scientific American*. Born in 1955 in Paris, he trained simultaneously as a physicochemist at the Ecole Supérieure de Physique et de Chimie Industrielles de Paris (ESPCI) and in literature at the University of Paris. In addition to his activities as a scientific journalist in *Pour la Science* and various radio programs, he is the editor of many scientific books. His first book was a textbook in mathematics and his second publication, *Les secrets de la casserole*, is a best seller in France. For his efforts in molecular gastronomy, he was recently appointed Chevalier de l'Ordre du mérite agricole and Honorary Member of the international club Les Toques Blanches (international club of culinary chefs). More recently, he was made Laureate of the Académie du Chocolat. Nicholas Kurti's collaboration with Hervé This-Benckhard began in 1988; in 1992, together with Harold McGee, they organized the First International Workshop on Molecular and Physical Gastronomy, in Erice (Sicily).

The Chemistry of Good Taste[a]

Anthony Blake[b]

When Adam was tempted by Eve with the forbidden fruit, the human curiosity for new flavor sensations was exploited for the first time. We alone among the animals do not simply eat the food we find; we blend, cook, and process the natural foodstuffs we have available in order to improve its nutrition and palatability: we have cuisine. For most of human history, this search for new eating sensations has been a process of serendipity, of trial and error, yet the flavor of our food is enormously important to most of us; whether it be the selection of a fine wine, the buying of vegetables by the housewife, or the choice of a meal in a restaurant, our appreciation of these are all greatly dependent on the flavor of the product—we are tempted and tantalized by our senses of taste and smell. It can be quite reasonably argued that during human existence, the importance placed on flavor has significantly changed and molded the course of the world's history.

In this article we will examine how flavor has played this key role, yet, although flavor has been a major contributor to the appreciation of food, it is only in the past hundred years that it has been studied chemically in any depth, and only very recently at a physiological level: we still know very little about why we appreciate flavor and close to nothing about how. We should at this point define flavor, a word used loosely and often with different meanings. For the purpose of this article, it is used in the sense that it is our combined impression of taste and smell.

Taste is perceived in the mouth and the smell or aroma in the nose. Both of these senses are in the first instance defenses for our body against the outside world and specifically against eating harmful materials; these senses also, however, direct us to eat desirable or dietetically necessary foods, and this is especially seen when we develop cravings for a particular type of foodstuff. The physiological basis for these phenomena is, however, poorly understood.

If we consider our sense of taste, it was for a long time accepted that there were four basic tastes of sweet, sour, salty, and bitter, but increasingly we now accept a quite separate savory or meaty taste, often given the name "umami," which is typified by the taste of MSG (monosodium glutamate). Whereas taste is limited in its specificities, in contrast aroma is multidimensional; we appear to have the ability to recognize many thousands of individual aroma notes, and much research has been aimed at unraveling the physiology and biochemistry that link our nose to our brain [1]. The precise details are still poorly understood, but we know that ability to detect odor can be trained and is therefore dependent on the smells we experience. We know, however, that some people are able to detect certain aroma chemicals that others cannot, which suggests that their genetic makeup controls their innate ability to detect these chemicals. Whatever the chemistry may be, the sense of smell gives us an enormous ability to detect, store, and remember the aroma of our foodstuffs with a high degree of acuity. This ability to detect and appreciate flavor and to take pleasure in certain combinations has long affected our society and organization.

In the beginning, humans were gatherers of food, picking or scavenging what was available, and then hunters. Eventually, there evolved a society which grew and harvested

[a]*Chemical Intelligencer* 1995(2), 50–55.

[b]Department of Food Science and Technology, Firmenich SA, P.O. Box 239, 1211 Geneva 8, Switzerland

B. Hargittai and I. Hargittai (eds.), *Culture of Chemistry: The Best Articles on the Human Side of 20th-Century Chemistry from the Archives of the Chemical Intelligencer*, DOI 10.1007/978-1-4899-7565-2_17, © Springer Science+Business Media New York 2015

crops and tended livestock. The principal food problem was often one of storage, of how to save food from times of surplus and have it in times of shortage; this was a particular problem with meat and fish, which rapidly deteriorate. The early techniques of drying and pickling often made use of salt. Today we take salt for granted, but it was not always so that salt could be obtained easily—we should remember that the word *salarium* comes from the salt ration given to the Roman legionnaires. Salt was important, not only because we need it in our diet to preserve our bodies' electrolyte balance—hence our craving for salt after sweating—but also because of its use in preserving foodstuffs. After the decline of Rome, European culture largely vanished except for notable exceptions. The Lombards, who occupied northern Italy, were renowned for their culinary skills, and the city of Venice became the Mediterranean center of the salt trade; Venice was famous and initially became wealthy from its salt industry and a trade in salted meat and fish. This key role of Venice as a trading center for foodstuffs was to be strengthened in later centuries by the role it was to play in the medieval flavor industry, the spice trade.

Whereas the salty taste is associated with the physiological need for salt, the sweet taste evolved as a pleasurable sensation and a good indicator of food that was safe and nutritious. For many millennia, the most important sweetener was honey; cave paintings made some 20,000 years ago at Arama in southern Spain show the gathering of honey. Sugar, native to India, did not receive much attention as a sweetener until the spread of the Greek empire under Alexander the Great into India—the home of sugar cane; however, it was not until 1506 that sugar was taken by the Spanish to be grown as a cash crop in the Caribbean. This initiative was followed by the Portuguese, Dutch, and British, and sugar finally replaced honey as the world's main sweetener. By the late seventeenth century, sugar was so important a crop that the Dutch traded New York as part of a deal with the British that allowed them to produce sugar in Surinam. Sugar had ceased to be an expensive novelty and fed the sweet tooth of the industrial revolution, and, incidentally, the bacteria that cause tooth decay. Sugar was also used as a preservative of food though on a smaller scale than salt. Chemists only came onto the sweetener scene by accident; in 1879 at the Johns Hopkins University in Baltimore a young researcher, Constantine Fahlberg from Leipzig, working in the laboratory of Professor Ira Remsen, was investigating the oxidation of o-toluenesulfonamide. Fahlberg unexpectedly produced a condensed heterocyclic molecule (o-sulfobenzoic acid imide), which he accidentally discovered to be sweet—far sweeter than sugar. He patented his product, called it saccharin, and created the first high-intensity sweetener.

Since that time, many other molecules with sweetening properties have been discovered (usually by accident), and some have been commercialized. However, sugar sweetness has always been the standard against which these have been judged, and they have not always been found to be as acceptable; for instance, saccharin has a bitter aftertaste that is perceived differently by different individuals. Yet another sweetener, sucralose, a derivative of sugar itself, is poised to become the latest high-intensity sweetener approved for use in food. Sucralose was discovered in the research laboratories of the sugar producer Tate & Lyle and developed jointly with Johnson & Johnson; after exhaustive safety testing to establish the absence of health risks, it looks likely to be a virtually perfect replacement for sugar with high sweetness intensity, stability to heat and acid conditions, and a flavor profile almost identical to that of sugar itself. Chemistry has and continues to play a key role in satisfying modern consumers' demand for sweetness.

Saccharin (o-sulfobenzoic acid imide)

We should now, however, return to Venice in order to trace another subject where chemistry comes into the kitchen and to consider the major impact that the need for flavor and variety in food had on world history. In the twelfth to fourteenth centuries, the diet in Europe was cereals, mainly bread, with cheese, vegetables, fruit, and occasionally salted or pickled meat. Ingredients which gave character to an otherwise dull diet were herbs and spices. Spices were products of mystery, brought from far-off lands and sold at high prices; the spice traders can fairly be said to have been the earliest forerunners to the flavor industry, and once again Venice played a key part in this trade. By the fourteenth century, the demand for spices was great enough and their costs high enough to provide a major part of the justification for seeking sea routes to the lands where they grew. Once the feasibility of sailing from European ports to the spice-producing countries had been established by the voyages of Columbus, Dias, da Gama, and Cabral, the European race to overseas colonization had begun. The search for the spice-growing countries of Asia had, of course, an unexpected spinoff in the discovery of the Americas, an event that totally changed the diet of the world and the flavors of it. By the nineteenth century, the traders in spices, herbs, and decoctions of these had become the suppliers of flavors, tinctures, and essences to the largely artisanal food industry. As the food industry grew, so too did the need for reproducible and reliable flavor systems, and thus the modern flavor industry was born.

With the improvement of analytical techniques and with increasingly sophisticated knowledge of organic chemistry, the nature of the flavor industry changed; it now became possible to isolate, identify, and synthesize the individual chemicals responsible for flavor that are present in our foods. After the 1950s, explosive growth in knowledge came from the new techniques of chromatography, mass spectrometry, and nuclear magnetic resonance spectrometry. Key components were identified at an increasingly rapid rate and synthesized in significant quantities. Many examples could be given, but the following will serve to illustrate some key aspects of this work and the relevance of it to the food and flavor industry.

Strawberry is one of the most popular fruit flavors worldwide, but until the 1970s it was difficult to produce a food product with the realistic taste of fresh strawberries. Even if the actual fruit was used, the freshly picked taste of strawberries was lost. The reason is that a chemical component in fresh strawberries that gives much of this character is unstable and rapidly modified when the fruit is structurally damaged; this is why frozen strawberries taste so differently from fresh. This chemical was identified as Furaneol [2], and since then strawberry flavors have improved substantially in quality and authenticity.

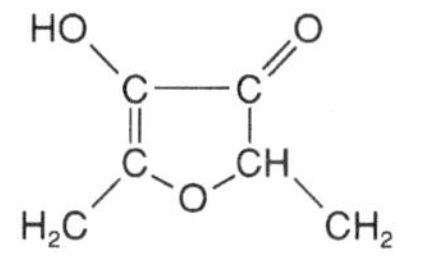

2,5 Dimethyl-4-hydroxy-3(2H)furanone
(Furaneol)

With the growing sophistication of analytical systems, other interesting chemical species that are responsible for characteristic flavors were identified in foodstuffs. Many chemical structures were found in nature for the first time; for example, a chemical that occurs in passion fruit was identified by Winter [3] to be from the family of oxathianes. A series of such compounds has since been identified, and these are particularly important in many tropical fruit flavors; two such chemicals are shown below.

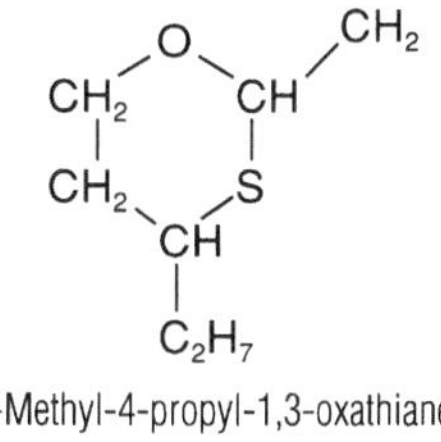

2-Methyl-4-propyl-1,3-oxathiane

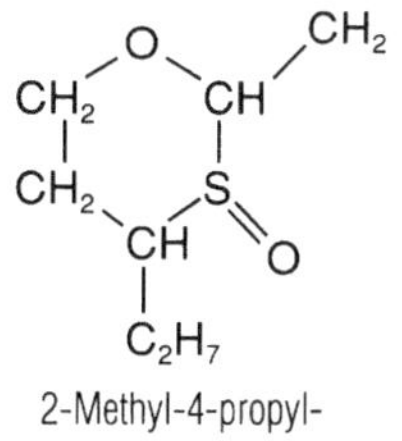

2-Methyl-4-propyl-
1,3-oxathiane-equatorial oxide

A characteristic of these compounds is that at high concentrations their odors are extremely unpleasant, and it is only at very low dosages that their odors become characteristic of the fruit and are appealing.

This phenomenon whereby odorous chemicals are perceived differently depending on concentration is not unusual but makes difficult the task of evaluating such materials. Another example of a very interesting chemical compound that is intensely unpleasant in its pure state but pleasant at low concentrations is the sulfur-containing terpene 1-p-menthene-8-thiol:

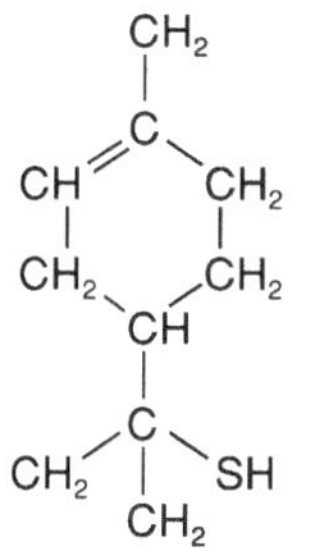

1-p-Menthene-8-thiol

The remarkable feature of this chemical is the low dosage at which it is detectable [4]; for example, its flavor threshold level is one microgram in 100 tonnes (roughly equivalent to a grain of salt in a small backyard swimming pool). Yet at this level in drinks it adds freshness, particularly to citrus flavors such as grapefruit.

Not only can flavor chemicals give detectable physiological responses at incredibly low levels, but synergistic effects can also occur which are impossible to predict. For example, a characteristic of cocoa is its mouth-filling bitterness, and for many years it has been known that the chemical theobromine is partly, but only partly, responsible for this.

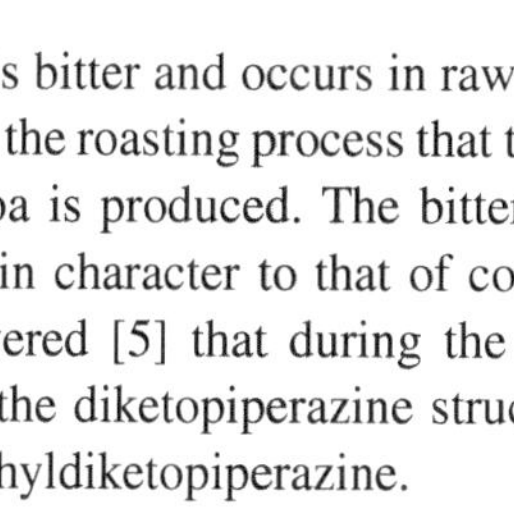

Theobromine

Theobromine is bitter and occurs in raw cocoa beans, but it is not until after the roasting process that the true flavor and bitterness of cocoa is produced. The bitterness of theobromine is different in character to that of cocoa. It has, however, been discovered [5] that during the roasting process compounds with the diketopiperazine structure are formed, such as 2,5-dimethyldiketopiperazine.

2,5-Dimethyldiketopiperazine

Such materials act synergistically with theobromine, and only when they are present together with theobromine is the typical bitterness of cocoa obtained. These diketopiperazines are thought to form during the roasting process from the proteins or peptides in the cocoa beans, and experiments involving the heating of model peptide solutions confirm the feasibility of this.

The search for more and more elusive chemical species continues, and components can be detected at lower and lower levels in the nanogram range. Likewise, the search for more sophisticated separation and analytical techniques goes on. As an example, we can cite the separation of materials of high volatility and lability using supercritical carbon dioxide [6]. At pressures above 74 atm and above 31 °C, carbondioxide becomes a supercritical fluid with the diffusivity and viscosity associated with a gas but the density and dissolving power of a liquid. This supercritical fluid can be used as the moving phase in high-pressure chromatography to effect the separation of components in flavorful extracts in a very gentle way, with no thermal degradation; it then allows the removal of the solvent at room temperature or below. Such techniques combined with more precise analytical systems add to our knowledge of chemical structures and their relevance to flavor.

However, in spite of these improvements in our analytical techniques, there are many foodstuffs, usually cooked ones, whose flavors are still too complex for complete resolution and reconstitution. At this point we should distinguish between two types of flavor. The flavors of raw foods such as fruit, vegetables, or uncooked meat are relatively simple; the chemicals that give rise to them are the products of a limited number of precise biochemical pathways, and, as such, these flavors may be referred to as primary flavors. When, however, we cook food, many chemical interactions can and do take place, and a very large number of secondary flavor components are produced; the number and low levels of these make their analysis very difficult yet they are essential to the final flavor. The human species is the only one that cooks its food, and it is here, in the kitchen or on the barbecue, that chemistry is unconsciously practiced and where the knowledge of the flavor chemist directly interrelates with the art of the chefs de cuisine.

As a specific example, let us discuss the cooking of meat. Meat is rarely eaten raw and in this state has little intrinsic flavor; however, with quite gentle and brief cooking there are major changes in both the character and the intensity of flavor, especially when temperatures over 100 °C are involved. The development of flavor is associated with color changes, and such flavor-producing processes are often called browning reactions. The chemistry of these has now been studied in considerable detail.

The book *The Curious Cook* by Harold McGee [7] provides a fascinating account of the early investigations into this subject. At the end of the eighteenth century, the French chemists Rouelle and Thouvenel investigated the cooking of meat and the generation of meat flavor. For several decades it was thought that a single component was responsible for the savory character of cooked meat, and the name osmazome was given to this material in 1806 by Louis Jacques Thenard at the University of Paris. We now know that the flavor of meat is very complex, and the chemistry of its development equally so. It was another French chemist, Louis-Camille Maillard, who laid the foundation of modern meat flavor research with his investigation of the heat-induced reactions between sugars and amino acids. His interest was primarily in the consequences of such reactions in living systems, but he also recognized the importance of such reactions in foodstuffs. However, as far as food technology was concerned, the research of Maillard remained an obscure piece of work until the 1950s, when Unilever filed a patent disclosing the generation of meat flavor by reacting cysteine and hydrolyzed proteins with pentose sugars such as ribose [8].

We now know that the flavor of meat is generated from relatively simple chemicals in the tissues of the raw meat: sugars and amino acids, especially cysteine, play a major role in the generation of non-specific meaty and roast meat character while fat plays a complex role in determining both the type of meat flavor (beef, chicken, lamb, etc.) and the characteristic notes associated with frying, roasting, and grilling. Although it is only since the 1950s that we have understood the role played by cysteine and sulfides in meat cookery, the use of onions and garlic in the cooking of meat has been practiced since the time of the Pharoahs. Important flavor precursors in raw alliaceous vegetables such as onions, garlic, and leeks are also cysteine derivatives, and the characteristic flavor of these when cut or chopped is due to disulfides; it is interesting to speculate that such compounds and their use in meat cookery actually aid the generation of meat flavor during cooking.

Reactants	Anethole H₂C—CH=CH—⬡—OCH₂	Anisaldehyde O=CH—⬡—OCH₂	Anisketone CH₂—C(=O)—CH₂—⬡—OCH₂
Methanethiol CH₂—SH	Sulfurous vegetable	Stewed radish	Stewed meat
Ethanethiol CH₂—CH₂—SH	As above, weak	Stewed tofu	Stewed meat
Propanethiol CH₂—CH₂—CH₂—SH	As above, weak	Stewed meat	Stewed meat
Butanethiol CH₂—CH₂—CH₂—CH₂—SH	As above, weak	As above, but weak	As above, but weak

Taken from the 1966 Annual Report of the Food Industry Research and Development Institute, Taiwan.

The development of meat flavor has been exhaustively studied, and many of the chemicals in it have been identified. Although no single chemical has been identified as being fully characteristic of meat, van den Ouweland et al. published a paper [9] which supports the view that specific chemical structures can be important to our overall impression of meat like odor. They showed by comparing a series of sulfur-containing molecules that those which possess the structural grouping shown below possess aroma reminiscent of cooked meat:

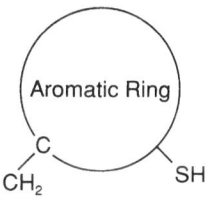

Gradually, we are learning more about the interactions that take place during the cooking of food and that give the flavors we enjoy. The unraveling of this chemistry is not restricted to Western styles of cooking. A publication from the Food Industry Research and Development Institute of Taiwan [10] reports a study of the flavor chemicals that are generated during cooking with five-spice powder, a blend of spices frequently used in Chinese cooking and which contains substantial quantities of aniseed. The results of this work are interesting in that they show that the heat-induced reactions between anisaldehyde and anisketone (components of aniseed oil) and short-chain thiol compounds, typical of those occurring in onions and leeks, can yield chemicals with meaty aromas. Interestingly, no such meatlike aromas are generated when anethole, the main component of aniseed, is used.

Commercial meat flavors improve all the time, and it is the part analytical, part pragmatic, and yet creative development which appears to give the best results.

Another example where kitchen practice preceded chemical understanding is seen when we consider the taste of savory products. Earlier, we discussed the basic taste sensations; for fruit products the key tastes are sweet and acid, and a correct balance of these is important to have a proper appreciation of the flavor of an apple pie or a fruit flan; in the case of savory cooking and specifically in the case of meat products, the taste sensations of salt and umami are important. Umami is a word used increasingly in the West and comes from the Japanese descriptions given to meat broths. Much emphasis is given in Japanese cuisine to subtlety of taste, and it is therefore not surprising that much of the research in this area has been carried out in Japan. The word umami roughly translates as "mouth filling," and the identification of the chemicals that give rise to this sensation were identified at the University of Tokyo [11, 12]. The umami taste is triggered by glutamate ions and also by 5′-ribonucleotides derived from the breakdown of RNA (ribonucleic acid); specifically, 5′-inosinate and 5′-guanylate anions have a pronounced umami effect, which is now recognized to be distinct from the other basic tastes of sweet, sour, salty, and bitter—umami [13] is our fifth taste.

Although umami has only been recently recognized by food scientists, it has been important to chefs for centuries. We see this when we look at the foods that contain it. Seaweed is a rich source of glutamate and has long been used in Japan as a condiment; tomatoes are rich in glutamate, but the chef who first created spaghetti sauce did not realize this, nor did he realize that when he added grated parmesan he further boosted its level; anchovy extract is a key component in Worcestershire sauce and is rich in umami taste. Today, when we go to our local Chinese restaurant, beef and vegetables cooked in oyster sauce is usually on the menu, and once again oyster sauce contains glutamate. We have always used mushrooms in our savory cooking, and the significance of truffles, shiitake, bolets, and other species of fungi becomes much more obvious when we understand that these often contain important quantities of the ribonucleotides that also contribute umami taste.

It is to be hoped that these few examples show how knowledge of flavor chemistry can help us understand and make better use of our food materials. The development of cuisine is alive, and never before have consumers been so interested in tasting and trying the foods of other countries. As the food industry develops new products and new techniques of manufacture, storage, and preservation, it is certain that the flavor industry will be called upon to develop the flavor systems to guarantee the flavor quality of the end products.

The aim of this article has been to give you some food for thought with some thoughts on food and to show that having a chemist in the kitchen can also be in very good taste.

References

1. Gibbons, B. *National Geographic Magazine* **1986,** *170*(3), 324–361.
2. Willhalm, B.; Thomas, A. F. German Patent DE-AS 1517023, Firmenich SA.
3. Winter, M. French Patent FR-A 7524121, Firmenich SA.
4. Demole, E.; Enggist, P. European Patent EP-A 54847, Firmenich SA.
5. Pickenhagen, W., et al. German Patent DE-AS 2445674. Firmenich SA.
6. Flament, I., et al. In *Supercritical Fluid Processing of Food and Biomaterials;* Rizvi, S. S. H., Ed.; Blackie Academic & Professional: London, U.K. 1994; Chapter 5.
7. McGee, H. *The Curious Cook* Macmillan: New York, 1990.
8. Morton, I. D.; Akroyd, P.; May. C. G. U.S. Patent 2,934,437, Lever Bros. Company NY.
9. Van den Ouweland, G., et al. In *Thermal Generation of Aromas;* ACS Symposium Series 409, Eds. Parliment, T. H. et al. American Chemical Society: Washington, D.C., 1989; Chapter 40.
10. Food Industry Research and Development Institute, Taiwan. Annual Report, 1988.
11. Ikeda, K *J. Chem. Soc. Tokyo* **1909,** *30,* 820.
12. Kodama, S. *J. Chem. Soc. Tokyo* **1913,** *34,* 751.
13. *Umami: A Basic Taste;* Kawamura, Y.; Kare, M. R. Eds.; Marcel Dekker. New York, 1987.

Anthony Blake has degrees in chemistry and biophysics from the University of Oxford. He has worked for the last 28 years in the food and flavor industries, 20 of these with the Unilever group and the last 8 years with the Geneva-based company Firmenich SA, where he is director of Food Science & Technology. He is a fellow of the British Society of Flavourists, and his hobbies include cooking (his specialities are Chinese and Indian) and the history and development of food.

The Cooking Chemist

Can a Cooked Egg White Be "Uncooked"?[a]

Hervé This-Benckhard[b]

Egg white is a solution of proteins in water, about 6 g of proteins to 34 g of water. When egg white is heated, it first becomes milky and then its consistency changes: it becomes more viscous and then turns into a gel, first soft and finally rubbery. The generally accepted explanation of this phenomenon is as follows. About 70% of the proteins in egg white are globular proteins. When egg white is heated, the globular proteins uncoil: the protein has become denatured. The hydrophobic parts, which previously were hidden inside the globules, are now exposed to water, and to avoid this contact they link up with similar groupings of neighboring molecules and form a network. The protein has coagulated.

In addition to the hydrophobic bonds, there are other types of links that could lead to coagulation. They are, in ascending order of strength: hydrogen bonds between a donor atom and a hydrogen atom on a lateral group of an amino acid, disulfur bridges between two cysteine molecules, and, finally, covalent bonds. The main purpose of this investigation was to determine which of the above bonds was responsible for the coagulation and to see whether coagulation could be reversed by breaking those bonds or, in simple language, whether a cooked egg white could be "uncooked."

The following experiment was carried out. An egg white that had been cooked was whisked. Since it is known that whisking breaks the hydrophobic and hydrogen bonds, the gel would have turned into a liquid if those bonds had been responsible for the coagulation. However, all that the whisking did was to break up the gel into tiny fragments.

To see whether the disulfur bridges were responsible for the coagulation, about 1 g of sodium borohydride, which is a strong reducing agent, was mixed with the broken-up egg white. A few seconds of whisking produced a foam containing no solid particles. After a few hours, this foam turned into a translucent liquid, which under the microscope looked identical with fresh egg white. This result proves that disulfur bonds are responsible for the coagulation of egg white and, as a corollary, that covalent bonds play no part in the cooking of egg white.

In the unlikely event that someone wants to uncook an egg white for culinary purposes, sodium borohydride **must not** be used since it is a poison. However, ascorbic acid, although a less powerful reducing agent, would do.

Finally, it should be emphasized that this experiment was not an exercise in molecular gastronomy. It was just a modest example of the endeavors to establish the scientific bases of culinary processes.

[a]*Chemical Intelligencer* 1996(4), 51.

[b]6 allée Georges, Versailles F-78530 Buc, France

Favorite Recipes[a]

Diversified Nutmeg

Marye Anne Fox[b]

I must admit that my husband (Jim Whitesell, also a chemist) is by far the better cook in our family. These two recipes reflect two uses of nutmeg, which we bring back every year from a family vacation in Jamaica.

LEEK AND MUSHROOM SOUP

8 oz fresh mushrooms, preferably found and identified during a Sunday afternoon forest walk
4 medium-sized leeks
4 T butter
1/4 cup dry sherry
3 cups chicken broth
1 cup fresh cream
Lots of fresh ground nutmeg
Parsley springs for garnish

Clean and coarsely chop the mushrooms. Coarsely chop the leeks after removing the roots and the green tops. Sauté the chopped leeks in the butter until they are tender and transparent. Add the sherry and cook for about one minute more, or until bubbles begin to form in the liquid. Stir in the broth and chopped mushrooms. Heat to boiling and then reduce heat to simmer for about 20 minutes or until the mushrooms are tender. Puree the resulting mixture in a food processor or blender, before adding the cream and nutmeg. Reheat to the desired temperature and serve, with a parsley sprig added as a garnish.

JAMAICAN SUGAR COOKIES

2 1/2 cups white baking flour
1 cup sugar or molasses
1 t. freshly ground nutmeg
1 1/2 T baking soda
Pinch of salt
1/2 cup butter
1/2 cup whole milk or buttermilk
1 large egg
Powdered sugar (optional)

Mix together dry ingredients before adding the butter, milk, and egg. Roll out the dough to a thickness of about 1/4 inch. Cut into triangles and bake at 375 °F for 10 minutes. Cookies can be further decorated by dusting with a nutmeg/powdered sugar mixture, if desired.

Jelly from Poison Elder

Guy Ourisson[c]

Sambucus racemosa is a wild shrub, growing in mountains up to 2000 m throughout central and eastern Europe, but

[a]*Chemical Intelligencer* 2000(3), 55 and 57.
[b]North Carolina State University, Raleigh, North Carolina 27695, USA

[c]Institut de France, Académie des Sciences, 23 quai de 53 Conti, F-75006 Paris, France (deceased)

B. Hargittai and I. Hargittai (eds.), *Culture of Chemistry: The Best Articles on the Human Side of 20th-Century Chemistry from the Archives of the Chemical Intelligencer*, DOI 10.1007/978-1-4899-7565-2_19, © Springer Science+Business Media New York 2015

more widespread: it is a panbo-real species. It is sometimes very abundant in open spaces (clearings) but can be present also in forests.

Despite one of its vernacular names[d] ("poison elder"), it appears to be innocuous, in a large measure certainly because its appetizing coral-red berries are quite unpalatable. Yet, they can not only be fermented into a schnapps, but can also be used to prepare a remarkable jelly.

These berries can be extremely abundant and conspicuous. They form palm-size racemes and are usually at 1.5–2 m height, and thus easily picked. A short hike through places invaded by *S. racemosa* can easily yield 5–10 kg of berries, mixed with stems, without any apparent depletion of the stock. The only problem for weekend pickers is that these beautiful berries may be just ripe one weekend and moldy the next, if it has rained. This part of the jelly-making process, on a sunny Sunday, is as pleasurable as the last part: eating the finished product.

In the kitchen, berries can quickly be separated from stems with a fork. In a large cooking pan, they are then brought to a boil without any addition of water and kept boiling for a few minutes. The resulting mash can be directly filtered from skins, seeds, and remaining woody stems in the setup sketched here. This is slow (one night is a good indication) and far from quantitative, but if you have brought back 10 kg of these easily collected berries, you will be satisfied with 2–3 kg of juice, which will yield nearly 10 jars of jelly.

The juice is then boiled again under constant supervision, heating the pan from one side so as to produce asymmetric convection currents, which progressively collect to one side a creamy, brightly yellow, soft foam (probably rich in β-carotene), which is carefully removed by scooping it up with a skimmer or even a simple wooden ladle (I use a small skimmer made of artistically woven grasses and rushes, bought on a roadside from an anonymous and sculptural Zulu lady—but an ordinary kitchen skimmer will do). The juice, initially orange and milky, becomes progressively clear and bright red. Let it cool down, refilter it through clean cloth using the same setup as before, measure it, and add sugar to your taste. Usually, recipes of jellies recommend using 1 kg of sugar for 1 liter of juice. I prefer using only two-thirds of that amount of sugar. Bring to a brisk boil, remove any possible foam, evaporate enough water to ensure that a hanging drop sets upon cooling, and pour into pots. Cover with a cellophane foil in the usual way. The resulting

A chair is inverted on a stool (or another chair). To the legs is firmly fixed a piece of clean cloth, hanging over a large enough receiving pan. The mash (see recipe) is poured into the cloth and filtered without disturbance. For the initial filtration, any cotton cloth will do. For the final one, it is highly preferable to use a linen cloth, sometimes sold commercially for the purpose of jelly making.

"poison elder jelly" is beautiful to look at, and its taste is quite unique: sweet with a slightly bitter aftertaste.

The easily fermented stem fragments, cooked seeds, and foam can be disposed of on your compost heap, if you have one.

Quince Recipes

Sir John Cornforth[e]

The most original preparations in our kitchen are made with quinces. Wien we came to live at Saxon Down, I brought with me a young quince tree. It is now around 30 years old and it bears a crop every year (up to 70 kg!).

Quinces are picked (or bought, if available) slightly unripe and are allowed to develop their aroma by storage for a week or two.

QUINCE PUREE

Ripe quinces (ca. 2 kg) are pricked with a fork and placed in a steamer for around 15 minutes or until they are soft enough to cut away the flesh from the core. Instead of steaming, the quinces can be cooked in an oven at 180–190 °C (like baked apples) for 25–50 minutes.

---

[d]*Engl.* dwarfelder, Hart's elder, poison elder, *Germ.* Traubenfliedter, Berghotder, Hirschholder; *Fr.* sureau de montagne, sureau à grappes; *Ital* Sambuco montano, Zambuco di montagna; *Flerm.* Bergvlier, Peterselievlier.

[e]School of Chemistry, Physics & Environmental Science, University of Sussex, Falmer, Brighton, East Sussex, BN1 9QJ U.K. (deceased)

Blemishes on the skins are cut out, then the skin and flesh are cut away from the core, weighed, and boiled with water (600 ml/kg) and sugar (250 g/kg) for 10–15 minutes. The mixture is then pureed (we use a blender). The puree can be frozen and kept indefinitely.

QUINCE FOOL

For 500 ml of puree, use 250 ml of double cream (or 125 ml of double cream and 200 g of thick yogurt). Whip the cream, add the yogurt if used, add the quince puree, mix, and refrigerate. Serve cold.

Part VI
A Chemist's Photo Album

Henry Eyring and Morris S. Kharasch[a]

John D. Roberts[b]

The next round of choices from my photo album are Henry Eyring and Morris Kharasch. The original pictures were taken with flash and a slow, but very fine grain, 35-mm Kodachrome slide film. Although almost a half-century old, the colors are still very vivid and fresh-looking.

Henry Eyring (1901–1981) was born in Mexico and moved in 1912 to Arizona, where his father started a farm. His early education was, to say the least, unusual for chemistry in that he received a B.S. in mining engineering at the University of Arizona in 1923 and a M.S. in metallurgy in 1924. He then turned to chemistry and received a Ph.D. in radiochemistry at Berkeley in 1927. Postdoctoral research with F. Daniels at Wisconsin on the decomposition of nitrogen pentoxide kindled his interest in reaction kinetics, and he spent 1929–30 as a National Research Council Fellow in Berlin at the Kaiser Wilhelm Institute. Here, he and Michael Polanyi developed a potential-energy surface for $H\bullet + H\text{---}H \rightarrow H\text{---}H + H\bullet$, which was to have an enormous influence on chemists and chemistry. Why? Because such energy surfaces provide a framework based on thermodynamics and quantum and statistical mechanics for discussion of comparative reaction rates, especially when coupled with the later idea of an activated complex that could be assigned a more-or-less definite structure and corresponding thermodynamic properties. This approach, along with its subsequent controversial development by Eyring and other chemical physicists, was of extraordinary, although mostly qualitative, value for physical organic chemists interested in the rates and mechanisms of reaction of organic compounds. After a subsequent year at Berkeley, Eyring spent 15 years at Princeton and then, as a

devout Mormon, was convinced to come to the University of Utah and help build up its research program.

Henry Eyring (left), Morris S. Kharasch (right)

Eyring was a wonderful kindly man who had an amazing breadth of research interests and who seemed to contribute more to ways of looking at how to solve problems than actual final solutions. The portrait was taken in his office at the University of Utah in the summer of 1953.

Morris Selig Kharasch (1895–1957) was born in the Ukraine and came to the United States in 1908. He received both B.S. (1917) and Ph.D. (1919) degrees from the University of Chicago and was subsequently a National Research Council Fellow at Chicago until 1922. After six years at the University of Maryland, where he worked primarily on organomercurials, he returned to Chicago and began his truly seminal research on free-radical chemistry. Nothing emphasizes the word "seminal" more than his discovery in 1933 with Frank Mayo of the "peroxide effect"

[a]*Chemical Intelligencer* 2000(3), 56–57.

[b]California Institute of Technology, Pasadena, CA 91125, USA

in the addition of hydrogen bromide to unsymmetrically substituted alkenes. Every student of organic chemistry of that era knew of Markovnikov's rule for the addition of hydrogen halides to such alkenes, which, however stated, resulted with the halogen winding up on the carbon with the least number of hydrogens. But hydrogen bromide was erratic and, even worse, with the same compound could add one way or the other in a seemingly aimless way. Kharasch and Mayo showed that it was peroxides that facilitated the anti-Markovnikov addition. Somewhat later (1936), they showed the discrepancies as being the result of the now well-known competition between free-radical and polar addition mechanisms. Around that time, when polar organic reactions were just becoming understood, Kharasch's many papers on unusual halogenations, copper- or chromium-influenced abnormal Grignard reactions, and such seeming oddities as addition of tetrachloromethane to alkenes were treated with almost incredulous disbelief by his more conservative colleagues. His many illustrious students and postdoctoral fellows included Frank Mayo, Herbert C. Brown, Wilbert H. Urry, Elwood Jensen, and Cheves Walling. My picture was taken about 1951 in Kharasch's office at the University of Chicago.

Part VII

Encounters with Chemistry

Charles Darwin[a]

William B. Jensen[b]

Many famous nonchemists have left behind accounts of their first encounter with chemistry. Whether the person in question was a psychologist, a writer, a critic, an artist, an economist, a mathematician, or a philosopher, whether the experience was brief or prolonged, whether it was pleasant or unpleasant, the purpose of this column is to record these encounters and to do so in the person's own words whenever possible.

The English naturalist Charles Darwin (1809–1882) needs no introduction to modern-day scientists. As a result of the publication in 1859 of his book *On the Origin of Species*, his name has become virtually synonymous with the concept of biological evolution. Among the more than 19 books that he wrote during his life, his travel journal, *The Voyage of the Beagle* (1839), and his study of human evolution, *The Descent of Man* (1871), have remained, along with *On the Origin of Species*, continuously in print and are still widely read.

In old age he wrote a short autobiography, which was published in an abridged form by his son Francis in 1887, along with a collection of his letters [1]. A fully restored edition of the autobiography was finally published by his granddaughter, Nora Barlow, in 1956 [2]. In the autobiography, Darwin revealed that, as a schoolboy, he had developed a keen interest in chemistry, largely at the instigation of his older brother Erasmus, who was being trained, like his father and grandfather before him, for a career in medicine [3]:

---

[a]*Chemical Intelligencer* 2000(2), 59.

[b]Department of Chemistry, University of Cincinnati, Cincinnati, OH 45221-0172, USA

B. Hargittai and I. Hargittai (eds.), *Culture of Chemistry: The Best Articles on the Human Side of 20th-Century Chemistry from the Archives of the Chemical Intelligencer,* DOI 10.1007/978-1-4899-7565-2_21, © Springer Science+Business Media New York 2015

Towards the close of my school life, my brother worked hard at chemistry, and made a fair laboratory with proper apparatus in the tool-house in the garden, and I was allowed to aid him as a servant in most of his experiments. He made all the gases and many compounds, and I read with great care several books on chemistry, such as Henry [4] and Parkes' *Chemical Catechism* [5]. The subject interested me greatly and we often used to go on working till rather late at night. This was the best part of my education at school, for it showed me practically the meaning of experimental science. The fact that we worked at chemistry somehow got known at school, and as it was an unprecedented fact, I was nicknamed "Gas." I was also once publicly rebuked by the head master, Dr. Butler, for thus wasting my time on such useless subjects; and he called me very unjustly a "poco curante" [6], and as I did not understand what he meant, it seemed to me a fearful reproach.

Given the enormous importance of "Gas" Darwin's later work in biology, chemists can perhaps forgive the fact that his early interest in chemistry waned as he grew older.

References and Notes

1. *The Life and Letters of Charles Darwin*; Darwin, F., Ed.; Appleton: New York, 1896.
2. *The Autobiography of Charles Darwin, 1809–1882, with Original Omissions Restored*; Barlow, N., Ed.; Collins: London, 1956.
3. Ref. 1, Vol. 1. p 32.
4. William Henry (1774–1836), British technical chemist and the discoverer of Henry's law of gas solubility. The book referred to is probably Henry's *The Elements of Experimental Chemistry*. Originally entitled *An Epitome of Chemistry*, it went through at least 11 editions between 1801 and 1829.
5. Samuel Parkes (1761–1825), British technical chemist and soapmaker. Author of the *Chemical Catechism*, which passed through 12 editions between 1806 and 1829.
6. Italian for "happy-go-lucky."

Part VIII

Book Reviews

Books

Stalin's Captive: Nikolaus Riehl and the Soviet Race for the Bomb[a]
Operation Epsilon: The Farm Hall Transcripts[b]
Hitler's Uranium Club: The Secret Recordings at Farm Hall[c,d]

Arnold Kramish[e]

Among the jacket photographs of *Stalin's Captive* is that of Paul Rosbaud, appropriately with those of Otto Hahn and Lise Meitner, the discoverers of nuclear fission, and with that of Igor Kurchatov and Yuli Khariton, the "fathers" of the Soviet nuclear weapons. For Rosbaud, a German scientific editor and secret agent, passed the German atomic secrets to the British during World War II. Somehow, those secrets also became known to the Soviets. And, although Rosbaud died in 1963, the three books here reviewed owe a great debt to him.

Even though, largely through Rosbaud, the British had known since 1943 of the lack of German atomic progress, General Leslie R. Groves, head of the Manhattan Project, the American atomic effort, had to know for himself. He directed his Alsos mission to follow the troops through Germany to collect documents and scientists. Ten of the most important German scientists, including Nobel laureates Max von Laue and Werner Heisenberg, were incarcerated for eight months in a Georgian mansion, Farm Hall, at Godmanchester, near Cambridge. In visiting Farm Hall's cellar, one sees traces of the wiring from every room to a secret "listening room," where every word of the scientists was recorded on tape. This was "Operation Epsilon," whose transcripts were released only in 1993, 30 years after Paul Rosbaud's death.

Interrogation was not the only purpose of "Operation Epsilon." General Groves feared that the German scientists would be captured by the Soviets and help them to make the atomic bomb. Little did Groves know that most of his "secrets" had already been acquired through espionage. Also, in a practical sense, none of the ten Farm Hall internees would have been as helpful to the Soviets as Nikolaus Riehl, the metallurgist who supplied the Germans with the uranium

---

[a]*Stalin's Captive: Nikolaus Riehl and the Soviet Race for the Bomb* By Nicolaus Riehl and Frederick Seitz (Rockefeller University). *History of Modern Chemical Sciences.* Series Editor Jeffrey L. Sturchio (Merck & Co. Inc.). American Chemical Society and the Chemical Heritage Foundations: Washington. D.C. 1996. XXII + 218 pp. $34.95. ISBN 0-8412-3310-1.

[b]*Operation Epsilon: The Farm Hall Transcripts* Introduced by Sir Charles Frank, OBE, FRS. Institute of Physics Publishing (U.K.): Bristol and Philadelphia. 1993. IX + 313 pp. ISBN 0-7503-0274-7. Also published by University of California Press: Berkeley. 1993. $30.00.

[c]*Hitler's Uranium Club: The Secret Recordings at Farm Hall* By Jeremy Bernstein (Aspen Center of Physics, Aspen, Colorado) American Institute of Physics: Woodbury, New York. 1996. XXX + 427pp. ISBN 1-56396-258-6.

[d]*Chemical Intelligencer* 1997(2), 59–60.

[e]2065 Whethersfield Court, Reston, Virginia 20191 (deceased)

B. Hargittai and I. Hargittai (eds.), *Culture of Chemistry: The Best Articles on the Human Side of 20th-Century Chemistry from the Archives of the Chemical Intelligencer,* DOI 10.1007/978-1-4899-7565-2_22, © Springer Science+Business Media New York 2015

metal for their pitifully small effort but then supervised manufacture for the Soviet nuclear program.

In *Stalin's Captive*, Frederick Seitz, himself a distinguished solid-state physicist, introduces a translation of Riehl's memoirs, "Ten Years in a Golden Cage." Riehl, born in Saint Petersburg in 1901, had obtained his doctorate in Otto Hahn's laboratory, under the supervision of Lise Meitner. As research director of the Auer Gesellschaft, he developed methods of extraction and purification of uranium. (Incidentally, he was also the inventor of the fluorescent electric lamp.) Until 1950, at Elektrostal, near Moscow, Riehl perfected his own process as well as others that he was confident had been learned through espionage. After those services were no longer required from him, Riehl was transferred beyond the Urals to an institute in Sungul, where the tasks involved fission product chemistry and radiation biology.

After a three-year "cooling-off" period on the Black Sea, to isolate him from current work, Riehl and family were allowed to return to Germany, having been highly honored by the Soviets. At a May 1996 symposium in Dubna, Russia, Riehl's contributions to the early Soviet nuclear program were given due recognition. Riehl died in Munich in 1990.

For the reader wishing a lively account of the German atomic program and some insight into the early Soviet program, the Riehl autobiography is recommended reading. Frederick Seitz's long introduction is itself worth the read. Indeed, much about Seitz's remarkable career is also illuminated. The book is short on the goings-on at Farm Hall, but for those desiring to eavesdrop on the 10 German scientists, minute-by-minute, day-by-day, one of the two "Operation Epsilon" accounts is the source. Each has its particular merit.

A unique merit of the edition published by the British Institute of Physics and the University of California Press is its introduction by Sir Charles Frank, who knew many of the 10 internees during the war and Rosbaud before that. Sir Charles was one of their rare visitors at Farm Hall.

The transcripts themselves are published unannotated, which has a certain appeal, since one is not distracted by scientific explanations, etc, which many readers will not require.

However, for those more comfortable with explanation and who believe they will not miss Sir Charles's introduction, Jeremy Bernstein's *Hitler's Uranium Club* is the book to read. (The title is something of a misnomer, for Hitler ignored the program.) Bernstein, a physicist and well-known science writer, makes much use of Max von Laue's significant and increasingly better-known *Lesart* letter of April 4, 1959, to his close friend and confidant Paul Rosbaud. Rosbaud had been a confidant of all of the Farm Hall scientists, as well as of Nikolaus Riehl. That is how he stole their wartime secrets. However, his relationship with von Laue, strongly anti-Nazi, was particularly close.

In his 1959 letter, von Laue asserted that after the Farm Hall internees heard about Hiroshima, a "version (*Lesart*) was developed that the German atomic physicists really had not wanted the atom bomb, either because it was impossible to achieve it during the expected duration of the war or simply they did not want it at all." The *Lesart* was bitterly contested by many after it was published in 1986 [Kramish, A. *The Griffin: The Great Untold Espionage Story of World War II*; Houghton Mifflin: Boston, 1986; pp 242–248].

The ultimate value of the full text of the Farm Hall transcripts is that it confirms the statements of von Laue's 1959 letter to Paul Rosbaud. For that, and for the general impressions of what 10 scientists talk and gossip about when they are locked up together, reading just one of the books of Farm Hall transcripts is enough for anyone.

Arnold Kramish served in the Manhattan Project and, later, with the U.S. Atomic Energy Commission. He is the author of many books and articles on nuclear history and serves as a consultant to government and industry, including the ANSER Corporation. He contributed this review on November 18. 1996, the 100th anniversary of the birth of Paul Rosbaud!

Books

Force of Nature: The Life of Linus Pauling[a]
Linus Pauling in His Own Words[b]

Zelek S. Herman[c,d]

By virtually any standard of measure, Linus Pauling ranks as one of the most influential and celebrated scientists of the twentieth century. His list of publications (*The Publications of Professor Linus Pauling*, compiled by Z.S. Herman and D.B. Munro; available for downloading on the Internet: http://charon.lpi.org/~zeke) fills 95 pages of fine print and contains over 1100 entries concerned with quantum mechanics, crystallography, molecular biology, medicine, nutrition, biostatistics, nuclear physics, and world peace, including Letters to the Editor and 13 books on subjects ranging from quantum mechanics to the achievement of human well-being. Of these 1100 publications, approximately 700 are scholarly ones, and approximately 300 represent original scientific ideas—a sum unequaled in number and variety by any other scientist, living or dead. Pauling's book *The Nature of the Chemical Bond*, in its three editions and numerous translations, is the most cited scientific book of all time. He is the only person to have been awarded two unshared Nobel prizes (Chemistry, 1954; Peace, 1962), and there are many knowledgeable people who argue that he should have also received the Nobel Prize for Physiology or Medicine for any of such subjects as the nature of the bonding of oxygen to hemoglobin, the alpha helix, the elucidation of the cause of sickle-cell anemia (the first "molecular disease"), or the Pauling oxygen meter, an invention that saved the sight of countless premature infants.

During the course of his long life, Pauling was awarded nearly 50 honorary degrees, including an honorary high school diploma. (Because he was not allowed to take two American history courses concurrently instead of sequentially, Pauling did not graduate from high school.) He also received the U.S. Presidential Award for Merit, the International Lenin Peace Prize, the Gandhi Peace Prize, and nearly every award of the American Chemical Society. In addition, he was awarded memberships in national academies and learned societies throughout the world. Pauling was acclaimed for his outstanding abilities as a lecturer and teacher, and the number of his students who went on to distinguished scientific careers of their own is significant in itself. According to one of his former students, Professor Alexander Rich of the Massachusetts Institute of Technology, Einstein remarked on hearing Pauling's name, "Ah, that man is a real genius!"

Owing to his controversial advocacy and analysis of the effects of large doses of vitamin C in mitigating illness and promoting well-being, and in spite of opposition from orthodox medical practitioners, Pauling almost single-handedly transformed the nutritional habits of people in the developed world regarding the use of supplemental vitamins. For this, his name became a household one and virtually synonymous with vitamin C.

[a]*Force of Nature: The Life of Linus Pauling*
By Thomas Hager. Simon and Schuster: New York. 1995. 721 pp. Hardback. $35. ISBN 0-684-80909-5.

[b]*Linus Pauling in His Own Words*
Edited by Barbara Marinacci. Simon and Schuster: New York. 1995. 326 pp. Hardback $35. ISBN 0-684-80749-1. Paperback $15. ISBN 0-684-81387-5.

[c]*Chemical Intelligencer* 1997(2), 60–62.

[d]521 Del Medio Avenue, #107, Mountain View, CA 94040, USA e-mail: zelek.herman@forsythe.stanford.edu

B. Hargittai and I. Hargittai (eds.), *Culture of Chemistry: The Best Articles on the Human Side of 20th-Century Chemistry from the Archives of the Chemical Intelligencer*, DOI 10.1007/978-1-4899-7565-2_23, © Springer Science+Business Media New York 2015

Thus, it is a difficult task for a biographer to portray this unique human being in such a way as to make him a living being—with his sparkling blue eyes, his incredible intelligence and knowledge, and his belief in the essential goodness of humanity—for those who did not have the unforgettable opportunity to interact with him.

From sociological and political points of view, Thomas Hager, a science writer living in Pauling's native Oregon, has produced, in his meticulously researched book, *Force of Nature*, an eminently readable, interesting, and sympathetic portrayal of Linus Pauling. His description of Pauling's difficult early life (Linus's pharmacist father, Herman Pauling, died when Linus was nine years old, leaving him as the mentor and part-time breadwinner for his sickly mother and his two younger sisters) in turn-of-the-century Oregon makes for fascinating reading, as do the accounts of his friendship with Lloyd Jeffries, his surreptitious and meticulous chemical experiments, and his romance with his wife-to-be, Ava Helen Miller. Hager's extensive narrative of Pauling's troubles with various governmental agencies and the administration of the California Institute of Technology over his peace work and his tireless campaigning for the cessation of atmospheric nuclear testing should be required reading for anybody interested in twentieth-century history. However, perhaps for want of time, space, or energy, Hager's description of Pauling's dealings in the latter part of his life with the two presidents of the Linus Pauling Institute of Science and Medicine, namely, Arthur B. Robinson and Emile Zuckerkandl, is not as satisfactorily written, and it is not easy for the reader to draw substantive conclusions regarding Pauling's role in these rather sordid matters.

Moreover, from the scientific perspective, Hager's accounting of Pauling's numerous scientific contributions is inadequate. In spite of its length, *Force of Nature* has not a single diagram (Pauling was noted for the quality of his scientific artwork), and there are numerous factual errors. For example, on p. 143, Hager states that carbon monoxide has carbon "double-bonded to a single oxygen atom." (Hager should have read Pauling's description of carbon monoxide on pp 289–290 of *General Chemistry*.) On p 192, Hager describes Pauling and Wilson's book *Introduction to Quantum Mechanics* as "staying in print for three decades" when, in fact, the original McGraw-Hill hard-bound book was in print for five decades (the record at McGraw-Hill), and the book is still in print in the Dover reprint. More egregious is the statement on p 220 that the tetrahedral angle is "precisely 104.67 degrees" (it is exactly arccos (−1/3), or approximately 109.47 degrees, as every chemist learns in his or her introductory chemistry course). Hager's book is also characterized by numerous misspellings (e.g., "natuer" in the Library of Congress Cataloging-in-Publication Data on the back of the title page and "*National Reivew*," p 562.) Pauling was a stickler for good proofreading and was an exemplary

proofreader himself, as any of his secretaries or editors can attest.

Linus Pauling was often involved in controversy; indeed, he thrived on it. He had supreme confidence in his scientific and intellectual abilities, which were considerable, and he enjoyed being in the limelight. He was a remarkable human being, and *Force of Nature* is a worthwhile portrayal of his life and times. For those readers interested in his scientific contributions, David Newton's short book *Linus Pauling: Scientist and Advocate* [Facts on File: New York, 1994; 136 pp; see review in *The Chemical Intelligencer* **1996**, *2*(4), 59] is a much better source of information.

Barbara Marinacci's work, *Linus Pauling in His Own Words*, a companion to *Force of Nature*, is a selection of readings from Pauling's works. Marinacci, the sister of Pauling's son-in-law, Caltech professor Barclay Kamb, was a consultant of the Linus Pauling Institute of Science and Medicine. Pauling wrote the introduction some months before he died, at the age of 93, on August 19, 1994, and this book is the closest thing to an autobiography. Marinacci's book is divided into four main sections: I. The Path of Learning 1901–1922; II. The Structure of Matter 1922–1954; III. The Nuclear Age 1945–1994; and IV. Nutritional Medicine 1954–1994. She has chosen "from more than one hundred separate sources, including publications and a number of unpublished manuscripts, notes, and interviews." Each chapter has an introduction by Marinacci, and her comments are interspersed between the selections. The book is very readable and gives an accurate portrait of Linus Pauling. Marinacci's comments are very revealing. I quote at length from her preface:

> Pauling had a changeable voice when he wrote or talked; words, contents, and tone all depended on his audience, the subject undertaken, his mental preoccupations in that period, and even his mood. (Though an eternal optimist and even-tempered, sometimes he sounded gloomy indeed; he also occasionally became noticeably angry, especially over any kind of injustice.) He liked to reminisce about events in his life, so favored stories abound in the archives. Compared with most other people's, his memory was prodigious; yet the same tale might be told differently each time—emphasizing particular educational contexts or dramatic subtexts, giving variable details, and even yielding disparate messages. And being only human, when remembering something, he occasionally distorted, dramatized (he relished well-told tales), made small errors, or simply forgot, especially in old age.
>
> Actually, Pauling rarely abandoned any keen interest; once taken up, it might lead him into a larger area that would incorporate it, or else it would be resumed periodically, as time and circumstance allowed. Pauling's scientific pursuits, convictions, and commitments had a remarkable consistency and logical progression, all the way from structural chemistry to biochemistry to nuclear cautions and peace to orthomolecular medicine.

Marinacci quotes at length (p 124ff) from an interesting manuscript in the Pauling Institute files written by Pauling in 1982 and entitled "The Discovery of the Alpha Helix." While this manuscript was unpublished at the time Marinacci was editing her book, thanks to the efforts of Dorothy Bruce Munro this manuscript was posthumously published in *The Chemical Intelligencer* [**1996,** *2*(1), 32–38].

In contrast to Hager's account, Marinacci's devotes much space to the latter part of Pauling's life and his involvement during the last 22 years of his life with the Linus Pauling Institute of Science and Medicine. It is ironic that Marinacci is so sanguine about the future of the Pauling Institute when she writes (pp 284–285) that "the Linus Pauling Institute now appears healthy enough to move into the twenty-first century, taking the founder's name along with it," for soon after writing this, Barbara Marinacci was dismissed from the Institute, along with most of the researchers and support staff, including this reviewer and Pauling's long-time assistant, Dorothy Bruce Munro. In a move orchestrated by Pauling's psychiatrist son, Linus Pauling, Jr., who took over his father's place as chairman of the board of trustees in 1992, the Linus Pauling Institute ceased operations as a California nonprofit organization at the end of July 1996 and moved its nameplate, endowment of nearly $1.5 million, three junior researchers, and the chief executive officer to Corvallis, Oregon, to become part of Oregon State University, Pauling's alma mater and the place housing the Linus and Ava Helen Pauling Archives.

Pauling's name and his work in chemistry, crystallography, mineralogy, biochemistry, molecular biology, and genetics will live on. Whether his institute and the vitamin research carried on there (as well as Pauling's work on nuclear structure) will be relegated to the dustbin of history remains to be seen. Pauling concluded his Introduction to *Linus Pauling in His Own Words* with "None of us knows how posterity will regard us once we are gone and our work is finished. We must leave it to others to evaluate the contributions we have made to knowledge of ourselves and our world. I hope you, the reader, will find these selected words of mine informative or provocative…and some of them, perhaps, even inspiring."

Books

Strange Brains and Genius: The Secret Lives of Eccentric Scientists and Madmen[a,b]

Walter Gratzer[c]

According to his friend Thomas Moore, Isaac Newton once invited a colleague to dinner in his rooms in Trinity College; the guest arrived to find a dinner for one on the table and Newton deep in contemplation. Unable to gain the philosopher's attention, he presently sat down and consumed the meal. As he rose to leave, Newton recovered himself and his eyes fell on the remains of his dinner. "Really," he observed "were it not for the proof before me, I could have sworn that I had not yet dined." The figure of the absent-minded professor from children's books probably has some basis in fact, for what arguably marks the man of genius is a capacity for intense concentration in pursuit of a long train of reasoning. Einstein, too, had the ability to exclude the world from his cogitations: as a young man, he would sit, mechanically rocking the cradle that contained his howling baby son, while writing equations with his free hand.

Does this trait then count as eccentricity? Clifford Pickover has collected some choice examples of unusual behavior by some of the odder personages of the last three centuries. Here is Nicola Tesla, the inventor who dreamed up the induction motor and gave the United States the alternating current mains supply; who asserted that he could construct a mechanical oscillator to split the Earth in half like an apple; who planned to ionize the upper atmosphere like a neon tube and eliminate the night; and who ended his days caring for sick pigeons in his hotel room in New York.

Then there is Oliver Heaviside, self-taught and reclusive, who furnished his house with granite blocks. The Hon. Henry Cavendish shunned company, especially that of women, and installed a separate staircase in his house to eliminate the risk of a chance encounter with his housekeeper. Francis Galton, who brought quantitative measurement to biology, is remembered here for his numerous original conceits, such as the Universal Patent Ventilating Hat, with its hinged crown to allow the brain to cool off after intense thought (though his many tips to travellers, such as keeping one's clothes dry in a storm by taking them off and sitting on them, or drinking a suspension of gunpowder in a warm soap solution to restore failing strength, do not get a mention). Geoffrey Pyke, whose inventions enlivened the Second World War, Theodore Kaczynski, mathematician and bomb-maker, and (unaccountably in this company) Samuel Johnson are accorded chapters to themselves, and there are briefer accounts of the careers of some also-rans.

[a]*Strange Brains and Genius—The Secret Lives of Eccentric Scientists and Madmen*
By Clifford A. Pickover (IBM T.J. Watson Research Center, Yorktown Heights, New York). Plenum Publishing Corporation: New York. 1998. XIV + 332 pp. $28.95. ISBN 0-306-45784-9.

[b]*Chemical Intelligencer* 1999(3), 56–57.

[c]Randall Institute, King's College, London WC2B 5RL, UK

B. Hargittai and I. Hargittai (eds.), *Culture of Chemistry: The Best Articles on the Human Side of 20th-Century Chemistry from the Archives of the Chemical Intelligencer*, DOI 10.1007/978-1-4899-7565-2_24, © Springer Science+Business Media New York 2015

Statue of Nicola Tesla in Niagara Falls. (Photo by I. Hargittai).

Many of the most celebrated eccentrics to have graced the history of science have not earned a place in Pickover's pages—the great mathematician Kurt Gödel, for instance, who starved because he believed his enemies were out to poison him, or, in an earlier era, the Bucklands, father and son, whose hobby it was to eat all animals that came their way. (The nastiest, Buckland fils recorded, was fried mole, at least until he tried the stewed bluebottles; but mice en croûte were delicious and frequently served up for dinner guests.) Perhaps these and others do not conform with Pickover's thesis that "most of the greatest minds through history believed in God and were interested in religion." (No

room here for the great figures of the Enlightenment.) Pickover also comments that "like many inventive minds ... Tesla never married." But the same surely must be true of many minds of no consequence at all. "Obsessive-compulsive behavior" (avoiding the cracks in paving stones for instance), Pickover believes, is a common attribute of geniuses. They tend also to have little concern with sex, and here Pickover gives St. Paul as his prime exemplar. I would not want to trade saints with Dr. Pickover, but surely one could offset his preference by St. Augustine's heartfelt prayer, "Make me chaste and continent, but not just yet." More, "many eminent people have had at least one child take his or her life." Galton, whose statistical investigations included such subjects as the efficacy of prayer, would have made short work of these assertions.

Read Pickover, then, for a collection of curious and often entertaining stories, and for some painless potted biography, but not as a serious contribution to the etiology of scientific genius. For scientist of his evident standing, he is also at times a mite credulous. Did Tesla's hair really turn white overnight—though transiently—when his mother died? Did the supposed deficiency of glial cells in area 39 in the left hemisphere of Einstein's brain indeed explain why he did not speak until he was 3? I confess I prefer the story, apocryphal as it may be, that Einstein's first utterance was a loud complaint about the scalding hot milk; and when his astonished and delighted parents asked why, if he could speak, he had not revealed it before, the sapient child replied: "because up to now everything was in order."

Books

Der Rücktritt Richard Willstätters 1924/25 und seine Hintergründe. Ein Münchener Universitätsskandal?[a,b]

H. Steffen Peiser[c]

The *Berliner Tageblatt* of June 27, 1924, had it right! While German universities after World War I felt discrimination from world science, they discriminated commonly against some of their own citizens because of their race or religion. That newspaper's correspondent, also correctly, linked the resignation of Richard Martin Willstätter to the exclusion of V. M. Goldschmidt's scientifically unsurpassed application for the mineralogy professorship at the renowned University of Munich. The cause, the newspaper alleged, was anti-Semitism. Indeed, Willstätter and Goldschmidt shared a Jewish background. Yet, only in his posthumously published autobiography did Willstätter clearly reveal that his resignation was a protest against the University's conduct, and especially its handling of the Goldschmidt affair. Litton analyzes this episode from all sides, presenting many relevant and fascinating details from distressing and often contradictory sources, but carefully leaves some ultimate conclusions to the reader. Here is a complex story worthy of a tragic plot for a play or film.

Nobel laureate (for chemistry in 1915) Richard Martin Willstätter (1872–1942) held leading professorial appointments at the University of Munich, the Eidgenössische Technische Hochschule in Zurich, and the Kaiser-Wilhelm Institut in Dahlem, Germany; he then returned to a full professorship at his old university in Munich. After his 1924 resignation, he continued research to some extent but never accepted another appointment. He had married in 1902, but his wife died in 1908, leaving a son, who died at the age of 11, and a daughter, Margarete (later Bruch), whom the author does not claim to have sought out for her viewpoint on her father's resignation. Willstätter was always popular with students, and they remained loyal to him after his resignation. Willstätter had endured anti-Semitism even in his boyhood. Much later, under the Nazis, he was deprived of his possessions, and he died in exile in Switzerland.

[a]*Der Rücktritt Richard Willstätters 1924/25 und seine Hintergründe. Ein Münchener Universitätsskandal?*
In German, *The Resignation of Richard Willstätter 1924/25 and Its Background. A Scandal at the University of Munich?*
By Freddy Litton. Institute für Geschichte der Naturwissenschaften: Munich. 1999. VII + 88 pp. DM 19.80 (EUR 10.10). ISBN 3-89241-033-X

[b]*Chemical Intelligencer* 2000(2), 62.

[c]403 Russell Avenue, #313, Gaithersburg, MD, USA (deceased)

B. Hargittai and I. Hargittai (eds.), *Culture of Chemistry: The Best Articles on the Human Side of 20th-Century Chemistry from the Archives of the Chemical Intelligencer,* DOI 10.1007/978-1-4899-7565-2_25, © Springer Science+Business Media New York 2015

This reviewer wishes that the author had described in more detail Wülstätter's innovative methods of isolating natural products, such as chlorophyll, alkaloids, plant pigments, scents, and enzymes. For these contributions, he is rightly celebrated for his leading role in the development of modern biochemistry. The science, however, is not part of Litton's story, but one may wonder whether the mind of a genius engages itself adequately to protect family and personal fortune. This is a reasonable question even in connection with Willstätter's resignation.

Whatever conclusions readers will reach from this little book, they will be shocked by the extent to which the State, politics, and public opinion influenced academic appointments. The Dean at the University, under pressure from these influences, was even able to override basic precepts of established procedures for making such appointments. Furthermore, we see here an all too typical example of how people hide their prejudices in meaningless verbosity, often pleading an insidious defense against any charge of anti-Semitism. We are also reminded that anti-Semitism in Germany was widespread, even life-threatening to prominent persons, much before Hitler's rise to power. One must ask oneself whether Jews did not foresee the dangers. Could Jews by common aggressive action have saved Germany from its dreadful history? Why was it conventional wisdom among Jews to accept bitterly felt indignities in silence and to keep a low profile in all threatening situations in preference to mounting an open and unified protest against maltreatment? Why did some Jews find the characteristics of some other Jews so offensive? And why did Willstätter, in letters about the Goldschmidt affair, on the one hand, express profuse gratitude to the University and, on the other hand, put a wordy veil over his true feelings rather than sharply accuse despicable colleagues on the Faculty? This reviewer asks himself these questions; he lived through those dreadful times. Litton, a true historian, wastes no words on hypothetical questions about what might have been a far happier life for a great scientist.

Part IX

Stamp Corner

Honoring Loschmidt[a]

Alfred Bader[b]

One of the most surprising and instructive stamps honoring chemists in this decade is the Josef Loschmidt stamp (1) issued by Austria in 1995 to commemorate the 100th anniversary of Loschmidt's death on July 8, 1895. In the nineteenth century, Loschmidt was known as an able physicist, the man who first calculated the Loschmidt/Avogadro number in 1865.

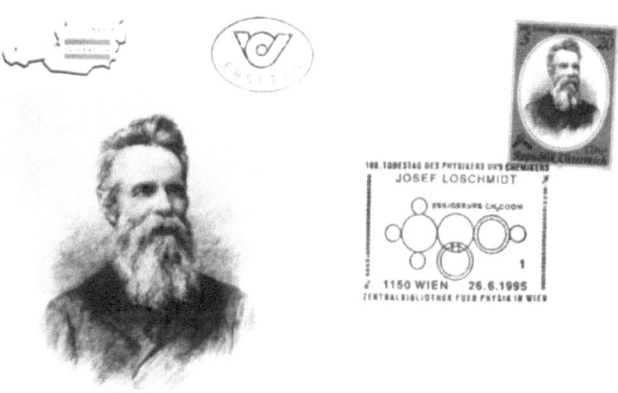

100. Todestag von Josef Loschmidt

First-day postmark (2).

Josef Loschmidt stamp (1).

In 1861, Loschmidt published a little book [1] (Fig. 1) giving the structural formulas of a great many compounds. Coming from an unknown high school teacher in Vienna,

a man without a Ph.D., the book was virtually ignored. It was mentioned only twice in rather derogatory footnotes written by August von Kekulé and in a brief abstract. Not a single Austrian chemist of the nineteenth century ever referred to this book.

Then around 1910, Richard Anschütz, Kekulé's former secretary and his successor as Professor of Organic Chemistry at the University of Bonn, discovered Loschmidt's book, which by then was already very rare. Anschütz was astounded to see that in 1861, the very year in which Kekulé had published his opinion that one could not depict structural formulas, Loschmidt had published several hundred structures, many of them correct and clear. Among these were structures for cyclopropane, mannose, benzene, toluene, phenol, aniline, benzidine, and acetic acid, which was shown on the first-day postmark (2). Anschütz published an article on

[a]*Chemical Intelligencer* 1997(1), 63–64.

[b]924 East Juneau Avenue, Suite 622, Milwaukee, WI 53202, USA

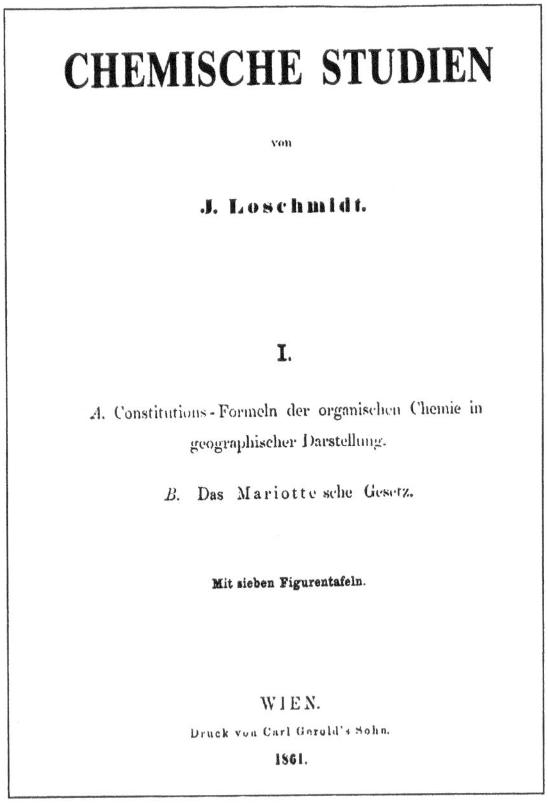

CHEMISCHE STUDIEN

von

J. Loschmidt.

I.

A. Constitutions-Formeln der organischen Chemie in
geographischer Darstellung.

B. Das Mariotte'sche Gesetz.

Mit sieben Figurentafeln.

WIEN.
Druck von Carl Gerold's Sohn.
1861.

Fig. 1 *Title page of Loschmidt's book (Aldrich reprint).*

OSTWALDS KLASSIKER
DER
EXAKTEN WISSENSCHAFTEN.
Nr.190.

KONSTITUTIONS-FORMELN
DER ORGANISCHEN CHEMIE IN
GRAPHISCHER DARSTELLUNG

von

J. Loschmidt

Mit 384 Figuren im Text und einem Bildnis

Herausgegeben von
Richard Anschütz

WILHELM ENGELMANN IN LEIPZIG

Fig. 2 *Loschmidt's book reprinted in* Ostwalds Klassiker der
exakten Wissenschaften *(Aldrich reprint).*

Loschmidt [2] and then went to the enormous trouble of taking Loschmidt's book, reformatting it so that it became very much more readable, and arranging for the publication of the reprint in *Ostwalds Klassiker der exakten Wissenschaften* [3] (Fig. 2).

In 1945, Moritz Kohn [4] published a long article on Loschmidt, essentially abstracting Anschütz's work. Apart from that, few chemists knew anything about Loschmidt's chemistry until William J. Wiswesser of the Wiswesser Line Notation published a startling article [5], "Johann Josef Loschmidt (1821–1895): A Forgotten Genius." The article ended with "all his contemporaries failed to realize that that tiny book of 1861 was really the masterpiece of the century in organic chemistry."

Since then, a good many papers have appeared describing Loschmidt's work as a chemist, and in June 1995, the University of Vienna held a symposium honoring Loschmidt's memory. Among the eminent chemists speaking there were Professor Max Perutz, the Nobel laureate, Professor Carl Djerassi from Stanford, Professor Ernest Eliel from the University of North Carolina, and Professor Albert Eschenmoser from the ETH. The lectures will be published by Plenum Press.

The postage stamp honoring Loschmidt shows beneath his portrait his structure of cinnamic acid. Chemists will note that this was published four years before Kekulé's circular structures of aromatic compounds and long before chemists were certain that the double bond in cinnamic acid is really *trans*. Surprisingly, the stamp does not show Loschmidt's greatest achievement in physics, the Loschmidt number. Unfortunately, at A.S. 20, the value of the stamp is very high, yet it is of such importance that it will surely interest every serious collector of stamps related to chemistry.

References

1. Loschmidt, J. *Chemische Studien;* Carl Gerold's Sohn: Vienna, 1861. Reprinted by the Aldrich Chemical Company, Milwaukee, WI, Cat. No. Z-18576-0, 1989.
2. Anschütz. R. *Chem. Ber.* **1912,** *45,* 539.
3. Anschütz, R. Ed. *Konstitutions-Formeln der organischen Chemie in graphischer Darstellung von J. Loschmidt;* Ostwalds Klassiker der exakten Wissenschaften No. 190; Wilhelm Engelmann: Leipzig, 1913. Reprinted by the Aldrich Chemical Company, Milwaukee, WI, Cat. No. Z-18577-9, 1989.
4. Kohn, M. *J. Chem. Educ.* **1945,** *22,* 381.
5. Wiswesser, W. J. *Aldrichimica Acta* **1989,** *22*(1). 17.

Russian Crystallography[a]

István Hargittai[b]

Crystallography has had a long tradition in Russian science. It was primarily the geological-mineralogical branches that produced crystallographers at the beginning, whereas later many of the leading crystallographers came from physics. Stamps commemorating Vernadskii, Fersman, and Baikov, all of whom were crystallographers, most notably the first two, were featured in these pages recently [1]. Here we present other aspects of Russian crystallography for which the stamp illustrations show mainly crystals and minerals rather than scientists. Although scientists often appeared on Soviet stamps, no crystallographers other than the three mentioned above did, and those three had added importance owing to their contributions to the Soviet nuclear program [1].

The history of Russian crystallography includes real giants. It is so rich in achievements and great individuals that singling out any small group of them is arbitrary. In this spirit, the aim of the present article is to give some flavor of Russian crystallography rather than to survey this branch of science in Russia. Perhaps the beginnings may be easier to characterize by considering the contributions of selected individuals than are the later times. Let us first mention three examples. A.V. Gadolin (1828–1892), a Russian artillery professor, a Finn, contributed to the classification of crystals into geometrical crystal classes, Evgraf S. Fedorov (1853–1919), among others, pioneered the system of 230 three-dimensional space groups, and G.V. Vul'ff

(1863–1925) made important contributions to the study of the structure and growth of crystals.

Evgraf S. Fedorov (1853–1919)

The *Historical Atlas of Crystallography* [2] may be as reliable a guide in singling out scientists as one can have, since it is the product of a concerted international effort. In addition to Gadolin, Fedorov, and Vul'ff, its picture gallery shows a number of other Russian crystallographers. There is the polyhistor Mikhail V. Lomonosov (1711–1765), who can be viewed as at least as much a representative of many other fields in science as of crystallography. There are quite a few stamps commemorating Lomonosov that are connected with the Russian Academy of Sciences, which he founded, and the premier institution of higher education, the Moscow State University, which was named after him. One of the Lomonosov stamps is shown here.

Continuing with the list of crystallographers presented in the picture gallery of the *Historical Atlas*, there is Aleksei Shubnikov (1887–1970), who is known for his work on classification, antisymmetry, and physical crystallography. In 1988, we republished three important papers by Shubnikov in a volume dedicated to his memory [3]. Shubnikov may be

Editor's Note: In addition to stamps, this section welcomes contributions on medals, coins, and other collectibles.

[a]*Chemical Intelligencer* 2000(4), 61–62.

[b]Department of Inorganic and Analytical Chemistry, Budapest University of Technology and Economics, Szt. Gellert ter 4, 1111 Budapest, Hungary
e-mail: istvan.hargittai@gmail.com

best known worldwide for his book on symmetry, which originally appeared in Russian in 1940 and was later revised and augmented jointly with one of his pupils. This expanded version was then published in English translation [4].

Other names on the list include Boris N. Delone (1890–1980), who made important contributions to geometrical crystallography, including three- and higher-dimensional tiling, and Nikolai V. Belov (1891–1982), whose most famous work was in the crystallography of silicates. The final name on the list is Aleksandr I. Kitaigorodskii (1914–1985), best known for his work on molecular packing [5]. The highly original Kitaigorodskii was a minor *enfant terrible* in Soviet science with his flamboyance and lack of respect for bureaucratic authority. Thus, it is no wonder that he was never made a member of the prestigious Science Academy and was more appreciated abroad than at home. Using a rudimentary so-called structure finder, he evaluated the efficiency of space utilization in the molecular packing for all 230 three-dimensional space groups and recognized the importance of complementarity in packing. His predictions of the relative frequencies of these groups have superbly withstood the test of time, having been corroborated by information on hundreds of thousands of crystal structures. Kitaigorodskii headed a large laboratory of chemical crystallography at the Institute of Element-organic Compounds of the Science Academy. Eventually, this laboratory was split, and a separate laboratory was formed that produced an enormous number of crystal-structure determinations. Its head, the late Yurii T. Struchkov, was, at one time, one of the world's 10 most prolific scientists, publishing a scientific paper something like every nine days.

One of the houses in which the Institute of Crystallography operated between 1941 and 1943 and the main building of the Institute of Crystallography in Moscow. (Photographs courtesy of V.V. Udalova).

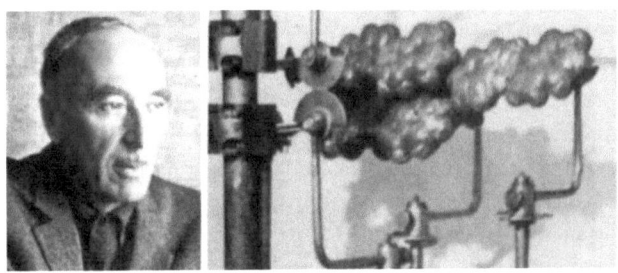

Aleksandr I. Kitaigorodskii (1914–1985) and his "structure finder."(Photographs courtesy of P.M. Zorkii).

An ultimately tragic fate befell another Russian crystallographer, Anatolii K. Bol'direv (1883–1946), Fedorov's pupil and successor at the Leningrad Institute of Petrology. Bol'direv was incarcerated in 1938 on fabricated charges, as was only too common at that time, and died in captivity in 1946 [6]. There may have been others, lesser-known scientists about whose fate we may never learn. Bol'direv's story is not unique either in having the ironic twist that he had spent years in prison under the Czar for revolutionary activities. The best source of what is available about Russian crystallography in the Soviet era was compiled by Ilarion I. Shafranovskii (1907–1994) [6], who can be considered to be Fedorov's scientific grandchild.

The Leningrad (formerly, and later again, St. Petersburg) Institute of Petrology was one of the early centers of Russian crystallography. A stamp displaying the building of the institute commemorated its 150th anniversary. There is a drawing in the background which, on closer inspection, illustrates the preparation of the stereographic projection of a crystal, shown here also in full [7]. The Department of Crystallography, which had a great tradition, was combined with two other departments and, as such, ceased to function as a crystallography institution, to the chagrin of many Russian crystallographers.

In modern times, the Institute of Crystallography of the Science Academy has been the most important institution of crystallography in the Soviet Union and in Russia [8]. It was originally established in Leningrad and moved to Moscow in 1934. During World War II, the staff was evacuated to the Ural region, where the institute was converted into a factory manufacturing crystals for defense purposes and thus helping the war effort. Shubnikov was the first director of the institute, and between 1962 and 1988 B.K. Vainshtein was in charge. The Institute became not only a first-class scientific center but also a showcase of Soviet science. At its peak, it employed about 1000 people, including 250 with Ph.D.-equivalent degrees. The Institute had its own Special Design Bureau for technology transfer of crystallographic instrumentation. Its staff comprised so many high-quality scientists that its scientific council had the right to confer Ph.D.-equivalent degrees, and well over 400 graduates can boast that they earned their degrees in this institute. The Institute has published book series and also has published the most prestigious Russian journal in the field, *Kristallografiya* (published in English translation formerly as *Soviet Physics—Crystallography* and currently as *Crystallography Reports*).

*Boris K. Vainshtein (1921–1996) and Margaret Thatcher
at the Institute of Crystallography in Moscow in 1987.
(Photograph courtesy of V.V. Udalova).*

Acknowledgments I thank Dr. Valentina V. Udalova (Institute of Crystallography, Russian Academy of Sciences), Ekaterina Udalova (Moscow), and Professor Petr M. Zorkii (Moscow State University) for their kind assistance in collecting the illustrative material for this article.

References

1. Hargittai, I. "More than Pure Science," *Chem. Intell.* **2000**, 6(1), 63–64.
2. *Historical Atlas of Crystallography*; Lima-de-Faria, J., Ed.; Kluwer: Dordrecht, 1990.
3. Shubnikov. A. V. "On the Works of Pierre Curie on Symmetry. "In *Crystal Symmetries: Shubnikov Centennial Papers*; Hargittai, I.; Vainshtein, B. K., Eds.; Pergamon Press: Oxford, 1988; pp 357–36; "Symmetry and Similarity," *Ibid.* pp 365–371; "Antisymmetry of Textures," *Ibid.* pp 373–377.
4. Shubnikov, A. V.; Koptsik, V. A. *Symmetry in Science and Art;* Plenum Press: New York, 1974.
5. See, for example, Kitaigorodskii, A. I. *Molecular Crystals and Molecules*; Academic Press: New York, 1973.
6. Shafranovskii, I. I. *Kristallographiya in the SSSR*, 1917–1991 (in Russian; *Crystallography in the USSR*, 1917–1991); Nauka: St. Petersburg, 1996.
7. Hargittai, I.; Hargittai, M. *Symmetry through the Eyes of a Chemist.* 2nd ed.; Plenum Press: New York, 1995; p 395.
8. Anon. "Pyatidesyatiletie Instituta Kristallo-graphii im. A.V. Shubnikova Rossiiskoi Akademii Nauk," *Kristallografiya* **1994**, *39*, 197 (in Russian; "The Fiftieth Anniversary of the A.V. Shubnikov Institute of Crystallography of the Russian Academy of Sciences").

*Stamps on the right:
M.V. Lomonosov (1986).
The 150th anniversary of the
Leningrad (St. Petersburg) Institute of Petrology. The stereographic
projection of a crystal seen in the background on the
stamp is shown separately in full [7].
Seventh World Congress of Crystallographers, held in 1966
in Moscow. There were 3000 participants, half of them from abroad,
and 1000 communications were presented at this meeting.
Three stamps of a set of six from 1963 displaying semiprecious
minerals from the Ural Mountains: 2k. topaz, 6k. amethyst,
and 10k. izumrud.
International Congress on the Enrichment
of Useful Minerals (1968).
Soviet–Polish space flight, 1978: Experiment "Sirena," growing
semiconductor crystals in the cosmos.*

Part X

Articles

Training of a Molecular Scientist, East and West[a]

Kozo Kuchitsu[b]

Prologue

The seed of the present article was an after-dinner talk given at a symposium on molecular structure [1]. In spite of the title, I did not make any clear-cut comparison between the "East" and the "West." It was my attempt to tell the audience, in a Japanese way of speaking, how I had been trained to be a molecular scientist under the influence of our cultural traditions.

My Parents

I was born in downtown Tokyo and was raised in a typical Japanese family. My parents (Fig. 1) were both born in Gumma Prefecture, a lovely district in the countryside located near the mountains, about 120 km north of Tokyo. It was a well-known district, producing good-quality silk and rice. As my father Shuji (1885–1959) was a second son, he had to leave his village and came to Tokyo to work when he was very young. I was told that my mother Hide (1892–1971) was born in a small local city and was the granddaughter of a samurai. In my parents' time, raw fish such as tuna was a precious treat for people who lived far from the ocean. Even after spending half a century in Tokyo, my mother would say whenever we had something to celebrate in our family, "Let's have raw fish for dinner!"

My parents did not speak a single word of English. They lived their entire lives in simplicity, quietness, and politeness. I regard these three qualities as the essence of our culture, mainly handed down with Buddhism and Confucianism. Shuji liked Kabuki plays and had memorized many songs from its plays. Hide also liked to listen to classical Japanese music, but European music was never harmonious to my parents' ears. They turned off the radio immediately whenever they heard a European opera. My mother told me, "Kozo, please don't marry a girl who sings soprano." I had good luck, because I met a girl who was an alto.

My first chance to visit a foreign country came in 1958, when I went to Iowa State University as a postdoctoral

Fig. 1 Shuji and Hide Kuchitsu, sitting at a charcoal brazier in the living room of my home in Tokyo (c. 1955).

Fulbright researcher. When I was leaving home, Hide said to me, "Kozo, I have no good advice to give you about your life in America, because I don't know anything about it. But I hear that American people like alcohol and when they drink they become very noisy. So, please don't drink strong liquor when you are at a dinner party."

Physical Chemistry: A Miracle

My interest in chemistry started when I was nine or ten. In Japanese, "chemistry" is written in two Chinese characters: 化学 (kagaku). The second character, 学, means "study." The first character, 化, means "an unpredictable change." We have another character, 変 (hen), to represent "a change that should be predictable if one is clever enough," such as a "change" in a time schedule. On the other hand, 化 often appears in fairy tales: say, a leaf of a tree suddenly "changes" into a gold coin, or a duck into a swan. So, to us "chemistry" is "a study of miracles".[1] I was first attracted by the exquisite color changes in chemical reactions; I saw magic in them. So when I entered the University of Tokyo, I decided to major in

---

[1]In our standard Japanese language, "science" 科学 (kagaku) and "chemistry" 化学 (kagaku) are pronounced identically. This is confusing, but it is not a rare exception; there are many words in Japanese that

[a]*Chemical Intelligencer* 1995(1), 6–10, 18.

[b]Department of Chemistry, Josai University, Sakado 350-02, Japan

Fig. 2 Chemistry building of the University of Tokyo (February 1994).

chemistry. My parents were very generous; though they had very little background in science and might have wanted me to pursue a career in business, they knew by intuition that science was "a good subject" for me to study.

Physical chemistry caught my interest, because the range of my vision had been widened remarkably when I started learning spectroscopy, from infrared to ultraviolet; thus, I could enjoy many more "colorful" reactions. I was taught in lectures and in reading that great explorers of physical chemistry had the "brain" and "hands" of a physicist and the "eyes" and "heart" of a chemist. I met Professor San-ichiro Mizushima in the Department of Chemistry (Figs. 2 and 3), who was indeed one of these distinguished physical chemists [2, 3]. He had discovered rotational isomerism in substituted ethanes by Raman spectroscopy in collaboration with Professor Yonezo Morino (Fig. 4) [4, 5]. I was so thankful that I could attend Professor Mizushima's lectures; I was illuminated by the idea that the water molecule was not

---

have different meanings and characters but are pronounced identically. The original Chinese words would be clearly distinguishable in speech because of their different intonations, but the intonations were lost completely when the characters were brought to Japan. For this reason, to refer to "science, not chemistry" we say "natural science" (schizen kagaku) and to refer "chemistry, not science" we say "study of 'miracles'" (bekegaku).

Fig. 3 Professor San-ichiro Mizushima (1899–1983).

Fig. 4 Professor Yonezo Morino (1908–).

merely a symbol H_2O but that it had a real shape, an isosceles triangle, with a precisely measurable O–H bond length and H–O–H angle, and it was even vibrating with three normal modes of precisely measurable frequencies.

Professor Morino

Being fascinated by the lectures of Professor Mizushima, I was attracted to theoretical and experimental studies of molecular structure. By the way, one of the commonest meanings of my first name, Kozo, is "structure." So I thought that I had been born to be a structural chemist. In 1948, the year of my entrance to the University of Tokyo, Professor Morino came back from Nagoya University; he had graduated from the Chemistry Department of the University of Tokyo in 1931 and had stayed there until 1943. He was already an established structural chemist and was actively doing research in this area. Furthermore, I was charmed by his fine personality. So I decided to enter his laboratory.

Professor Morino was among the youngest faculty members, but he was known to be one of the strictest teachers in the department. There was a story told among the students about how one of his junior colleagues, Professor Takehiko Shimanouchi (1916–1980), had started his world-famous calculations on normal vibrations using the Urey-Bradley force field [6]. One day, when TS was a young student, he

was making a spectroscopic measurement. Dr. Morino at that time held the position of "assistant" and was the immediate supervisor of TS. He was sitting next to TS and analyzing his spectroscopic data with a slide rule, which was one of his favorite computational tools. TS made a terrible mistake, nobody knows what, and Dr. Morino got angry. TS ran out of the room, and Dr. Morino chased him with his slide rule all the way to the end of the floor. So TS was unable to return to his spectrometer and went into theoretical calculations.

When I first came to Professor Morino's office as a senior undergraduate student, in April 1950, he gave me two important instructions. They were the first and the last academic instructions that I received from him. He often gave me good personal advice at other times during our close collaboration of about 20 years, but this advice was only remotely related to science. His first instruction was the following: "I will give you 'gas electron diffraction,' and I will tell you why. I am interested in internal rotation and rotational isomerism." He did not say anything further; at least, I do not remember any of his further remarks. His second instruction was, "Read the *Journal of Chemical Physics*. I will tell you why. You will meet your future teachers and friends there." He was right; I met many of my excellent teachers and lifelong friends in science abroad first through this journal.

However, it was not easy to follow his instructions. I had a hard enough time starting my thesis work in gas electron diffraction, mainly because we were still suffering from the serious destruction caused by World War II. However, it was even harder to "meet my teachers and friends in the *Journal of Chemical Physics*". It was not only because of the severe language barrier. Tokyo was a very different city in 1950. There were very few places in Tokyo where we could find recent issues of this popular journal. One such place, located in the city center (Hibiya), was the CIE Library (Civil Information and Education as I recall), attached to the U.S. Occupation Forces. I had to take a crowded train to get there. At that time there were many "pushers" in each train station. Many of them were students, paid by the national railway system. Every morning I was packed into the train that I took to get to the library. So I had received good physical training before I came to my study in chemical physics.

Professor Morino and I did not spend much time engaged in scientific discussions, but curiously we could understand each other. He knew what I was doing and planning, and I understood what he wanted me to do even without him saying much. One example was our joint effort for a scientific paper. When we were preparing a manuscript, I usually started the rough draft. Naturally, there were many details with which I was not satisfied, but I did not know what to do. When I brought the draft to him, he discovered these points immediately and said, "We should make changes in these places. Let me do the work for you." A few days later, he returned the manuscript to me with the changes that he had made, which were exactly as I had wanted. I did not know why our communication was like this, but I took it for granted and did not give it much thought for 20 years.

Then came the day of his retirement from the University of Tokyo. I went to his office on the morning of March 31, 1969. After short greetings, he said to me, "Kozo, I will give you a book. Here is my science." It was a book of Zen [7].

Zen

I knew that both Professor Morino and his thesis adviser, Professor Masao Katayama (1877–1961), had been well informed in the scriptures of Zen. Though I had never studied Zen seriously, I realized at that moment that I had also been trained in this tradition.

There are two well-known sayings which come from Zen. If I try to interpret them literally in my own words, they are "Do not rely on words" and "from heart to heart." I think they represent similar ideas. One is puzzled by an apparent contradiction: one is not supposed to "rely on" written words, and yet one is given a book. A simple-minded answer to this dilemma is "symbolism." It is hard to understand all that is written in the book given to me by Professor Morino, and I am still "working" on it.

I am told that Zen temples are open to the common people; many people, young and old, men and women, even foreigners, visit these places for quiet meditation and stay as long as they wish. One day, my wife and I wanted to invite to our home a professor from New England who was visiting our department. He said, "I will be happy to come, but I am afraid my wife will not be with me. She is now visiting a temple in Kyoto and won't be back until the end of next week." According to one of my friends who has visited a Zen temple, he was given a very difficult problem for meditation. I do not remember what his problem was, but one famous problem that I know is, "What had you been before your parents were born?" He sat quietly seeking his answer. On the first day, he thought mostly about fishing. On the second day, he thought mostly about his family. On the third day, he became hungry and sleepy. When he started nodding, a priest hit him on the shoulder with a soft bamboo cane. A big noise. He bowed quietly and returned to his meditation. A few days later, he thought he had found an answer, and he brought it to his master. The master, listening for the approaching footsteps, felt that his pupil was not ready. The door of his room was not opened, and "the manuscript was returned to the author by the referee." After a few more days, he found another answer, and this time the door was opened. The master and the pupil met and discussed, probably without words, and the pupil's manuscript was now accepted for publication in the *Journal of Metaphysics*.

Japanese Archery

When I was 12 years old, I was deeply impressed by a booklet on Japanese archery, based on a speech made by Professor Eugen Herrigel (1884–1955) in relation to Zen [8]. When I gave a talk in Texas, I told the following "fiction," being inspired by my fond, but very faint, memory of his remarkable story.[2]

A German philosopher who visited Japan wanted to study the typical Japanese way of thinking, and somebody advised him to study archery (Fig. 5). He was happy, because he was a good rifle and pistol marksman, and he thought that the technique would be much the same. When he came to his master, the first lesson puzzled him. He was told to breathe with his stomach, and he started his argument on physiology. The second lesson was even more puzzling. He was told not to use his muscles. He argued with the master using his knowledge of physics. The master took the strongest bow, stretched it, and asked the German to put a glass of water on his elbow. The water was not disturbed, and the German found that the master's muscle was completely relaxed. Day after day, he was given harder lessons. The most important and the hardest lesson was the following: "Do not try to shoot the target. Invite your target, and when it comes to you shoot yourself." After some time, the German became skeptical. He requested the master to show him the technique, but the master kept on saying that the German had to be patient. Finally, the German decided that this idea was totally unrealistic. So he challenged the master, "I will leave you unless you show it to me today." The master said, "Very good. Come to the archery field this evening." It was a very dark evening. He set a tiny piece of glowing incense near the target, and he shot two arrows from a distance in the dark. The German found that the first arrow hit the center of the target, and the second arrow split the first arrow.

[2] Professor Herrigel was an eminent German philosopher, who taught first in Heidelberg and later in Erlangen for many years. He stayed in Sendai (the central city in northeast Honshu) for more than five years (1924–1929), teaching philosophy and classical languages (mainly Latin and Greek) at Tohoku University. As he was very interested in mysticism, he and his wife wanted to learn about traditional Japanese spiritual life, particularly Zen, by studying Japanese arts. They both started studying archery, and his wife studied flower arrangement and brush painting as well. With the help of a Japanese professor, he received intensive training under the strict guidance of a great master, Kenzo Awa. Owing to his incredible struggle and devotion, he was finally given the fifth grade in archery. His first booklet [8] was based on a speech made before the Germany–Japan Society in Berlin in 1936. After World War II, he published a completely revised and extended version describing his remarkable experience in Sendai 20 years before and his deep philosophical insight into archery and Zen [9].

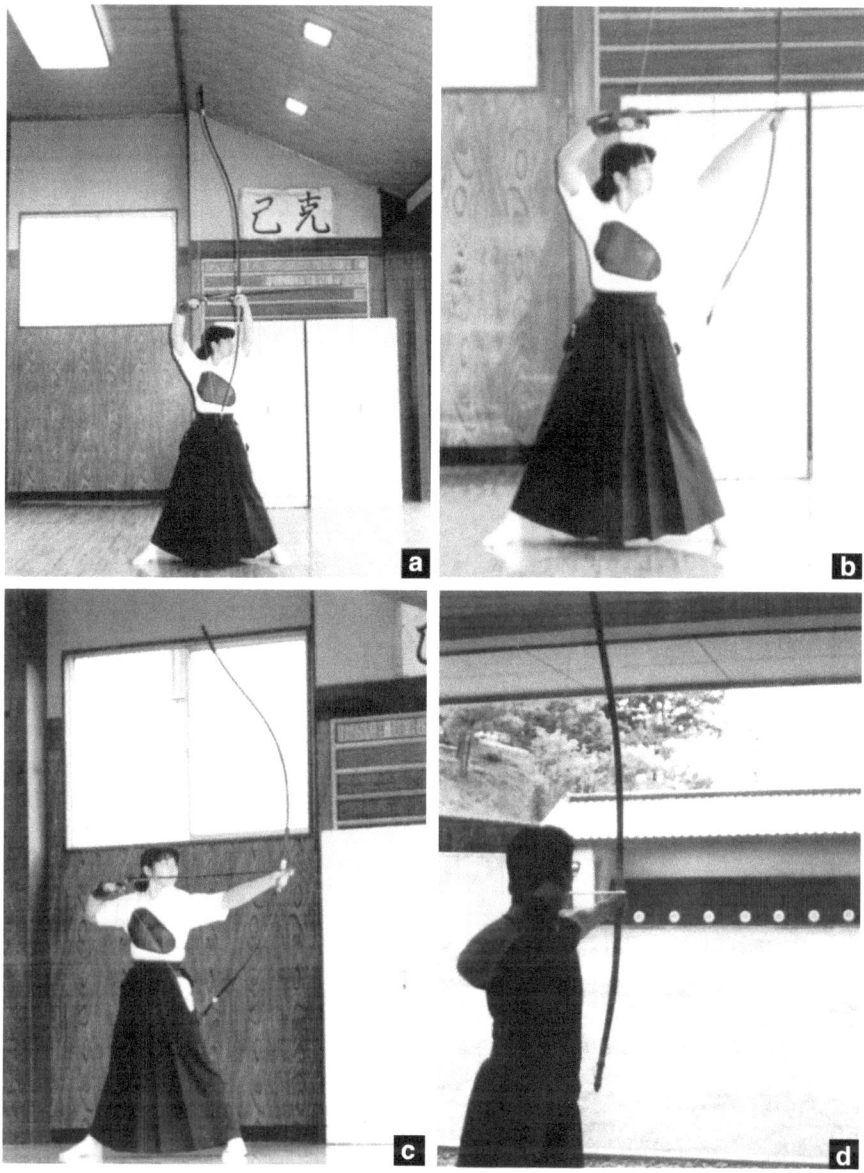

Fig. 5 Japanese archery. A national female student champion, from Josai University, is shown practicing archery: (a) starting, (b) middle, and (c) final positions. The display on the wall, "克己," made with a brush in a classic style (i.e., reading from right to left), means "Be a master of yourself." (d) An archery field at Josai University.

Assistant as an Arbitrator

The reader may wonder how we can train our young scientists in such a way. A scientist must be trained logically and technically instead of metaphysically. Somebody has to be present between an unrealistic professor and a realistic student. Fortunately, our university system has young, capable "assistants." They are typically 25 to 35 years of age, and most of them have Ph.D. degrees. They are ready to train our students chemically, and even physically, because every assistant has a fancy personal computer, which should be more "powerful" than "Dr. Morino's slide rule." On the other hand, it is hard to be a good assistant. An assistant must be very clever, because this is not an independent position, unlike that of an American "assistant professor." An assistant can be 50 years old and still be one of the senior members of his professor's laboratory. In order to obtain an independent research position, he must train himself constantly, so that he may become a master of himself, his students, and his professor.

Castle of Molecular Science, a Modern Style of Training

In olden times, we had beautiful castles in our Japanese cities. Just as in many European cities, farmers, technicians, and merchants were living together, with the castle as their spiritual and political center. I often regard myself as living in such a "gemeinschaft" of molecular science, made up of a group of spectroscopists, crystallographers, theoreticians, etc. I was "born" in this community and am still living in it.

The Japanese molecular scientists in our "gemeinschaft" needed a castle, and one was built in 1975—the Institute for Molecular Science in Okazaki, a middle-sized city near Nagoya. Our leaders had spent more than 15 years in building this "castle." Professor Morino was among these active leaders.

When our leaders were making their plans, they invited Dr. Gerhard Herzberg of the National Research Council of Canada as one of the councillors. I heard that his advice was the following: "Get the best people and let them do whatever they want. Then you will have a nice castle." The Institute has invited the best "samurais," from all over Japan and from overseas, in the past 20 years, so that Dr. Herzberg's advice has been confirmed [10]. Besides being extremely active in research, the Institute is engaged in the training of senior graduate students, jointly with eight national research institutes in Japan. This graduate school, called the Graduate University for Advanced Studies, was opened in October 1988. The graduate students who enter this Institute receive excellent opportunities for training in molecular science by a group of active "professors" and "assistants" with first-class equipment and supercomputers.

Epilogue

Here is my question: "Did my students feel that they could communicate with me without much verbal discussion while they were working in my laboratory?" If a majority of them answer "Yes," then this would indicate that Japan is still under the influence of the cultural tradition of my professors and parents, in spite of the drastic change from silk and rice to cameras and cars, or from a slide rule to a supercomputer. If the students' answers are "No!", then it might be because I was so busy reading the *Journal of Chemical Physics* in the CIE Library that 1 missed the chance to visit the temples and archery fields.

Acknowledgment The author is grateful to Dr. Jon T. Hougen, National Institute of Standards and Technology, for his helpful comments on the original manuscript.

References

1. Kuchitsu, K. After-dinner talk at the Eighth Austin Symposium on Gas Phase Molecular Structure, University of Texas at Austin, March 4, 1980.
2. Mizushima, S. *Annu. Rev. Phys. Chem.* **1972**, *23*, 1.
3. Mizushima, S. *Proc. Am. Philos. Soc.* **1979**, *123*, 164.
4. Mizushima, S. *Structure of Molecules and Internal Rotation;* Academic: New York, 1954.
5. Morino, Y. *J. Mol. Struct.* **1985**, *126*, 1.
6. Shimanouchi, T. In *Physical Chemistry,* Vol. 4; Eyring, H., Henderson, D., Jost, W., Eds.; Academic: New York, 1970; pp 233–306.
7. *Shōbōgenzō-Zuimonki* (正法眼蔵随聞記); narrated by Priest Dōgen and recorded by Ejō, annotated by Ōkubo, D.; Sankibō-Busshorin: Tokyo, 1968. [I am told that this book has been translated into English: Cleary, T. *Record of Things Heard;* Boulder, CO, 1980.]
8. Speech by Herrigel, E. *Die ritterliche Kunst des Bogenschiessens,* Berlin, 1936. Japanese translation: Shibata, J.; Iwanami: Tokyo, 1940.
9. Herrigel, E. *Zen in der Kunst des Bogenschiessens;* Kurt Weller Verlag: Konstanz, 1948. English translation: *Zen in the Art of Archery,* translation by Hall, R. F. C.; Routledge: London, 1953. Japanese translation: Inatomi. E.; Ueda, T.; Fukumura: Tokyo, 1981.
10. Annual Review, Institute for Molecular Science, 1993.

Kozo Kuchitsu was born in September 1927 in Tokyo. He graduated from the Department of Chemistry of the University of Tokyo in March 1951. He became Assistant in 1956, Lecturer in 1961, Associate Professor in 1962, and Professor in April 1969, succeeding Professor Yonezo Morino. His main field of research is physical chemistry, in particular, experimental studies of the geometric and electronic structures of molecules by gas-phase electron diffraction and spectroscopy and of the dynamics of gas-phase molecules and molecular clusters, such as dissociation, ionization, and photoemission caused by collisions with electrons and excited atoms and by photoexcitation. He has taught physical chemistry, quantum chemistry, and structural chemistry to graduate and undergraduate students. He was Dean of the Faculty of Science prior to his compulsory retirement from the University of Tokyo in March 1988. He became Professor of Nagaoka University of Technology in April 1988. He was Chairman of the Chemistry Department when he left Nagaoka in March 1993, again because of compulsory retirement. He is currently Professor of Chemistry at Josai University. He is President of the Physical Chemistry Division of IUPAC (International Union of Pure and Applied Chemistry) and is on the Board of Councillors of the Institute for Molecular Science.

The Lab[a]

Alan L. Mackay[b]

Each venture
Is a new beginning, a raid on the inarticulate
With shabby equipment always deteriorating
In the general mess of imprecision of feeling.

T. S. ELIOT, *EAST COKER*

"The lab" is the frame for one's working life. Joseph Needham extrapolated from *laborare est orare* to *laboratorium est oratorium*. The lab is where one puts questions to nature in dialogue with one's colleagues and struggles with the physical and administrative machinery.[1]

The laboratory founded by J.D. Bernal (1901–1971) [1] at Birkbeck College has, since 1947, been one of the major world centers of crystallography and a focus for intellectual activity. With the arrival in 1977 of T.L. Blundell as head of the department, the lab has moved from small science into the world of big science in which it operates today.

In 1937, J.D. Bernal, at the age of 36 and newly elected Fellow of the Royal Society (FRS), moved from Cambridge, where he had been Director of Research in Crystallography, to become Professor of Physics (and head of the Department of Physics) at Birkbeck College, University of London, in succession to P.M.S. Blackett (later Nobel Prize winner and President of the Royal Society).[2] However, Bernal was almost immediately sucked into wartime activities (for which he received a Royal Medal from the Royal Society) and did not return to Birkbeck until 1946.

Bernal's ideas on how he saw, around 1934, the development of molecular biology can be seen in fictional form in C.P. Snow's novel *The Search* ("Constantine" is Bernal). His general and specific ideas about science were set out in his masterpiece *The Social Function of Science* [2] of 1939.

After the war, Bernal assembled his team for crystallographic research and left the actual physics to the care of his staff, particularly to Werner Ehrenberg and Reinhold Fürth. The physics department was a small one, and physics research was located in the cellars of the old building in Breams Buildings, in the City near Chancery Lane, which had suffered severely in the bombing of London. This laboratory should have been kept as a museum to show the dreadful conditions under which science was then carried out. Research included cosmic rays, with laboratories on the top of the Senate House and down in Holborn Underground Station.

The crystallographic work started in 1946 in two old Georgian houses, also bombed, at 21 and 22 Torrington Square, just north of the Senate House of London University (then still housing the Ministry of Information, which had been Orwell's model for *1984*) and behind the steel frame of the future buildings of Birkbeck College (which had been put up just before the war).

Eventually, two Nobel Prize winners and six FRSs appeared among the research workers emerging from these undistinguished buildings.

One of the two later Nobel Prize winners was Derek Barton of the chemistry department. His laboratory was on the floor above that of crystallography, and on several occasions water flooded through the building. The chemists kept stocks of solvents under the wooden stairs, and the laboratories in the new building were equipped by using the insurance money received after the old laboratories were burned out in a fire.

[a]*Chemical Intelligencer* 1995(1), 12–18.

[b]Department of Crystallography, Birkbeck College (University of London), Malet Street, London WC1E 7HX, UK

[1]The best recent account of the working life in science is by Steven Rose, *The Making of Memory*: Bantam Press: London, 1993.

[2]Collaboration with Blackett's group on rock magnetism continued while Blackett was at Imperial College.

B. Hargittai and I. Hargittai (eds.), *Culture of Chemistry: The Best Articles on the Human Side of 20th-Century Chemistry from the Archives of the Chemical Intelligencer*, DOI 10.1007/978-1-4899-7565-2_29, © Springer Science+Business Media New York 2015

J. D. Bernal and an early model of a monoatomic liquid.

J. D. Bernal about 1960 in 21/22 Torrington Square.

Bernal carefully chose the themes of the research groups to carry out the revolution in molecular biology and materials science that the application of the methods of X-ray diffraction promised. Harry Carlisle took on the structure determination of large-molecular-weight biological compounds, including viruses; Werner Ehrenberg was to make a high-intensity microfocus X-ray source; Jim Jeffery acted as clerk of works, made the former dwelling houses into laboratories, and started the work on the structure of cement that has continued until the present day; Donald Booth started to design and build computers for crystallographic calculations (he made the third computer in Britain, and Booth's unit eventually became a separate department and the foundation of the present computer science department); Sam Levine was to apply mathematics to the understanding of colloidal systems; and several research students began work on the relatively simpler biological molecules that were the essential bricks for building molecular biology. Stan Lenton, demobilized from the Eighth Army, was steward, and Anita Rimel became departmental secretary. When things were in working order, the lab was formally opened on July 1, 1948, by Sir Lawrence Bragg himself, the founder of X-ray crystallography.

In general outlook, the laboratory was to be a materialization of the institute for molecular biology that had been unsuccessfully proposed by Joseph Needham in 1935 to the Rockefeller Foundation. These plans had developed from discussions of the Theoretical Biology Club in Cambridge, of which Bernal was a member.

The lab was very socially and politically conscious, a wide spectrum of views being expressed every teatime. Bernal had also thought about the social organization and had brought the institution of the tea club from Cambridge so that everyone met twice a day with considerable regularity. When I arrived as a part-time research student in 1948, the cold war was in full swing, and the lab was convulsed by the news that Yugoslavia had been expelled from the Cominform and the first break in the communist system had appeared. Several people had worked in student brigades in the reconstruction of Yugoslavia and had firsthand experience.[3] Everyone had his or her own experiences of the war, some in combatant roles, and there was, as in the whole of Britain (which in 1945 had voted the Conservatives out by a huge majority), a conviction that things would have to be done differently. Many people worked for a variety of political and social organizations, and the wartime sense of social purpose persisted for many years after the war itself.

Besides undertaking the creation of molecular biology, the lab set out to solve the scientific problems of cement and other materials. These research interests came from the good connections Bernal had with the Ministry of Works and the Building Research Station. Later, Rosalind Franklin worked on the graphitization of coal, as well as on viruses. That is, besides intellectual exploration, science was seen as the basis for technology, and, reciprocally, applied technology was considered a good place in which to look for fundamental problems. Work on cement has continued from the immediate postwar period to the present day.

Bernal was asked whether he ran his lab on communist lines. He replied no, he had advanced only as far as the stage of feudalism, and the policy was that people should plough the lord's land for half the time and for the other half they could cultivate their own patches. This worked very well. For his own work, Bernal developed a beautiful combination of theoretical and experimental geometry that is the basis of our present view of the structures of liquids. In spite of his writing about the planning of science, Bernal was always wanting instant experiments. If you followed all suggestions, you would get a life's work every week. Sometimes he would ask every morning "what's new?", but at other times he would not be seen for weeks. In many ways it was "guerrilla research," trying to advance wherever you could get a foothold, stretching problems looking for solutions to meet solutions looking for problems. Anne Sayre wrote about the atmosphere of crystallography that "the idea of competition didn't seem to emerge very strongly until the 1960s or so.

---

[3] Edward Thompson who died in 1993, had been leader of the British group who worked on the Youth Railway in Bosnia, and he was the first to start "the new left," with his journal *The New Reasoner*. Yugoslavia has had a significant influence on British politics, and the present events are particularly tragic for this "new left."

Uncompetitive societies tend to be good for women." Certainly we were shocked by a colleague who returned from a year at Purdue and reported that one Ph.D. student would not help another with his vacuum leaks in case he lost his own advantage. Visiting research workers arrived. An early one was Sven Furberg, who rapidly found the way in which nucleotide residues joined in pairs, knowledge which was an essential preparation for the double helix. Our first Japanese, Indian, and Chilean students arrived, starting a network of lifelong friendships.

There has always been a steady flow of visitors. Being 400 yards from the British Museum, the lab was at the center of gravity of the universe, mostly scientific of course, but also political, with an actress and a general or two as well. Paul Robeson sang in the computer room, and Picasso scribbled a large drawing on the wall. Linus Pauling, Buckminster Fuller, Avram Ioffe, André Lwoff, Donald Coxeter, and dozens of others came.

Molecular biology became respectable when, in 1953,[4] the double helix was solved by Crick and Watson in Cambridge (*Nature*, April 25, 1953) and, about the same time, Stalin (and Prokofiev) died.[5] Besides supervising research and working on liquid structure, Bernal had been producing his book *Science in History,* published in 1954 [3]. His final book *The Origin of Life*, published in 1967 [4], was written in the period of his decline and did not have the percipience of his younger period. After his stroke, he was not able to keep up with the runaway progress of molecular biology.

From 1954 to 1961, Aaron Klug worked on virus structures at Birkbeck with Rosalind Franklin (who had come from Kings College in 1953), Ken Holmes, John Finch, and others. In his 1982 Nobel Lecture [5] (and in his careful account of Rosalind Franklin), Klug has described the work that he and his colleagues did in the old slum of 21 Torrington Square when they were creating molecular biology.

For example, there was great excitement when the news came that poliovirus had been crystallized. There was a lab meeting to discuss whether the virus should be brought into the building and how it should be handled. Rosalind Franklin said simply that she would do it. As it turned out, she took the X-ray diffraction pictures at the Royal Institution. I have wondered whether she knew then that she was suffering from a fatal cancer. She died in 1958, having worked until a few weeks before.

For the first period of small science, the lab was desperately underfunded. Money came first mainly from the Nuffield Foundation and from the Rockefeller Foundation. There was great waste of effort in trying to reuse obsolete or cast-off or war-surplus or homemade apparatus. John Jennings had a huge electromechanical calculator captured from the Germans. Much could be bought cheaply in Lisle Street (Soho) from war-surplus stores so that we had a searchlight reflector, optical components from bombsights, radar sets, etc. Workshop training was part of everyone's education.

There was a constant struggle with computing since the determination of crystal structures by X-ray diffraction demanded huge and tedious computation (for three-dimensional Fourier syntheses). All kinds of analog machines were tried. We began with Beevers–Lipson strips and shop cash register machines. Jim Jeffery built a Fourier synthesizer from post office telephone relays and George Parry tried the optical diffraction analog—"the fly's eye." Harry Carlisle's group spent their time man-handling thousands of IBM cards through sorting and tabulating machines. The next generation struggled with punched paper tape and then punched cards for the early main frames. The present power of personal computers is bliss which was then unimaginable. People can now get on with the real science.

The general poverty and squalor reflected the general lack of scientific education in the British ruling class, which is simply not equipped to manage a modern technological society. There was a period of great optimism in 1965, when R.H.S. Crossman visited Birkbeck College to speak about the "white-hot technological revolution" that the coming Labour government (of Harold Wilson) would accomplish. Bernal, myself, and many other academics took part in the Bonnington Group, which planned the introduction of more science into government, but little resulted.[6] Several academics who took the problems of the college to Crossman were reprimanded by the Master, who was a conservative, a classicist, and hostile to science and especially to crystallography. In contrast, Bernal once gave a couple of lectures for the Arts Faculty on Tennyson and science. He wrote with illumination on G.B. Shaw, on architecture, and on dozens of other such topics, his earliest essays being the sharpest [6]. It is a characteristic of crystallographers that they are not content to be a kind of spectroscopic service where they wait for problems to be brought to them. Half of a problem may involve crystallography, and in most cases crystallographers have learned the other half of the problem to retain command of the topic as a whole.

---

[4] 1953 was a tense year. The Russians exploded a hydrogen bomb; the Rosenbergs were executed; and the Korean War was fought to an armistice.

[5] A reviewer of Watson's book asked, "I wonder what else happened that week?"

---

[6] In the *Guardian*, February 14, 1994, David Marquand wrote, confirming this impression: "[The Labour party] had captured the public imagination and seized the initiative well before the campaign had begun. It had done these things because the Wilsonian story of a marriage between socialism and science seemed both convincing and inspiring. Since then the mainstream left has had no project."

Computing in Britain largely developed from the needs of crystallography, and even machine translation of natural language began in the crystallographic lab. The world of computing is staffed by many ex-crystallographers. In Bernal's heyday, people in the department did not hesitate to concern themselves with sociology, economics, politics, architecture, and many other topics. The proper word is "polytropic"—the epithet applied to Odysseus when he is introduced in the first lines of the Odyssey. It is translated as "of many wiles," active in many directions, ingenious, etc. It was an agreeable atmosphere where science was manifestly a total way of life, not compartmentalized for administrative convenience.

An M.Sc. course in crystallography, at first in collaboration with Kathleen Lonsdale of University College, was begun in 1948 and has continued ever since as the staple teaching activity. The national organization for crystallographers, at first a subset of the Institute of Physics, has continued from that time and has engendered a real feeling of community among crystallographers.

I think that Bernal was enormously impressed with what I have called the laser effect—simply that if people all pushed together for a particular objective, then the resultant adding of the amplitudes rather than the intensities produced far greater results than if they pushed individually with random phases. On this account he was able to overlook some of the excesses of regimentation in the Soviet Union, although he is believed to have made representations over the Hungarian events of 1956 [7]. It was a matter of "rendering unto Caesar what is Caesar's and unto God what is God's" in the formulation with which Jesus escaped taking sides when he was asked whether it was lawful to pay taxes to the regime. Science, properly applied and if people all pushed together, would leave them with enough free time in which to plough their own land. However, his personal administration was rather chaotic. If Bernal had not been preoccupied with politics, science would have progressed more but, as he wrote, "the science I could do will be done by others, but unless the political work is done there may be no science at all." If he had been single-minded about clear scientific objectives, there is no doubt but that Bernal would have got the double helix and much else.

It is difficult to recall the tense atmosphere of the cold war period. In October 1962, at the height of the Cuban missile crisis, I went to a big toy shop (Gamage's—now long gone) to buy a birthday present for our son, aged 5. I found that I could not bring myself to do this but instead joined some ad hoc demonstration which marched round the Soviet and American embassies. There might well have been no "next week."

Bernal was very active in antiwar movements following the Stockholm Peace appeal. From 1954 he was president of the World Peace Congress, succeeding Joliot Curie, and traveled all over the place on its affairs. He met Khrushchev and

Mao Zedong. This did not go down well with the Master of the College, but Bernal was scrupulous in keeping his academic appointments. In particular, arriving back from New York on July 22, 1963, he went straight to an academic board meeting but had to leave in the middle. That night he had his first stroke, and effectively the great Bernal days ended. His faculties for communication tragically faded. There was a decade with a kind of interregnum with divided collective leadership until the appointment of Tom Blundell (then aged 35) in 1977, which inaugurated the new period of big science. Tom's slogan was "expand out of the crisis," and it worked. Modernization painfully took place.

Opening of the Crystallographic Laboratory, July 1, 1948.
W. L. Bragg, Gordon Cox, J.D. Bernal.

Today, Tom Blundell is Chief Executive of the Biotechnology and Biological Sciences Research Council, but manages also the scientific leadership of the Imperial Cancer Fund Unit for Structural Molecular Biology embedded in the Department of Crystallography at Birkbeck College. The Department of Crystallography is a world center for the structure of proteins, dealing with the macromolecules involved in Alzheimer's disease and AIDS and other molecules of great importance such as the proteins of the eye lens. There are other topics, especially the structure of DNA, molecular dynamics, and electron microscopy. Outside "big science" facilities such as the synchrotron and neutron scattering are widely used. Work associated with drug design has long been a major line. Substantial grants now come in for other areas of crystallography, and the Department of Crystallography has about 120 full-time workers. Private companies and foundations now provide funding not available from the state. Pharmaceutical companies, in particular, now understand that their business is grounded on the structure of molecules. The fortunes of Britain's largest company rest on the properties of half a dozen molecules. A recent patent case, with implications in the billion dollar range, turned on the difference between two crystalline forms of the same material, and evidence was provided from work done in the lab. The cosy teatimes of the Bernal period have gone, to be replaced by more structured discussion of the strategy of research. The social fabric too has changed and is less collective, politics having largely dropped out, although some people work on environmental issues. This year the material fabric of the building itself is turned upside down for rebuilding, financed partly by gifts from the pharmaceutical companies who employ many of our graduates.

Bernal was a seer or "see-er" rather than a sage and could explain what science could do, both as science itself and in its contribution to society. He inspired people with his coherent vision. Both science and politics continue but are done in different ways. "Money" is added to the dimensions of mass, length, and time which form the framework of science. However, small money guerrilla research can also continue in new forms in the interstices of "big science," using the facilities of electronic mail, personal computing, data banks, etc.

In the recent period of the Thatcher regime, the antiscience attitude has been perpetuated with the absurd belief that wealth is not created by industry and agriculture but by people selling insurance to each other. Science coexists unhappily with other modes of conceiving of reality and is assailed from both the right and the left and by "fundamentalists." It is a sad decline from the industrial revolution, when Britain was the workshop of the world. Things could have been managed so much better both for crystallography at Birkbeck and for Britain in general. The asset strippers are in command, and I hate them for what they have done to Britain.

The old lab has long ago been demolished, but I could still find my way round it in the dark.

References

1. Hodgkin. D.M.C. "John Desmond Bernal (1901–1971)" (obituary). *Biogr. Memoirs of Fellows of the Royal Society* **1980**, *26* (December), 17–84.
2. Bernal, J.D. *The Social Function of Science;* Routledge, London, 1939.
3. Bernal, J.D. *Science in History;* Watts, London, 1954.
4. Bernal, J .D. *The Origin of Life;* Weidenfeld & Nicolson, London, 1967.
5. Klug, A. Nobel Lecture, December 8, 1982.
6. Bernal, J.D. *The Freedom of Necessity;* Routledge & Kegan Paul, London, 1949.
7. Goldsmith, M. *Sage: A Life of J. D. Bernal;* Hutchinson: London, 1980.

This photo of A. L. Mackay is from 1988, taken during his inaugural lecture, Birkbeck College, London.

Alan L. Mackay, elected a Fellow of the Royal Society in 1988, retired from "the lab" in 1991.

He graduated in physics from Trinity College, Cambridge in 1947 and, after a spell in industry, and on completing a Ph.D. in "the lab" of Birkbeck College, part of the University of London specializing in continuing education, he became an assistant lecturer in 1951. He is now Professor Emeritus, retaining part-time connections which permit the continuation of productive research, chiefly in what he calls "flexi-crystallography" or nonconformist crystallography—the geometry of structures outside the framework of the classical space groups in having molecules arrayed in curved rather than planar sheets.

In spite of the paucity of sabbatical leave, he has traveled widely and is an enthusiast especially for Asia and the Far East, having been a visiting professor in Japan (1969 and 1980), in Korea (1987), and in India (1992). He was an exchange fellow in Moscow in 1962.

A major retirement activity is cultivating the friendships which have resulted from the various visits and in continuing research collaborations. The problems of "computing for the over 65s" engage much of his attention (his e-mail address is UBCG04M@CCS.BBK. AC.UK).

He continues to be a cosmopolitan, polymath, socialist, atheist, and skeptic.

His publications, besides scientific papers and miscellanea, include *A Dictionary of Scientific Quotations* (Adam Hilger, Bristol, 1991; also now translated into Spanish).

When Resonance Made Waves[a]

István Hargittai[b]

I was appointed University Professor in 1991. The Diploma of University Professor was signed by the President of Hungary, and it was handed to me by the Secretary of State for Education and Culture. The brief formal ceremony was followed by an informal reception.

It was obvious that the Secretary, himself a former college professor, had read my curriculum vitae. He knew that I had done years of study in Moscow and later spent years at various U.S. universities as well. He asked me whether it was depressing to look back to the darkness of my Moscow years as compared with the freedom I must have experienced in the American laboratories. It was a leading question. In 1991, so soon after the great political changes of 1989–90, many people tried to dissociate themselves from past Soviet connections. This made me feel a little defiant. I told the Secretary that, regardless of the external political system, inside the laboratory I did not feel any difference. In fact, my four years at Moscow State University, 1961–65, were during Khrushchev's "thaw." That period carried the promise of change. Also, Moscow State University was a great school with excellent teachers and students and with often vibrant discussions in the laboratory. Besides, I was in my early twenties at the time.

All that I told the Secretary was true. Eventually, however, I have felt that my response to his question was not quite complete. Even during the sixties, ideology was penetrating the laboratory. A case in point was the nonacceptance of the concept of electronegativity. It was nothing like what had happened with regard to the theory of resonance, though, during the early fifties in the Soviet Union.

An Episode

Some time in the late eighties during a molecular structure meeting in Austin, Texas, we were having a late-night conversation, and the theory of resonance came up. A Soviet colleague and I quite surprised our Western colleagues when we mentioned the ostensibly grave ideological implications of this theory. In the fifties this theory was heresy in the eyes of Soviet officialdom. For me it was only history, but I remembered that I had read a whole volume about it, containing the minutes of a big meeting against resonance. I mentioned this book and painted a rather gloomy picture of the whole affair. The Soviet colleague tried to soften our Western colleagues' surprise. He told us that none of the proponents of the resonance theory had lost his or her life for advocating this theory. At most, some may have lost their jobs. The atmosphere froze immediately in that warm Texas night.

A little while later, I received two letters about that Texas encounter. An American friend wrote that he found the joke cruel and not funny at all. Obviously, he did not think that the story was true. The Soviet colleague also sent a letter. He apologized for having sounded as if he were belittling the situation, but, he continued, surely I must have known that ideological differences did cost lives in other branches of science, so chemistry did not fare so badly, after all.

There seems to be a widespread ignorance of the resonance controversy in Soviet chemistry among most in the West and among the younger generation of chemists in Russia as well. I don't even know whether the story can teach us anything today. They say, however, that "those who cannot remember the past are condemned to repeat it" [1]. Having this in mind and in view of the inherent historical interest of the story, I thought it worthwhile to bring it up in these pages. Hence, I will present here a brief review of the minutes of that ominous meeting I had referred to in our Texas conversation. I have inserted only a few of my own comments in braces.

The Minutes of the Meeting

The minutes of the meeting were published in *Sostoyanie Teorii Khimicheskogo Stroeniya v Organicheskoi Khimii* (*The State of Affairs of Chemical Structure Theory in Organic Chemistry*). Izdatel'stvo Akademii Nauk SSSR (Publishing House of the Soviet Academy of Sciences): Moscow, 1952; 440 pp. This volume is not a light booklet. It is a heavy, hardbound volume of densely printed pages. (It makes heavy reading, too.) The title page is shown in Fig. 1.

The Chemistry Division of the U.S.S.R. Academy of Sciences held a four-day, all-union conference between

[a]*Chemical Intelligencer* 1995(1), 34–37.

[b]Department of Inorganic and Analytical Chemistry, Budapest University of Technology and Economics, Szt. Gellért tér 4, 1111 Budapest, Hungary
e-mail: istvan.hargittai@gmail.com

Fig. 1 Title page of the hardcover volume published in 1952 containing the minutes of the 1951 meeting.

Fig. 2 Statue of A.M. Butlerov (1828–1886) in front of the Chemistry Department of Moscow State University. (Photo courtesy of Dr. A.A. Ivanov, Moscow) The term "chemical structure" was introduced by Butlerov in 1861 (cf. O. Bertrand Ramsay, Stereochemistry. Heyden: London, 1981; pp. 55–57).

June 11 and June 14, 1951, in Moscow. The subject of the meeting was the structure theories of organic chemistry. Four hundred and fifty chemists, physicists, and philosophers attended. They represented major centers of scientific research and higher education all over the Soviet Union. A report entitled "The State of Affairs of Chemical Structure Theory in Organic Chemistry," compiled by a special commission of the Chemistry Division, was presented, followed by 43 oral contributions. An additional 12 contributions were submitted in writing. The conference adopted a resolution and sent a letter to I.V. Stalin.

The letter to Stalin expressed self-criticism for past deficiencies in appreciating the role of theory and theoretical generalizations in chemical research. This has resulted, the letter added, in the spreading of the foreign concept of "resonance" among some Soviet scientists. This concept was an attempt to liquidate the materialistic foundations of structure theory. However, the letter continued, Soviet chemists have already started their struggle against the ideological concepts of bourgeois science. They have unmasked the falseness of the so-called "theory of resonance" and would cleanse Soviet chemical sciences from the remnants of this concept, the letter concluded.

During the meeting, there were repeated references to Stalin's teachings on the importance of the struggle between differing opinions and of the freedom of criticism. [George Orwell's doublespeak of *1984* pales by comparison.]

The report of the Chemical Division was submitted to the meeting by Academician A.N.Terenin on behalf of the special commission. [The names of the members of this commission and many of the names of the speakers participating in the subsequent discussions read like a who's who of Soviet chemistry. There were many academicians among them and also many future academicians.] The report consisted of the following chapters:

1. Butlerov's teaching and its role in the development of chemistry (Fig. 2)
2. The development of structure theory during the second half of the nineteenth century and the first half of the twentieth century
3. Advances of Soviet organic chemistry
4. Quantum chemistry and structure theory
5. About the so-called "theory of resonance"
6. On the mistakes of some Soviet chemists
7. Current state of Butlerov–Markovnikov's teaching on the intramolecular interactions of atoms and on reactivity
8. Perspectives on further development of structure theory

These are telling titles and I mention a few points from Chapter 6 only. Here we learn that Professor G.V. Chelintsev had criticized actively the concept of "resonance" in the press. It was mainly owing to him that Soviet scientific society had turned to this question. It was noted, however, that the basis of his criticism was his own "new structure theory," which completely contradicted the modern theory of chemical structure and was contrary to the experimental facts and theoretical foundations of quantum physics.

Ya.K. Syrkin and M.E. Dyatkina were named as the main culprits in disseminating the theory of resonance in the

Soviet Union. They were accused of having even further developed the erroneous concepts of Pauling and Wheland, of ignoring the works of Soviet and Russian scientists, of idolizing foreign authorities, and of quoting even works of secondary importance by American and English authors.

Others were also mentioned, among them the organic chemist A.N. Nesmeyanov (Fig. 3). He, along with R.H. Freidlina, interpreted the diverse reactivity of chlorovinyl compounds of mercury and other quasicomplex compounds in terms of the "resonance" between their covalent and ionic structures. These lesser sinners, along with many others, however, had eventually repented and had themselves become critics of the application of "resonance."

The report, as well as the subsequent contributions, was followed by questions and answers. Perhaps the most important question was about the idealism of the concept of resonance. The answer to this question started with a quotation from V.I. Lenin, according to which philosophical idealism was a one-sided exaggeration of an insignificant feature of the cognitive process. Such a feature was then detached from the matter and from nature and made into something absolute. The answer then turned to the concept of resonance, where the insignificant features of the cognitive process were specified to be the individual components of the approximate computational techniques employed in the calculation of the molecular wave function. They were made into something of primary importance, as if objectively existing in the molecule, and as if determining a priori the molecular properties. In reality, it was further explained, the resonance structures and their resonance was torn from the matter, and the theory of resonance became an absolute above the matter.

[If it sounds complicated, it is complicated. The sentences were formulated very carefully as the matter was extremely sensitive. The atmosphere was like that of a trial rather than that of a scientific discussion. The problem was not that the theory of resonance was criticized. There are chemists who do not like the description of a molecular structure by a series of resonance structures. What is frightening and mind-boggling is that such a dislike was made into an official dogma with philosophical justification.]

Only a small, though very vocal, group of chemists attacked blindly the theory of resonance and even more those whom they declared to be its proponents. They also attacked quantum chemistry and all of the science of the West. They advocated a return to historical Russian results and suggested that their own theories be used. These theories, however, had been shown to be worthless nonsense by many. However, all those present painstakingly dissociated themselves from the theory of resonance. At times the self-criticism of some excellent scientists was humiliating in the extreme.

It was characteristic of the atmosphere of the meeting that a philosopher (B.M. Kedrov) declared Schrödinger to be a representative of modern "physical" idealism. This made Schrödinger a relative of Pauling's. Furthermore, he stated that Dirac's superposition principle was as idealistic as Heisenberg's complementarity principle and even more idealistic than Pauling's theory of resonance.

Another speaker, a writer (V.E. L'vov), criticized the report for a serious political error, namely, that the protagonists of the theory of resonance were equated with the greatest Soviet scientists. These protagonists had been unmasked as spokesmen of the Anglo-American bourgeois pseudoscience by the press and Soviet society. According to L'vov, the report was vague about the main thrust of the ideological struggle taking place in theoretical chemistry. He also quoted, as a positive example, the criticism of Mendel by T.D. Lysenko, who proved that Mendel's work had nothing to do with the science of biology. Furthermore, L'vov attacked fiercely the theories of Heisenberg as well as those of Heitler and London. He protested the report's view of quantum mechanics as a development of Butlerov's teaching. The most important political task of Soviet chemistry, he declared, was the isolation and capitulation of the insignificant group of unrepenting proponents of the ideology of resonance.

[To me it is a great puzzle why a concept as innocent as the resonance of chemical structures triggered a reaction of such enormous proportions. I cannot offer any rational explanation. An important contributing factor, though, may have been the fear of foreign ideas. The story of resonance should not be viewed in isolation from the rest of Soviet life in the early fifties, the last years of Stalin's reign. To me a question that is most telling was that asked of M.E. Dyatkina: "How do you explain that you are so conspicuously familiar with the teachings of foreign scientists? May it be that you, along with Professor Syrkin, are intentionally bowing to foreign scientists?"

Reading this accusation brings back some personal memories. In 1965, as a Master's student at Moscow State University,

Fig. 3 A.N.Nesmeyanov (1899–1980) on a Soviet stamp. He was Rector of Moscow State University, 1948–1951, and President of the Academy of Sciences of the U.S.S.R., 1951–1961.

Fig. 4 Photo of the late Professor M.E. Dyatkina (1914–1972) by an unknown photographer. I am grateful to professor L.V. Vilkov (Moscow) for acquiring this photo.

I traveled every week to the Institute of Inorganic Chemistry of the Soviet Academy, where Professor Dyatkina (Fig. 4) was giving a not-for-credit course, something like Structural Inorganic Chemistry. The large auditorium was always packed with research workers and graduate students. Professor Dyatkina would basically tell us about her readings of the previous week and her interpretation of recent literature. She was not a colorful lecturer, yet she held our attention fully. She was also our living library, library facilities being scarce.]

The last entry in the minutes of the meeting is a dissenting opinion in the form of a short letter by E.A. Shilov, member of the Ukrainian Academy of Sciences. He was critical of the report and the resolution of the meeting for looking so much backward rather than forward. He suggested concentrating on new results and new teachings instead of conducting scholastic debates about questions such as where does resonance end and mesomerism begin and how does the "healthy" mesomerism of Soviet authors differ from Ingold's "erroneous" mesomerism and how can ideal structures be considered real at the same time. The result of ending such debates would be, Shilov added, that tire efforts and time of Soviet organic chemists could be

devoted to valid and productive work. This contribution was not delivered as an oral presentation during the meeting.

A Final Note

It is an irony of history that in 1952 the already world-famous Linus Pauling was denied a passport by the U.S. State Department, based on his leftist politics [2]. This happened exactly at the time when the theory of resonance, associated primarily with Pauling's name, was vehemently attacked in the Soviet Union. The attackers might have asked Professor Pauling himself about the ideological implications of his chemical resonance theory. He could have explained that "the several structures that are used in the description of a molecule such as benzene by application of the theory of resonance are idealizations, and do not have existence in reality" [3]. Such an interpretation would have taken the ostensibly alien and harmful edge off the theory of resonance, had the accusers been interested in his explanation. According to a later Soviet evaluation of the affair, "The minutes of the meeting is one of the most shameful documents ever created by a collective effort of scientists, and God only knows when it will be possible to wash off this disgrace" [4].

Acknowledgment Professor J. E. Boggs's (Austin, Texas) comments on this manuscript were very helpful.

References

1. *Santayana, Life of Reason.* Quoted in *The Pocket Book of Quotations.* Davidoff, H., Ed.; Pocket Books: New York, 1951.
2. Serafini, A. *Linus Pauling. A Man and His Science.* Paragon House: New York, 1989.
3. Pauling, L. *The Nature of the Chemical Bond,* 3rd ed.; Cornell University Press: Ithaca, NY, 1960; p 287.
4. Okhlobystin, O. *Yu. Zhizn'i Smert' khimicheskikh idei (Life and Death of Chemical Ideas)*; Nauka: Moscow, 1989; p 184.

Delayed Radical

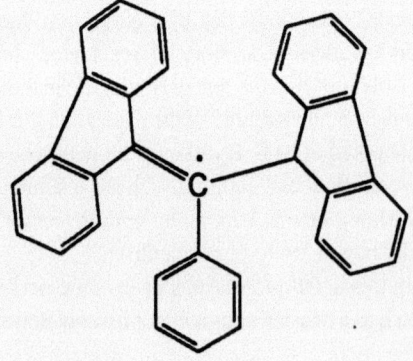

In 1932 C.F. Koelsch submitted a manuscript to the *Journal of the American Chemical Society* describing a free radical, and the manuscript was rejected. The paper was finally published in 1957 (79, 4439). The published version differed from the original manuscript only in some footnotes about quantum mechanical calculations and ESR. The sample itself proved rather stable even during the quarter of a century period. The Koelsch radical is also called the "delayed radical."

Source: Leffler, J.E., *An Introduction to Free Radical.* Wiley, New York, 1993 (PP. 201–2).

Wit and Wisdom of Albert Szent-Györgyi: A Recollection[a]

Irving M. Klotz[b]

During this gathering, we heard almost every known anecdote (true or fictitious) illustrating Albert Szent-Györgyi's humor and whimsy, so in a sense there is little that I can add. Nevertheless, as a self-proclaimed expert on the facets of temperament, personality, and social environment that generate scientific anecdotes, 1 may have some additional insights into "Prof" and his talent for captivating people.

Let me start with probably the most famous story about Szent-Györgyi, his experience with the editor of the journal to which he submitted his early renowned manuscript describing the original isolation and characterization of vitamin C. Recognizing that this substance had the properties of a sugar, but not having determined its detailed molecular structure, Prof decided to name it "ignose." The editor considered this name too flippant and insisted that Szent-Györgyi change it if he wanted the manuscript to be accepted for publication. Prof promptly responded, proposing as an alternative the name "godnose."

I was not at all surprised by Prof's linguistic talents, revealed at that early stage in his scientific career and on other occasions, for he had as an ancestor the greatest linguist of all times, our primordial ancestor—Adam who was—a Hungarian. I shall document that assertion presently. Let me first just remind the reader that although God created the world, it was Adam who invented language. Surely that was the greatest intellectual achievement of all times. Some young people may ask me how I know that Adam invented language. Since the classical education of the modern generation has been so sadly neglected, I must insert a few verses from Genesis:

> And out of the ground God formed every beast of the field and every fowl of the air, and brought them to Adam to see what he [Adam] would call them; and whatsoever Adam called every living creature, that was the name thereof. And Adam gave names to all the cattle, and to the fowl of the air, and to every beast of the field … GENESIS 2:19–20.

So we see that Adam was the first great linguist. And he had no dictionary to consult nor any role model to follow.

Now we must turn our attention to the crucial question: what language did Adam speak? The answer is Hungarian! This was proved by a great scholar of the past century (by coincidence, also a Hungarian), who based his argument on a fundamental principle enunciated by Plato, and by the Stoic philosophers, that a name expresses the inner nature of a thing, that there is an intimate relation between the appearance of a word and its meaning. For example, write out the letters "ollo." This is the Hungarian word for scissors; doesn't the word obviously look like the object it denotes? With examples like this, Hungarian scholars proved that Hungarian was the language spoken in the Garden of Eden.

For balance, I should point out that other cultures do not have the same high regard for the Hungarian language. For example, the Spanish Emperor Charles V said that

> Spanish should be spoken to the gods,
> French to men
> Italian to ladies
> German to soldiers
> and
> Hungarian to horses.

It is my impression that Hungarian jokes, especially the really filthy ones, are replete with references to horses.

This memoir was presented at the conclusion of a memorial symposium for Albert Szent-Györgyi held at the Marine Biological Laboratory at Woods Hole, Massachusetts, in July 1988. A brief abstract appeared in the Biological Bulletin, Vol. 174, p. 228 (1988).

[a]*Chemical Intelligencer* 1995(1), 40–43.

[b]Department of Chemistry, Northwestern University, Evanston, IL 60208-3113, USA (deceased)

Convocation in 1959 at Four Winds (Szent-Györgyi home) in Woods Hole, Massachusetts, to discuss energy transfer and water structure. From left to right, top row: Zoltan Bay, Koloman Laki, Andrew Szent-Györgyi, Irwin Isenberg, Guy Williams-Ashman, T. Fujimori, William McElroy, Sidney Velick; middle row: Irving Klotz, William Arnold, Robert Livingston, Bernard Pullman, Michael Kasha; bottom row: Richard Steele, Henry Linschitz, Marta Szent-Györgyi, Theodor Förster, Alberte Pullman, Albert Szent-Györgyi, Hugo Theorell.

Actually, in regard to what language was spoken in the Garden of Eden, I am most convinced by the contention of the English linguist John Webb that it was Chinese. His argument is very straightforward. What is the first word of any newborn baby, of any race or culture? It is "yä," an unequivocal Chinese word.

Prof was not a modest man as we know from his famous comment about "fishing with a big hook." Likewise, he also said, "research [needs] egotists ... [giant] egotists." However, there was a critical self-insight within him also, as in fact is obvious if one reads the full "big hook" statement as he actually put it in print (in 1962, in *Perspectives in Biology and Medicine*), not just the abbreviated quote popular with journalists and the public media:

"When I took up fishing I always used an enormous hook ... [it is] more exciting not to catch a big fish than not to catch a small one ... I have reduced since the size of my hook."

There are other bits of science folk wisdom to which he could give a refreshing flip; for example, "There is but one safe way to avoid mistakes; to do nothing.... This, however, may be the greatest mistake of all."

Or let me present Prof's own story of how he became a scientist:

"I must have been a very dull child....I read no books and needed private tutoring to pass my exams. Around puberty, something changed and I became a voracious reader and decided to become a scientist. My uncle, a noted histologist (M. Lenhossek), violently protested, seeing no future for

such a dull youngster in science. When his opinion gradually improved, he consented to my going into cosmetics. Later, he even considered my becoming a dentist. When I finished high school with top marks, he admitted the possibility of my becoming a proctologist (specialist of anus and rectum; he had haemorrhoids). So my first scientific paper, written in the first year of my medical studies, dealt with the epithelium of the anus. I started science on the wrong end."

All who knew him remember Szent-Györgyi as an intellectually as well as physically restless individual, dynamic, romantic in outlook, full of life and enthusiasm, even wild. He himself recognized these features of his personality. As he said in one of his autobiographical essays, "If 1 look at myself objectively the first thing I notice is that I find myself running every morning, at an early hour, very impatiently, to my laboratory. Nor does my work end when I return from my workbench in the afternoon. I go on thinking about my problems all the time; and my brain must be going on thinking about them even when I sleep. My brain must have done as the Hungarian laxative which was advertised by saying: 'While you sleep it does the work.'"

But there was also subtlety in his manner. I recall a story he told the students at Woods Hole in the early 1950s about one of his experiences in World War I. He had been sent to the Italian front somewhere in the Tyrol to a high ridge overlooking the front-line trenches. Although all was quiet on the Italian front the first days after his arrival, he did notice that at (precisely) noon each day one of the sergeants ordered a simple cannon to be shot off. When Szent-Györgyi inquired about this practice, his sergeant said, "Excellency, in this way all the regimental officers coordinate and set their wristwatches so that bombardments of the enemy stop precisely just before our men are scheduled to move forward out of their trenches." That made good sense. So Szent-Györgyi continued, "And how do you know when it is precisely noon?" "Oh, Excellency, every day at about 11:30 A.M., I go down to the village behind us over to the shop of the clockmaker, Willi Zwingli, who has a magnificent precise clock in his shop window, and I set my wristwatch by the reading on it." So Officer Szent-Györgyi complimented the orderly on his astuteness. A few days later, Szent-Györgyi had to walk down to the village to buy some cigarettes, and he saw the magnificent clock in the window and decided to walk in to compliment Herr Zwingli. In the course of the conversation, he asked the clock maker whether the clock kept good time. To which the craftsman answered, "Oh yes, Excellency, why everyday at noon when the Army cannon goes off up on the ridge, my clock reads exactly noon!"

Even his moral insights were expressed in striking images. In the post-World War II period, Szent-Györgyi was deeply tormented by the threats posed by the atomic bomb, and he wrote passionately about the dangers of its continued development. One of his most touching appeals was expressed as follows:

"We pray for peace but heap up H-bombs for safety…. The world is symbolized for me by the colossal statue of Christ standing on a hill in Spain, stretching out His arms to [all] mankind and wearing on His head an enormous lightning rod to protect Him, should the Almighty Father try to smite Him with lightning."

To illustrate to students that the commutative law of algebra does not hold in biology, Prof once remarked: "If one woman can produce a child in nine months, then nine women should produce it in one."

To my knowledge, only once in the history of the Marine Biological Laboratory have there been as many perspiring and aspiring Nobel laureates gathered for a scientific convocation as there were at this Szent-Györgyi memorial assembly. That other event took place in 1950 (give or take a year), the first time after World War II that Otto Warburg visited the United States and shortly after Szent-Györgyi had settled in Woods Hole. Incidentally, it has come as a great shock to me in recent years to realize that young life scientists today do not know who Otto Warburg was and have never even seen, let alone used, a Warburg apparatus. As the Romans said, *sic transit gloria mundi*. I can convey Warburg's personality by telling you that on a scale of arrogance from 1 to 10, if it is a logarithmic scale, Warburg rated 20. The subject of that Woods Hole meeting was "Photosynthesis" and particularly the question of the quantum yield of photosynthesis, a subject that Warburg pointed out he had settled definitively with a series of experiments in the 1930s. However, a large array of very competent researchers had accumulated strong evidence contrary to Warburg's reports. At the meeting, besides Warburg, the other Nobel laureates present were James Franck, Otto Loewi, of course Albert Szent-Györgyi, and, I believe, Carl and Gerty Cori. Incidentally, this was such a noteworthy cosmic congruence that *Time* or *Life* magazine sent photographers to Woods Hole to take a group picture of these eminent scientists. Warburg refused to sit for a picture with the others since they were below his Olympian level. So, ultimately, two pictures appeared, one of him alone and one of the others as a group.

To get back to my theme, a vigorous argument developed between Warburg and Franck. (Franck was one of the towering physicists of the first half of this century, being, among other things, the founder of the field of quantum photophysics.) Warburg was shouting at Franck, "You are wronk," and Franck was responding "You are wronk." Prof defused the confrontation by getting up and saying, "But gentlemen, please don't argue; you are both right!"

At about that same time—in 1949—my university was trying very actively to change its image from that of a school that provided a genteel education for the sons and daughters of wealthy families in the midwestern United States to that of a distinguished scholarly and research institution of international stature. So when I mentioned to my department head and then to the dean that a very famous Nobel prizewinner had arrived recently in this country and was not yet attached to any university, I was promptly told to invite him to visit. Those who knew Prof only in his later years can picture what a magnificent lecturer he was in his prime, and how charming he could be as an individual. So almost immediately after Prof's arrival on our campus, the dean opened discussions on an offer. Prof did not want to talk with the dean but asked to see the president of the university, and a meeting was promptly arranged. He was offered a salary three times that of the highest paid full professor, a whole floor of the chemistry department, and all kinds of other perquisites. He went back to Woods Hole to think the matter over. About a week later, he called me, and said "Irv, you and your colleagues have been most warm and generous, but I am going to decline your offer." So I asked whether there was anything else we could do to attract him, and he said, "No, Irv; the only objection I really have to Northwestern is that it has no one there who even thinks that he has been cheated out of a Nobel prize!"

On another occasion I was a minor intermediary for Prof in establishing one of his industrial connections. Starting with his work on vitamin C, I suppose, he had a fascination for trying to isolate pure, defined molecules from tissue extracts, be it a thymus hormone to cure myotonia or retine and promine to modulate cell growth. About the former, he wrote:

"I started hunting for fluorescent substances and soon discovered a substance in my extract which, if illuminated with near-ultraviolet, showed a splendid fluorescence. It was present in traces only. The isolation of this substance in crystals was the only brilliant piece of chemical work I ever produced. The crystals were sent for the analysis of their constitution to Merck & Company, whose report was expected with great excitement. I did not have to wait long for it. It told me that what I [had] isolated was a substance [a plasticizer] which I [had] extracted from my rubber tubing."

At that time he was also forced to abandon his thymus myotonia research for other reasons. For this work, he had to use goats as the animal model and he had a colony at the Marine Biological Laboratory, but as he said in his Hopkins lecture, "I work in a marine biological laboratory and the smell of goats dearly identifies them as non-marine organisms."

From thymus extracts he also isolated the two growth factors that he christened retine and promine. However, he could obtain only minute quantities from this gland. At the suggestion of Charles Huggins, he looked at urine and indeed found tiny quantities in it. So he decided he needed a large-scale collection and isolation facility. At that time, I was a consultant for Abbott Laboratories, a very large pharmaceutical firm about 40 miles north of Chicago. Abbott is very near the Great Lakes Naval Training Station and also Fort Sheridan, where tens of thousands of recruits for the Navy and for the Army were being trained continuously. So let me continue on with Prof's own words at that time:

"Abbott Laboratories in Chicago offered to collect and crudely extract for us quantities of several thousand gallons of urine weekly. This was a wonderful godsend. We blessed the armed forces, which were the final source. The U.S. Army is urinating now for me, and it is comforting to know that there is at least one army in this world which does something useful."

This incident reminded me of another of Prof's famous "thanks" for help, which he wrote in 1946 just after he had left Hungary. In a preface to his book *Chemistry of Muscular Contraction*, he expresses gratitude to the Josiah Macy Foundation and others and then concludes with the quip "... my thanks are [also] due to Professor J. W. McBain of Stanford University for giving me his fountain pen to write this book."

Since there are so many widely circulated anecdotes and stories about Szent-Györgyi, as a historian I must interject a small note of caution. The store of items we have accumulated is called nowadays "oral history." Some years ago I had a long conversation with Samuel Goudsmit, the discoverer, with George Uhlenbeck, of electron spin, during Goudsmit's visit to Northwestern to receive an honorary degree. I was trying to extract an oral history of the origins of wave mechanics in the heady period of the mid-1920s, when the young Goudsmit was in the vortex of events. At the outset of our conversation, Goudsmit said: "oral history?—*all* lies!" I

feel he was exaggerating. However, it is my impression that only about 50% of what we hear from people who are reminiscing is not true—the problem is to figure out which 50%.

Nevertheless, of one thing I am 100% certain. It will be a long, long time before another like Albert Szent-Györgyi appears again in the scientific world.

Irving M. Klotz received his undergraduate and Ph.D. degrees (Chemistry) from the University of Chicago. He joined the faculty of Northwestern University as an instructor and became Professor of Chemistry in 1950 and Morrison Professor of Chemistry and Biochemistry in 1963. In addition to his long-term interest in solvent water effects on protein structure and behavior, Professor Klotz has devoted substantial efforts to investigations of ligand–receptor interactions, structure and function of nonheme oxygen-carrying proteins, chemical modifications of proteins, and the construction of polymers with enzymelike properties. At present, he is also trying to understand some of the interpenetrations of science with the humanities. Professor Klotz is a member of the National Academy of Sciences (USA), Fellow of the American Academy of Arts and Sciences, Fellow of the Royal Society of Medicine, and recipient of the 1949 Eli Lilly Award of the American Chemical Society and the 1993 William C. Rose Award of the American Society for Biochemistry and Molecular Biology.

Gilbert Newton Lewis[a]

Some Personal Recollections of a Chemical Giant

Glenn T. Seaborg[b]

I had the extraordinary good fortune to serve at Berkeley as the personal research assistant to the great physical chemist Gilbert Newton Lewis during the first two years following my receipt there of my Ph.D. in chemistry. I have often been asked to identify the ablest and greatest scientist that I have known personally during my career as a scientist (now extending over 60 years). I unhesitatingly designate Lewis as one of the two best that I have known (the other being the extraordinary physicist Enrico Fermi). Yet Lewis is relatively unknown to the present generation of scientists, including chemists. And, somehow, the Nobel Foundation made one of their rare mistakes by not awarding him the Nobel Prize in chemistry.

Although in this essay I shall focus on my personal contacts and impressions, I begin with a short account of the remarkable College of Chemistry that he developed at the University of California at Berkeley. (His influence transcended the College of Chemistry—he played an important role in turning the Berkeley campus into a world-class university.)

In the fall of 1912, Gilbert Newton Lewis (Figs. 1 and 2), then at the Massachusetts Institute of Technology, accepted the position of Dean of the College of Chemistry and moved to Berkeley. When Lewis arrived, the chemistry faculty already had four members: Edward Booth, who served until he died in 1917; Edmond O'Neill, who retired in 1925; Walter C. Blasdale, who retired in 1940; and Henry C. Biddle, who left Berkeley in 1916. From MIT, Lewis brought with him William C. Bray, Merle Randall, and Richard C. Tolman, together with several graduate students. Bray and Randall were to stay at Berkeley, but Tolman left in 1916. George Ernest Gibson from England and Germany and Joel H. Hildebrand from the University of Pennsylvania joined the faculty in 1913. These proved to be the last non-Berkeley

Fig. 1 Gilbert Newton Lewis at the Massachusetts Institute of Technology, circa 1910, two years before he accepted the position of Dean of the College of Chemistry, University of California, Berkeley.

Ph.D.'s appointed to the faculty until Melvin Calvin's appointment in 1937 (see Table 1). Of the permanent chemistry faculty from 1912 to the present, I have known all but William C. Argo, Booth, O'Neill, and Biddle.

The photograph in Fig. 3 is one of the few of the members of the College of Chemistry and was taken in front of Gilman Hall in the fall of 1917, at about the time this building was completed. This photograph includes faculty members Ermon D. Eastman, Blasdale, Bray, Randall, Gibson, C. Walter Porter, T. Dale Stewart, O'Neill, Argo, and Lewis (Gerald E.K. Branch was away in the Canadian Armed Services, and Hildebrand was apparently out of town); Lewis's secretary, Constance Gray, and clerk M.J. Fisher; graduate students Esther Kittredge, Esther Branch (wife of Gerald Branch), Charles S. Bisson, Wendell M. Latimer, Charles C. Scalione, Hoy F. Newton, William G. Horsch, William H. Hampton, John M. McGee, George S. Parks, Parry Borgstrom, Albert G. Loomis, Angier H. Foster, and Axel R. Olson; undergraduate students Carl Iddings, William D. Ramage, Willard G. Babcock, and Reginald B. Rule; assistant George A. Linhart; glassblower William

This article was supported by the U.S. Department of Energy under contract No. DE-AC03-765700098.

[a]*Chemical Intelligencer* 1995(3), 27–37.

[b]Department of Chemistry, and Lawrence Berkeley Laboratory, University of California, Berkeley, Berkeley, CA 94720, USA (deceased)

B. Hargittai and I. Hargittai (eds.), *Culture of Chemistry: The Best Articles on the Human Side of 20th-Century Chemistry from the Archives of the Chemical Intelligencer*, DOI 10.1007/978-1-4899-7565-2_32, © Springer Science+Business Media New York 2015

Fig. 2 College of Chemistry Dean Gilbert N. Lewis, circa 1925.

J. Cummings and woodworker James T. Rattray and curator Harry N. Cooper.

This photograph was taken at about the time when Lewis published his famous concept of the electron pair for the covalent bond [1]. This was one of the most important ideas of twentieth century chemistry.

I started my graduate work in the College of Chemistry at Berkeley in the fall of 1934. As an undergraduate at UCLA, I had become acquainted with Lewis's book *Valence and the Structure of Atoms and Molecules* [2], published in 1923, and was fascinated by it. This book described and elaborated the concept of the electron-pair bond and also his famous and useful "electron dot" depiction of the structure of atoms and molecules. I wanted to meet and become acquainted with this remarkable man, but I could not then have envisioned that I would be working with him on a daily basis.

I was drawn to Berkeley by my admiration for Lewis and by the presence there of Ernest Orlando Lawrence and his cyclotron, for I was intrigued by the relatively new field of nuclear science. When I arrived and started my classes and research, I found the atmosphere and surroundings exciting to an extent that defies description. It was as if I were living in a sort of world of magic with continual stimulation. In addition to Lewis, I met the authors of most of the chemistry textbooks I had used at UCLA—Hildebrand, Latimer, Bray, Blasdale, and Porter. I took classes from Olson, Branch, and William F. Giauque, and I opted to do my graduate research in the nuclear field under Gibson in a laboratory situated in Lawrence's nearby Radiation Laboratory. In my

thermodynamics class with Olson, I was introduced to the classic book *Thermodynamics and the Free Energy of Chemical Substances* [3], by Lewis and Randall. This book, also published in 1923, was a monumental contribution to placing chemical thermodynamics on an understandable theoretical and practical basis for chemists and chemical engineers. This book was also used, although augmented by more recent material, in Giauque's more advanced thermo-dynamics course that I took during the second semester of my graduate work.

Nearly everyone who participated as a member of the College of Chemistry in the Lewis era recalls and comments on the Research Conference presided over by Lewis in his own inimitable style. This was held each Tuesday afternoon during the school year, starting at 4:10 p.m. and lasting until about 5:30 p.m. in Room 102 at the extreme south end on the first floor of Gilman Hall (a building which, miraculously, is still there—the only surviving building of Lewis's day). Lewis's office was only a few doors away in Room 108, with its door usually open. His and the College of Chemistry's secretary, Mabel Kittredge (Mrs. Wilson), was located next door to him in Room 110 (Fig. 4). At the Research Conference, Lewis always occupied the same place at the central table—the first chair on the right side facing the speaker and the blackboard. Members of the faculty sat at the table, and the others (graduate students, postdocs, research fellows, etc.) sat in chairs set at two levels at the two sides and back of the room. Lewis always had one of his Alhambra Casino cigars in his hand or mouth and several more in his upper coat pocket (Fig. 5). The first of the two speakers, a graduate student giving a report from the literature, started when Lewis gave his inevitable signal, "Shall we begin!" The second speaker—a faculty member, research fellow, or advanced or finishing graduate student—then reported on research that had been conducted in the College. Although Lewis dominated the scene through sheer intellectual brilliance, no matter what the topic, anyone was free to ask questions or speak his piece; in the latter instance, prudence suggested that the comment had best not be foolish or ill-informed. If Lewis had any weakness, it was that he did not suffer fools gladly—in fact, his tolerance level here was close to zero.

During my three years as a graduate student and the subsequent years until the war, Lewis always attended the Nuclear Seminar held on Wednesday evenings in Room 102, Gilman Hall. This seminar was run by Willard F. Libby, together with Robert Fowler (until he left Berkeley in 1936), and was attended regularly by Latimer, Bray, and Eastman. Lewis also conducted some research with neutrons in 1936 and 1937. He was always highly supportive of my nuclear research, some of which was conducted in my spare time during the period in which I served as his personal research assistant.

Table 1 University of California, Berkeley—Chemistry Faculty

YEAR JOINED	NAME	DEGREE DATE	WHERE TAKEN: WITH WHOM
Faculty on hand at the time Gilbert Newton Lewis arrived in Berkeley			
	Booth, Edward	1877	UC Berkeley
	O'Neill, Edmond	1879	UC Berkeley
	Blasdale, Walter C.	1892	UC Berkeley
	Biddle, Henry C.	1900	University of Chicago
1912	Lewis, Gilbert Newton	1899	Harvard: T.W. Richards
	Tolman, Richard C.	1910	MIT
	Bray, William C.	1905	Leipzig: Luther
	Randall, Merle	1912	MIT: G.N. Lewis
1913	Hildebrand, Joel C.	1906	Pennsylvania: Edgar Fahs Smith
	Gibson, G. Ernest	1911	Breslau: Lummer
1915	Branch, Gerald E.K.	1915	UC Berkeley: Lewis
	Argo, William C.	1915	UC Berkeley: Lewis
1917	Porter, C. Walter	1915	UC Berkeley: Biddle
	Eastman, Ermon D.	1917	UC Berkeley: Lewis
	Latimer, Wendell M.	1917	UC Berkeley: Gibson
	Stewart, T. Dale	1916	UC Berkeley: Tolman
1921	Olson, Axel R.	1918	UC Berkeley: Lewis
	Hogness, Thorfin R.	1921	UC Berkeley: Hildebrand
1922	Giauque, William F (1949 Nobel Prize)	1922	UC Berkeley: Gibson
1923	Rollefson, Gerhard K.	1923	UC Berkeley: Lewis
1933	Libby, Willard F. (1960 Nobel Prize)	1933	UC Berkeley: Latimer
1937	Pilzer, Kenneth S.	1937	UC Berkeley: Latimer
	Calvin, Melvin (1961 Nobel Prize)	1935	Minnesota: Glockler (UC Berkeley. 1923, Gibson)
1938	Ruben, Samuel C.	1938	UC Berkeley: Latimer/Libby
1939	Seaborg, Glenn T. (1951 Nobel Prize)	1937	UC Berkeley: Gibson

Fig. 3 Members of the University of California at Berkeley College of Chemistry. Photo was taken in the fall of 1917, in front of the newly constructed Gilman Hall. FRONT ROW (from the left to right): M. J. Fisher (bookkeeper), Esther Branch, Esther Kittredge, Constance Gray, Gilbert N. Lewis, William L. Argo, Edmond O'Neill, T. Dale Stewart, C. Walter Porter, G. Ernest Gibson, Merle Randall, William C. Bray, Walter C. Blasdale, and Ermon D. Eastman. ASCENDING STAIRS (from left to right): Charles S. Bisson, Wendell M. Latimer, William J. Cummings (glassblower), Carl Iddings, Reginald B. Rule, J.T. Rattray (woodworker), Charles C. Scalione, Hal D. Draper, William G. Horsch, William H. Hampton, Willard G. Babcock, John M. McGee, George S. Parks, Parry Borgstrom, Albert G. Loomis, George A. Linhart, William D. Ramage, and Harry N. Cooper. SEATED (from left to right): Axel R. Olson and Angier H. Foster.

Fig. 4 Entrance door to Room 110, Gilman Hall, University of California, Berkeley, Mabel Kittredge's (Lewis's secretary) office, and the official entrance to Gilbert N. Lewis's office in Room 108.

Fig. 5 Gilbert N. Lewis with his cigar as he was typically seen at the Tuesday afternoon Research Conference in Room 102, Gilman Hall, University of California, Berkeley.

With this background in mind, let me now proceed to a description of my work with Lewis as a research associate. I'll never forget how this got started. I had completed my graduate research in the spring of 1937, my Ph.D. degree had been awarded, and it was time for me to go and find a job someplace. Lewis didn't recommend me for a position anywhere, which I could have regarded as a bad sign. Actually, in this case, it was a good sign. That meant that I still had a chance to stay at Berkeley in some capacity—which, of course, was my objective. One day in July after the next academic year had actually started (so I was technically without any salary), Lewis called me into his office and asked me if I would like to be his research assistant. Lewis was unique in having a personal research assistant, whose salary at that time was $1,800 per year. Although I was fervently hoping to stay in some capacity, I was flabbergasted to find he thought me qualified for this role, and I expressed my doubts to him. He smiled and indicated that if he didn't think I could do the job, he wouldn't have offered it to me. My acceptance of the position he offered was enthusiastic, and thus our two-year intimate association began.

Lewis had suffered some disappointment in his previous research with neutrons. In fact, I had played a role in advising him frankly where he was going wrong, an act that took some courage on my part, and this may have influenced him in his decision to undertake the risk of having me as his research assistant. He told me that he had decided to forgo research for a time, during which I would be free to continue the nuclear research that I had under way. As I have already indicated, I continued a rather substantial effort in the nuclear field, with his blessing, during the entire two-year period that I was associated with him.

In the late fall of 1937, Lewis resumed his research. He decided to try to separate the rare earths praseodymium and neodymium using a system involving repetitive exchange between the aqueous ions and their hydroxide precipitates. He employed a long, tubular, glass column extending from the third floor to the basement at the south end of Gilman Hall. The column was constructed with the help of Bill Cummings, the long-time glassblower in the college, and erected with the help of George Nelson, the irascible head of the machine shop. (He was irascible from the standpoint of graduate students but very polite to Lewis and now to me in my prestigious role as the assistant to the "Chief.") The long column was serviced by a machine-driven system for agitation in order to keep the hydroxide precipitates suspended along the column's length. It was my duty to keep this operating, which I did with only limited success. Lewis, with no help from me, measured the degree of separation of the praseodymium from neodymium with the spectroscope in the darkroom off Room 301, Gilman Hall. For whatever reason, including possible shortcomings in my performance, no detectable separation of praseodymium from neodymium was achieved.

In the early spring of 1938, Lewis returned to his former interest in acids and bases. If I recall correctly (this was 57 years ago!), he was, at least in part, motivated by the need for an interesting topic, supported by feasible experimental demonstrations, for a talk that he was scheduled to give at the Franklin Institute in Philadelphia in May on the occasion of his receiving a Doctor of Science degree and honorary membership in the Franklin Institute, in connection with the dedication of the Benjamin Franklin Memorial (i.e., the large new building housing the Institute's activities, including the science museum). In any case, much of our first work in this area was concerned with such demonstration experiments.

Our experiments were directed toward his generalized concept of acids and bases. In his 1923 book *Valence and the Structure of Atoms and Molecules*, Lewis had proposed a very general definition of acids and bases. According to that definition, a basic molecule is one that has an electron pair which may enter the valence shell of another atom to consummate the electron-pair bond, and an acid molecule is one which is capable of receiving such an electron pair into the shell of one of its atoms. Lewis wanted, with my help, to find a broad base of experimental evidence for this concept.

We worked in Room 119 (Fig. 6), at the north end of the first floor of Gilman Hall, a laboratory that Lewis had used for a number of years previously. It was here that he did his trailblazing work with Ronald McDonald and others during the period 1933–1935 on the isolation of deuterium by the electrolysis of water and the determination of a number of its properties [4]. The apparatus used for this work was still there in the east side of the room, a part of the room that we didn't use at this time. We used the laboratory bench extending along the west side of the room, flanked in the back by a row of windows. The sink, at which I washed and cleaned our glassware each evening (Fig. 7), was at the extreme right (north) end of the bench, and our writing desk adjoined the opposite end of the laboratory bench against the south wall.

Fig. 6 Entrance to Room 119, Gilman Hall, University of California, Berkeley, where Seaborg served from 1937 to 1939 as research associate to Gilbert N. Lewis.

Our indicator experiments were performed on the laboratory bench top at the ambient room temperature in ordinary test tubes. For later, more sophisticated (but still basically simple) experiments, which I shall describe presently, we used a low temperature bath that consisted of a large, wide-mouth Dewar filled with acetone which was cooled by the addition of chunks of dry ice. Our vacuum bench, used in later experiments, was in the center of the room, opposite and parallel to the laboratory bench.

I was immediately struck by the combination of simplicity and power in the Lewis research style, and this impression grew during the entire period of my work with him. He disdained complex apparatus and measurements. He reveled in uncomplicated but highly meaningful experiments. And he had the capability to deduce a maximum of information, including equilibrium and heat of activation data, from our elementary experiments. I never ceased to marvel at his reasoning power and ability to plan the next logical step toward our goal. I learned from him habits of thought that were to aid continuously my subsequent scientific career. And, of course, working—and apparently holding my own—with him boosted my self-confidence, which was not at a very high level at this stage of my life.

Starting at this time, I worked with Lewis on a daily basis, interspersed with intervals when he was otherwise occupied and during which I pursued my nuclear research. He would arrive each day between 10 and 11 a.m. in his car, a green Dodge, which he would park on the road, South Drive (Fig. 8), between the chemistry buildings and The Men's Faculty Club (where I was living at that time). When I spotted his car, I knew that it was time to join him in Room 119. We then usually would work together until about noon or 1 p.m.,

Fig. 7 Sink area in Room 119, Gilman Hall, University of California, Berkeley, where Seaborg serviced the experiments he performed with Gilbert N. Lewis in between moonlighting as a nuclear chemist.

Fig. 8 The Campanile, Le Conte Hall, Gilman Hall, and Chemistry Building on the campus of the university of California, Berkeley, circa 1941. South Drive, where Lewis would park his green Dodge every morning, is in the foreground.

when he went to The Faculty Club to play cards with his friends (he didn't eat any lunch) while I went to lunch. He usually returned to our laboratory at about 2 p.m., and we would work together until late afternoon. This gave me time to work on my other research projects before he came, during the noon break, and after he left. However, he often gave me assignments to assemble materials, prepare solutions, etc., over the noon hour, or overnight, or when he left town for a day or two. These assignments were usually unrealistically demanding for such a time scale, and I had to scramble to meet his demands. This was done not for the purpose of keeping me busy, but because he underestimated the size of the tasks. Sometimes we worked in the laboratory during the evening after dinner, often on Saturday morning, and occasionally on Sunday. We did most of the writing up of our work for publication on Sunday afternoons.

Lewis gave his talk at the Franklin Institute in Philadelphia on Friday morning, May 20, 1938, as scheduled. During his talk, he performed the demonstration experiments that we had developed. So far as I know, his talk was well received. However, the main impact came from his publication, based on the talk, which appeared in the September issue of the *Journal of the Franklin Institute* [5]. In the preparation of this paper, which was written entirely by Lewis without my help, he used additional data that we developed in subsequent experiments. However, the main thrust of the paper was his beautiful exposition of his concept of generalized acids and bases, which had a worldwide impact and became the "bible" for workers in this field. His primary acids and bases are characterized by their instantaneous neutralization reactions, which occur without any heat of activation. In this paper, he also introduced his concept of secondary acids and

bases, whose neutralization requires a heat of activation. I soon found that I was destined to work with him on a program of experimental verification of this idea.

I helped Lewis pack his equipment for his travel by train to Philadelphia for his demonstration lecture at the Franklin Institute. I was pleased to see him bring into our laboratory and place on the bench two suitcases because I felt this would give me ample room to pack the material for his demonstration experiments. However, he told me that he would need much of this space for his cigar boxes. He filled one entire suitcase and part of the other with cigar boxes, which meant that I had to exercise some ingenuity in order to get the equipment, chemicals, etc. into the remaining space.

Lewis and I resumed our experiments on generalized acids and bases during June and early July 1938, after he returned from his trip to Philadelphia. We found many cases where, with one solvent and one indicator, the colors obtained seemed to be dependent only upon the acid or basic condition of the solution and not at all upon the particular acid or base. By means of the color changes, the solutions could be titrated back and forth as in aqueous solution. For example, with thymol blue dissolved in acetone, the color was yellow with either pyridine or triethylamine, while the acids $SnCl_4$, BCl_3, SO_2 and $AgClO_4$ gave an apparently identical red color. With crystal violet in acetone, the color changed successively from violet to green to yellow upon the gradual addition of $SnCl_4$ or BCl_3, after which the original violet color could be restored upon the addition of an excess of triethylamine.

Because similar effects could also be obtained with HCl, and since we had been working in the open with reagents that had not been especially dried, we were afraid that some of

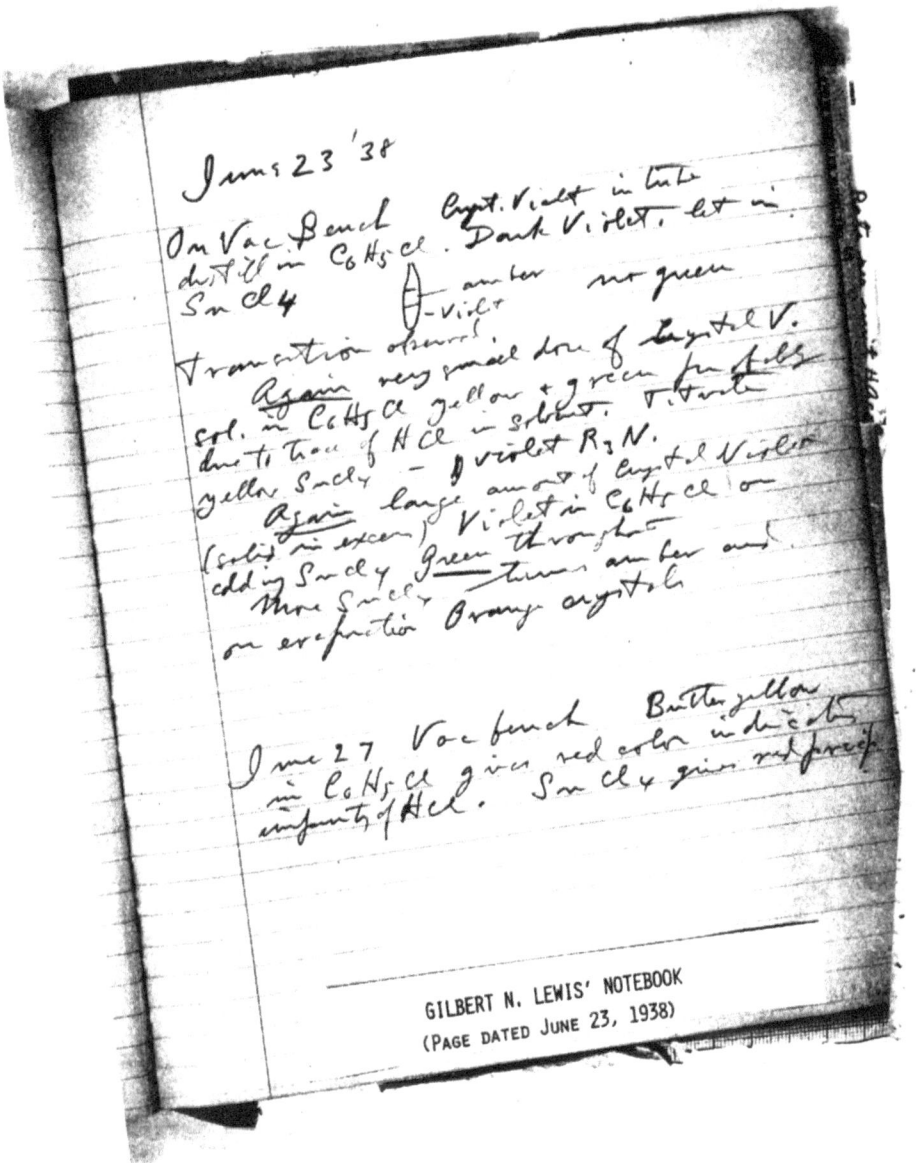

Fig. 9 Entry in Gilbert N. Lewis's handwriting in Seaborg's laboratory notebook describing experiments an acid/base systems carried out in Seaborg's absence. June 23, 1938.

the similarities in color produced by the different acids could be due to small impurities of H-acids in the reagents. We therefore conducted experiments with very dry solvents, given to us by Dr. C.H. Li, with indicators that themselves contain no labile hydrogen, such as butter yellow, cyanin, and crystal violet, and upon the vacuum bench to prevent the pickup of water. These experiments gave the same results as those performed in the open with ordinary reagents.

Toward the end of June, Lewis gave me leave to go to San Diego to give a talk on my nuclear work at a meeting of the American Physical Society. During my absence, he conducted vacuum bench experiments to observe the color changes when $SnCl_4$ and triethylamine were added to a solution of crystal violet in thoroughly dried chlorobenzene, when $SnCl_4$ or HC1 were added to a solution of butter yellow in chlorobenzene, etc. I reproduce in Fig. 9 his notes covering one of these experiments as he recorded them in my notebook.

In addition to taking some vacation during the summer of 1938 with his family at their cottage in Inverness, he spent a good deal of time on his paper "Acids and Bases," which he was getting ready to send to the *Journal of the Franklin Institute*. The process of formulating his thoughts and setting them down on paper suggested to him many little confirmatory experiments, which we then performed. I reproduce in Fig. 10 a sample page from my journal (notebook) of this period.

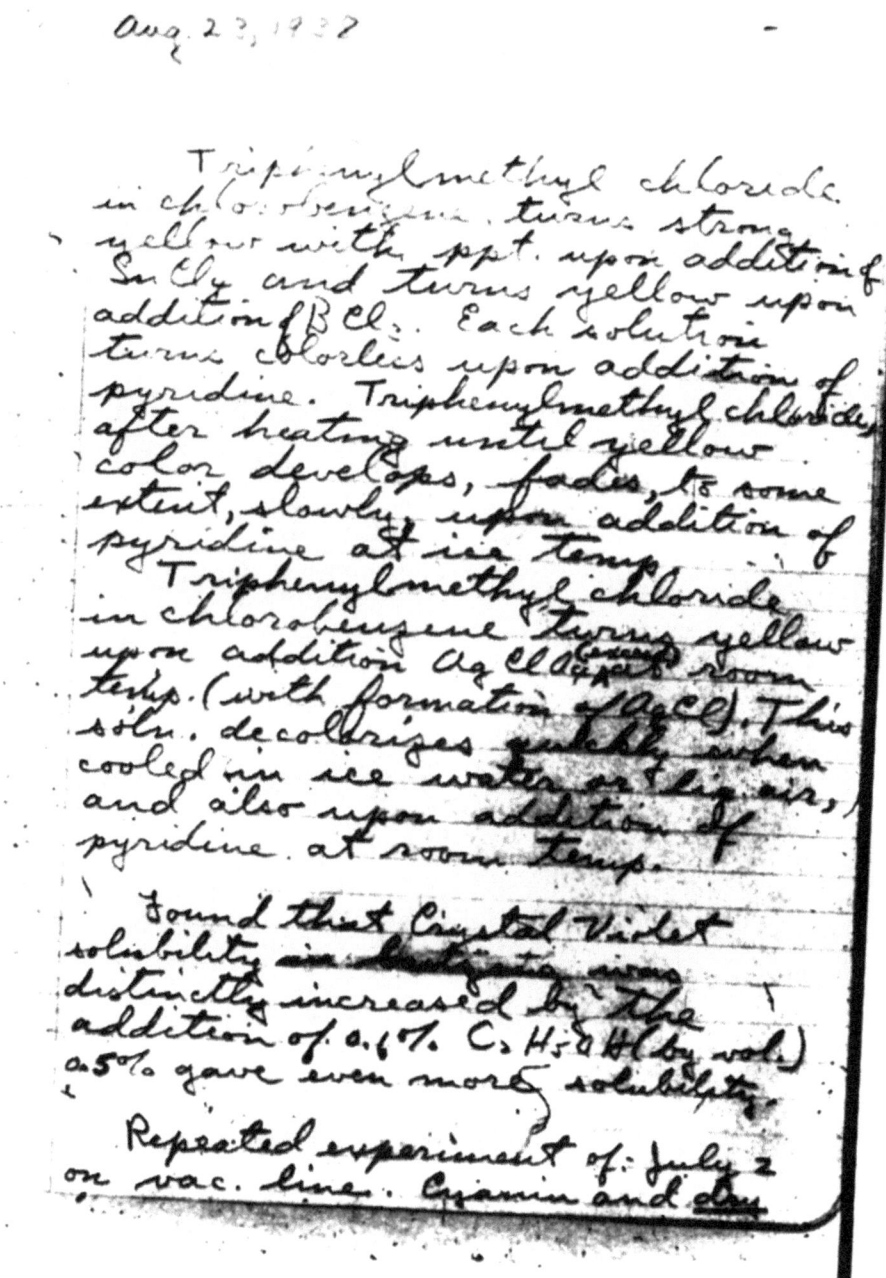

Fig. 10 Sample page from Seaborg's laboratory notebook during his collaboration with Gilbert N. Lewis. August 23, 1938.

In September, Lewis turned to his next project—experiments related to his concept of secondary acids and bases—and from the latter part of September until Christmastime, I worked with him on a daily basis on much the same schedule as I outlined earlier. We did some broadly based experiments, which led to the publication of our background paper "Primary and Secondary Acids and Bases" [6], and a detailed investigation of a specific secondary and primary base, which was published as a companion paper entitled "Trinitrotriphenylmethide Ion as a Secondary and Primary Base" [7].

It was in the course of this detailed investigation of this secondary and primary base that I was to see firsthand a master researcher at work and to be privileged to be a participant in his work. Here was a prime example of simple experiments leading to interesting and fascinating interpretations, and in my description I shall do my best to capture the flavor of the process. As background for understanding these experiments, we should recall that Lewis had suggested that there is a large group of acids and bases, called primary acids and bases, which require no energy activation in their mutual neutralization, and there is another group, called secondary acids and bases, which do not combine with each other (nor does a secondary base combine with a primary acid nor a secondary acid with a primary base) except when energy, and frequently a large energy, of activation is provided.

As the experiments that we conducted are illustrative of the Lewis method, I shall describe them in some detail. Based on the results of some preliminary experiments and Lewis's intuition and analysis, we decided that the intensely blue 4,4′,4″-trinitrotriphenylmethide ion should be a base that could exist in the primary and secondary forms and be a good material for experimentation to give support for and information on this concept.

Lewis soon deduced that our first interest should be in the secondary base B_s^- in the blue form that requires a heat of activation to be converted to the primary form $\left(B_p^-\right)$ in which it reacts instantaneously with acetic acid. Thus, he deduced that the two forms would have the formulas shown below.

$$B_p^- \qquad\qquad B_s^-$$

We launched into a series of kinetic experiments to measure the rate of the fading of the blue B_s^- upon the addition of acetic acid or other acids which combined instantaneously with the small proportion of B_p^- that was present. This mechanism, for any acid HY, can be summarized as follows:

$$B_s^- \rightleftarrows B_p^- \qquad\qquad (1)$$

$$B_p^- + HY \rightarrow BHY^- \qquad\qquad (2)$$

$$BHY^- \rightarrow HB + Y^- \qquad\qquad (3)$$

Lewis suggested that reaction (2) is the rate-determining step and that the concentration of B_p^- depends upon the concentration of B_s^-, the temperature, and the difference in

energy between B_p^- and $B_s^- \cdot$ On this basis, the reaction should be bimolecular, and the measured heat of activation should be the same with all acids (HY) of sufficient strength.

To test this, we measured the rates of reactions (rates of fading of the blue color) over a range of temperatures in order to determine the heat of activation. The experimental method was simplicity itself. The first experiments were performed in open test tubes, but it was found that trinitrotriphenylmethane was sensitive to oxygen under the conditions used, and therefore the reaction vessels were evacuated. Our solvent was 85 % ethyl alcohol and 15 % toluene, and our first series of experiments were with acetic acid. The reaction vessel, in the form of an inverted Y, with the alkaline blue methide ion solution in one limb and the acid in the other, was placed in the low-temperature bath (of acetone cooled with dry ice). When temperature equilibrium was attained, the vessel was tipped rapidly back and forth until the contents were thoroughly mixed. The reaction (rate of fading of the blue color) was then followed by comparing the color with a set of standard color tubes. (The set of standard color tubes consisted of solutions of crystal violet, which had blue colors nearly identical to those of the blue methide ion, made by successive two fold dilutions to cover the entire range of diminishing blue color.) After the experiments had indicated that the reaction was always of first order with respect to the colored ion, the procedure was simplified further. The time was taken merely between the mixing and the matching of a single color standard, which corresponded to 1/16th of the original concentration of the blue methide ion (i.e., the color standard was made by four twofold dilutions of the original matching crystal violet solution).

We made measurements with acetic acid at four temperatures: −53 °C, −63 °C, −76 °C, and −82 °C. From these measurements, we could calculate that the reaction was first-order with respect to the acid and that the heat of activation for the reaction of fading the blue methide ion was 8.6 kcal. According to our interpretation, then, this is the energy difference between the secondary form B_s and the primary form B_p of the methide ion. We next measured the heat of activation for the same reaction for five additional acids, for which the reaction also proved to be bimolecular, and found the same value for the heat of activation within the limits of our experimental uncertainty—an average of 9.1 kcal. Such a result is to be expected from our interpretation that the heat of activation should be equal to the difference in energy between the primary and secondary forms of the base. If the activation occurred only at the moment of collision between the reacting molecules, it would be hard to explain why the heat of activation, or, in other words, the potential barrier in the activated complex, should be the same for such very different substances as alcohol (for which we also measured the heat of activation, indirectly, as described below) and our other acids—chloroacetic, furoic, α-naphthoic, lactic, and benzoic—as well as acetic acid.

I have recounted here in some detail only the central conclusions from this research. Lewis made many other deductions that are too involved to be easily described here, but which can be enjoyed by reading the paper reporting this work. I shall merely sketch some—by no means all—of these conclusions. From some of our other measurements, he was able to deduce the equilibrium constant for the reaction in which the blue methide ion is formed from the reaction of the hydroxide (or ethylate) with the trinitrotriphenylmethane and the heat of activation, from which he found that the heat of activation for the reverse reaction (B_s^- plus ethyl alcohol), corresponding to the difference in energy between the primary and secondary forms of the base, is 8.9 kcal, in good agreement with our direct determination for the six acids (9.1 kcal). He could deduce from our measurements that only one-eighth of the trinitrotriphenylmethane was in the form of the blue methide ion under the conditions of the kinetic experiments. He also concluded that our kinetic measurements with such weak acids as phenol and boric acid suggest that these displace the solvent alcohol from the nitro groups in the blue methide ion to an extent depending upon their concentration and that the ion with the phenol attached is less reactive than the corresponding alcohol compound.

We found that an orange color was produced immediately upon the addition of the strong acid HCl to a solution of the blue methide ion. We also found this upon the addition of the relatively strong trichloroacetic acid. Lewis found a ready explanation for this. When the blue ion has been formed and the central carbon has lost its power of acting immediately as a base, the basic power has, in a certain sense, been transferred to the three nitro groups. Therefore, a sufficiently strong acid should attach itself at one or more of the nitro groups, and in this process the blue ion should act as a primary base.

We finished these experiments just before Christmas time in 1938. After a diversion in January to test another of his ideas experimentally, we began in February the process of writing our two papers [6, 7] on primary and secondary acids and bases for publication in the *Journal of the American Chemical Society*. Writing a paper with Lewis was a very interesting process. We did most of our work on these papers, sporadically over several months, on Sunday afternoons in our laboratory, Room 119 in Gilman Hall. The process consisted of Lewis, pacing back and forth with cigar in hand or mouth, dictating to me. I recorded his thoughts in longhand. However, his output was interspersed with discussions with me and even with experimental work when he wanted to check a point or simply wanted a break. His sentences were

carefully composed, and the result was always a beautiful and articulate composition.

After we had finished with the two papers up to the point of the summary of the second paper, he said to me that he was tired of this process and suggested that I write this summary by myself. By this time, I was familiar enough with his thought processes to make this feasible. I wrote the following, which he accepted after no more than a glance at it and without changing a word:

Trinitrotriphenylmethide ion was expected and has proved to be a secondary base. In alcohol when this blue ion is added to any weak acid at temperatures between −30 and −80° the formation of the corresponding methane is slow and can be followed colorimetrically. The rate of neutralization was studied with numerous acids and under like conditions the rates diminish with diminishing acid strength. With the weakest acids the rates are not proportional to the concentration of acid, and this fact is explained. With the six acids of intermediate strength the rates were found proportional to the concentrations of blue ion and of unionized acid, and unaffected by neutral salts. In these cases the heat of activation was calculated from the temperature coefficient of the rates and was found approximately constant with a mean value of 9.1 kcal. By indirect methods the rate of neutralization by alcohol itself was determined. Here the heat of activation is found to be 8.9 kcal. The constancy of the heat of activation over the great range from chloroacetic acid to alcohol can hardly be explained by the theory of an activated complex. The value obtained is taken as a measure of the difference in energy between the primary and secondary forms of the base. The small departures from this constant value are attributed in part to experimental error, but especially to differences in the actual composition of the reacting ion. Several kinds of evidence are adduced to show that the actual composition of the blue ion depends not only upon the solvent but in several cases upon the presence of other solutes.

While the trinitrotriphenylmethide ion is a secondary base with respect to addition of acid to the central carbon, it is a primary base with respect to addition of acid to the nitro groups. In the presence of strong acids an orange substance is thus formed which contains more than one free hydrogen ion per molecule. The very slow rate of fading of the orange compound is studied, and an explanation is suggested for the large catalytic effect of water. Mono- and dichloroacetic acids give mixtures of the orange and blue substances and the rate of fading in these solutions leads to some of the conclusions already mentioned.

During January 1939, Lewis and I worked to make an experimental test of an old, rather far-out, idea of his. This is far afield from acids and bases but is, I believe, worth mentioning as a further illustration of the breadth of his intellect and interests. A number of years before (1930), he had published an article in *Science* magazine on the "The Symmetry of Time in Physics" [8]. He had already revealed this idea in his third book, *The Anatomy of Science*, the published account of his philosophical Silliman Memorial Lectures, published in 1926 by the Yale University Press [9], A consequence of this theory, as it applies to radiation, is that we must assign to the emitting and the absorbing atom equal and coordinate roles with respect to the act of transmission of light. A consequence of this, Lewis told me, is that the receiver or observer of the light (for example, the apparatus used for this purpose) is of importance equal to that of the emitter of the light and exerts its own influence upon how the light manifests itself.

Lewis told me he wanted to test this hypothesis by setting up a Michelson interferometer to detect the interference fringes with different receivers or detectors of radiation and to thus determine if some properties of the radiation depend on the receiver or detector as it should if it conformed with his theory on the symmetry of time. He asked me to set up a Michelson interferometer in the dark room off Room 301 at the southwest corner of the third (attic) floor of Gilman Hall. This room contained a spectrograph with which Lewis had made his spectrographic measurements mentioned earlier on rare-earth samples.

I went to the Department of Physics and borrowed a Michelson interferometer that was ordinarily used for demonstration experiments in some of the physics lecture courses. In order to make this operate correctly, I had to prepare some "half-silvered" surfaces on glass with a silver layer of such thickness that about one-half of the incident light would be reflected and the other half transmitted through the layer. Since Professor Axel Olson had some experience with this "half-silvering" process, I enlisted his help. Lewis and I detected the interference fringes with each of a number of different types of photographic film in order to see if we could detect any gross differences in the way the films reacted. We found some peculiar effects, which excited Lewis for a time, but my skepticism prevailed when I was able to explain these as due to rather prosaic failures in our techniques and to show how we could eliminate the effects by correcting our techniques. These negative results then convinced Lewis to go on to something else.

During the period from January to June 1939, Lewis and I did scouting experiments with a wide range of indicators, acids, and bases. Many interesting observations were made that are not susceptible to summarization in a reasonably brief fashion. As always, there were moments of excitement. I recall a series of experiments, conducted with test tubes immersed in our acetone-carbon dioxide bath, on the development of color when trinitrobenzene and sodium phenollate were reacted in absolute ethyl alcohol over a range of temperatures below room temperature. We found that large excesses of NaOH were needed to produce the indicator color. This elicited some bizarre interpretations from Lewis. However, when these experiments were repeated on the vacuum line, the action of NaOH was more reasonable. Apparently, in our open test tube experiments, large amounts

Table 2 Color Production in Mixing Several Bases with Nitro Compounds

	DNB	TNB	TNT	TNX	TNM
NH_3	+	+	+	+	
NH_2R	+	+	+	+	
NHR_2		+	+		
NR_3		+	+		
OH^- or OR^-		+	+	?	

of CO_2 were absorbed in the alcoholic solution from our CO_2-cooled acetone bath!

Our research during this period did result in one coordinated project from which some interesting conclusions could be drawn. We reacted each of the bases ammonia, methylamine, dimethylamine, triethylamine, and hydroxide with each of the acids m-dinitrobenzene (DNB) and symmetrical trinitrobenzene (TNB), trinitrotoluene (TNT), trinitroxylene (TNX), and trinitromesitylene (TNM)—25 combinations in all—and made observations on the degree of development of color (a measure of the degree of reaction between these acids and bases). At any point in Table 2 corresponding to a given base and a given nitro compound, the sign + indicates the formation of color.

We found that with trinitrobenzene the intensity of color is least with triethylamine, greater with dimethylamine, and still greater with methylamine and ammonia. For the direct addition of the base to one of the ring carbons that is not attached to a nitro group, there is the possibility of double chelation of hydrogen atoms to nitro groups in the case of methylamine and ammonia, thus strengthening the acid-base combination. With the weaker acid m-dinitrobenzene, methylamine and ammonia—which are capable of double chelation—give good colors, while the two stronger bases dimethylamine—which is capable of only one chelation—and triethylamine—where no chelation is possible—give no color at all. Thus, our conclusion was that the stability of the colored compounds is greatly enhanced by chelation, in which the hydrogens of an aliphatic amine are attached to oxygens of the nitro groups. Similarly, we could deduce that the chief effect of introducing methyl groups into symmetrical trinitrobenzene is to diminish resonance between the nitro groups and the ring and that this effect, which is very strong when the nitro group is ortho to two methyl groups, as in symmetrical trinitroxylene, becomes weak when only an ortho methyl is present, as in symmetrical trinitrotoluene. Trinitromesitylene, in which each nitro group lies between two methyl groups, showed no color with any base.

Lewis and I didn't write up this work for publication until about a year later owing to the press of our other activities.

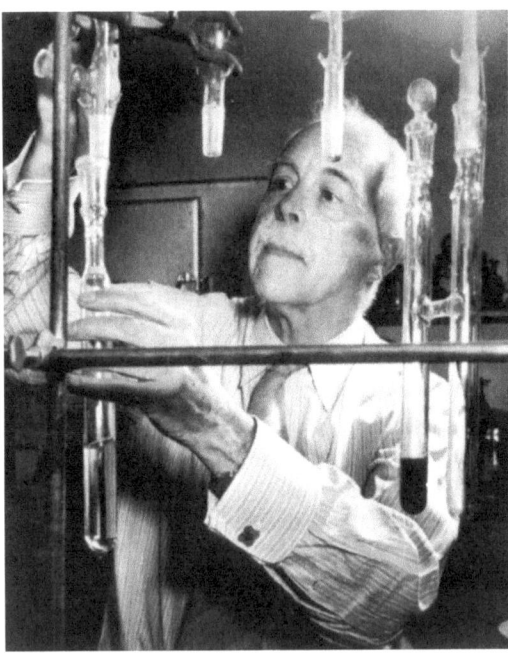

Fig. 11 Gilbert N. Lewis at vacuum line during work
on spectroscopic observations.

Fig. 12 Gilbert N. Lewis at work in his Gilman Hall office.
Room 108, University of California, Berkeley.

When we did, of course, it was done by the same method of dictation with me serving as a scribe. Our publication, which included explanations for all of our observations, was entitled "The Acidity of Aromatic Nitro Compounds toward Amines. The Effect of Double Chelation" [10].

During my last months with Lewis, April, May, and June 1939, he turned part of his attention toward spectroscopic observations on light absorption and the observations of fluorescence and phosphorescence in various colored organic substances. For this we used the spectrograph in Room 310, Gilman Hall, where Ted Magel, then a graduate student, was working. Lewis was now beginning his experimentation on the relation of energy levels in molecules to their emission of light and was already beginning to think in terms of the triplet state. Besides Magel, Otto Goldschmid, a volunteer research fellow, and Ed Meehan, an instructor in the College of Chemistry helped us in these measurements. Melvin Calvin, David Lipkin, Jacob Bigeleisen, and Michael Kasha were active in this program (Fig. 11).

Also during this time, Lewis was working with Calvin, putting the finishing touches on their review paper "The Color of Organic Substances," which they mailed in August for publication in *Chemical Reviews* [11]. Lewis had been interested in the color of chemical substances for a long time, and, in fact, this was the subject of his acceptance address in New York on May 6, 1921, when he received the Nichols Medal of the New York Section of the American Chemical Society. He had been working with Calvin, off and on, during much of the last year. I can recall looking in on them in Room 102, where they had their writing sessions, and finding them totally immersed in their piles of reference journals and notes.

During all of the time that I was working with Lewis, he was, of course, serving as dean of the College of Chemistry and chairman of the Department of Chemistry. These positions would ordinarily entail heavy administrative duties, but he did not allow himself to be burdened by them. Nevertheless, I believe, he discharged his responsibilities very well (Fig. 12). He was efficient and decisive, highly respected by the faculty members in the college, and eminently fair in his dealings with them. To a large extent, he ran the college from his laboratory. I recall that his efficient secretary, Mabel Kittredge, would come into our laboratory, stand poised with her notebook until she commanded his attention, and then describe clearly and briefly the matter that required his attention or decision. Lewis would either give his answer immediately or ask her to come back in a little while, after he had given the matter some more thought. This system worked very well in those days but might not be adequate today, and, in any event, certainly could only function with a person of Lewis's ability.

Lewis never addressed me with a harsh word although there were times when he might have been justified in so doing. I recall one occasion when, together with friends,

I had overindulged in alcohol the previous evening to the extent that on the following morning, I had to steady my right hand with my left hand in order to turn a stopcock on our vacuum line. With a grin on his face, he recommended that I should return to my room in The Faculty Club "to rest" awhile; I took his advice and was able to return to my duties in the afternoon.

Sometime in June 1939, Lewis told me that he was putting me on the faculty of the College of Chemistry as an instructor. In his whimsical way, he expressed the opinion that he had been taking up "too much of my time." This was a revealing comment considering that I was supposed to be serving as his full-time research assistant. However, I have good reason to believe that he was not at all unhappy with my additional research and writing projects. He told me my salary would be $2,200 per year, that of a third-year instructor. Thus, to my delight, he was giving me full credit for my two years in the capacity of his research assistant.

In conclusion, I want to say that I regard it as extraordinarily good fortune that I was granted the privilege of spending this time working so closely with Gilbert Newton Lewis, an extraordinary scientist of the twentieth century.

References

1. Lewis, G.N. *J. Am. Chem. Soc.* **1916**, *38*, 762–785.
2. Lewis, G.N. *Valence and the Structure of Atoms and Molecules*; American Chemical Society Monograph Series; Chemical Catalog: New York, 1923.
3. Lewis, G.N.; Randall, M. *Thermodynamics and the free Energy of Chemical Substances*; McGraw-Hill: New York, 1923.
4. Lewis, G.N.; McDonald, R.T. *J. Am. Chem. Soc.* **1933**, *55*, 3057–3059.
5. Lewis, G.N. *J. Franklin Inst.* **1938**, *226*, 293–313.
6. Lewis, G.N.; Seaborg, G.T. *J. Am. Chem. Soc.* **1939**, *61*, 1886–1894.
7. Lewis, G.N.; Seaborg. G.T. *J. Am. Chem. Soc.* **1939**, *61*, 1894–1900.
8. Lewis, G.N. *Science* **1930**, *71*, 569–577.
9. Lewis, G.N. *The Anatomy of Science*; Yale University Press: New Haven, Connecticut, 1926.
10. Lewis, G.N.; Seaborg, G.T. *J. Am. Chem. Soc.* **1940**, *62*, 2122–2124.
11. Lewis, G.N.; Calvin, M. *Chem. Rev.* **1939**, *25*, 273–328.

Glenn T. Seaborg is currently University Professor of Chemistry, associate director-at-large of the Lawrence Berkeley Laboratory, and chairman of the Lawrence Hall of Science at the University of California, Berkeley.

From 1937–1939, following receipt of his Ph.D. in chemistry from the University of California at Berkeley, Seaborg served as the personal research assistant to the great physical chemist Gilbert Newton Lewis, Dean of the College of Chemistry at the University of California at Berkeley.

Winner of the 1951 Nobel Prize in chemistry (with Edwin M. McMillan) for his work on the chemistry of the transuranium elements, Seaborg is one of the codiscoverers of plutonium (element 94). During World War II he headed the group at the University of Chicago's Metallurgical Laboratory that devised the chemical extraction processes used in the production of plutonium for the Manhattan Project. He and his co-workers have since discovered nine more transuranium elements. In March 1994 he was honored with the recommendation by the other codiscoverers of element 106 that it be named "seaborgium," with the chemical symbol Sg.

In 1944 Dr. Seaborg formulated the actinide concept of heavy-element electronic structure, which accurately predicted that the heaviest naturally occurring elements, together with synthetic transuranium elements, would form a transition series of actinide elements, analogous to the rare-earth series of lanthanide elements. This concept is one of the most significant changes in the periodic table since Dmitri I. Mendeleev's nineteenth century design.

His codiscoveries also include many isotopes that have practical applications in research, medicine, and industry (such as iodine-131, technetium-99m, cobalt-57, cobalt-60, iron-55, iron-59, zinc-65, cesium-137, manganese-54, antimony-124, catifornium-252, americium-241, and plutonium-238), as well as the fissile isotopes plutonium-239 and uranium-233.

Molecular Craftwork with DNA[a]

Nadrian C. Seeman[b]

DNA isn't just a double-helical molecule. With specially designed synthetic sequences, DNA can be made to form specific branches, knots, catenanes, and polyhedra. The future may see DNA used as a material to make macroscopic objects, including crystals. A woodcut by M.C. Escher, the techniques of genetic engineers, and the minimization of sequence symmetry have produced this approach to nanotechnology.

We all know that DNA is the genetic material of living organisms. DNA molecules consist of two antiparallel polyanionic sugar-phosphate chains that are held together in a double helix by complementary interactions between purine and pyrimidine bases. The two purines found in DNA are adenine (A) and guanine (G), and the two pyrimidines are thymine (T) and cytosine (C). The rules for pairing them are known to schoolchildren all over the Earth: A always pairs with T, and G always pairs with C. The two ends of a DNA single strand are called the 5' end and the 3' end; sequences are written in the $5' \rightarrow 3'$ direction. The repeating unit of the polymer, a base, plus a single sugar-phosphate backbone component is called a "nucleotide." When we think about the chemical properties of the DNA molecule, it is usually with regard to its biological context, because the sequence of the bases defines the contents of genes: Many scientists throughout the world are engaged in devising improved means to modify, isolate, or characterize DNA derived from living species. They seek to understand genes, genomic organization, and gene expression and to formulate therapies for diseases whose origins can be traced to the genetic level.

The chemical features that make DNA such a suitable molecule to serve as genetic material also make it a molecule that can be used as an effective construction medium on the nanometer scale. When we build any object or device on the macroscopic scale, we know ahead of time which components to stick together and how they are to be placed relative to one another. We are all familiar with structures that are assembled according to design: Carpenters construct furniture, bricklayers build houses, plumbers build water systems, and tailors assemble garments on the macroscopic scale by linking the materials of their trades. When we are dealing with molecules, the situation is different. Even though we usually know what molecules we want to join, it is often difficult to direct associations between them: We can't just grab molecule 1 and graft it onto molecule 2. As much as we would like to, we cannot act on the microscopic scale in the straightforward way that is possible on the macroscopic scale.

However, just because we can't do it doesn't mean we can't think about doing it. It is a valid goal of chemistry to pursue control of molecular structure comparable to the structural control enjoyed by craft workers on the macroscopic scale. This approach to chemistry is often called "nanotechnology" [1]. The ability to join, cement, couple, or weave two molecules together, with the same certainty enjoyed by a carpenter, a bricklayer, a plumber, or a garment worker, would increase greatly the efficiency of chemists, materials scientists, and molecular biologists. Of course, there are means of ligating or bonding molecules together, but there are no nails, cement, screws, or threads to act as arbitrary fasteners between molecules, nor are there vices to clamp them together while they are being annealed.

A close chemical equivalent to macroscopic construction is self-assembly. The structural components in biological systems often self-assemble spontaneously. In self-assembling systems, complementary surfaces form cohesive structures. Microtubules and microfilaments are examples of systems that self-assemble to grow in one dimension, and the capsid proteins of simple viruses form self-assembled two-dimensional (icosahedral) surfaces. Many biological macromolecules can be induced to self-assemble into three-dimensional crystals, although such structures are not common within the cell. Molecular surfaces that are complementary in shape and charge are the primary elements in biological self-assembly.

Why Use DNA?

I wasn't thinking about molecular construction at all in the fall of 1980, when I realized that DNA offered special advantages in this area and that it could be used to build nanoscale objects. I was a crystallographer in the Department of Biological Sciences at SUNY/Albany.

[a]*Chemical Intelligencer* 1995(3), 38–47.

[b]Department of Chemistry, New York University, New York, NY 10003, USA

B. Hargittai and I. Hargittai (eds.), *Culture of Chemistry: The Best Articles on the Human Side of 20th-Century Chemistry from the Archives of the Chemical Intelligencer*, DOI 10.1007/978-1-4899-7565-2_33, © Springer Science+Business Media New York 2015

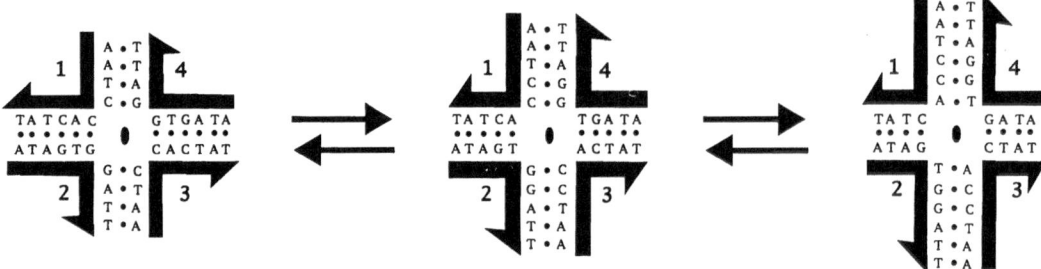

Fig. 1 Branch migration in a twofold symmetric 4-arm junction. The thick black lines correspond to strand backbones, with the half-arrowheads pointing in the $5' \rightarrow 3'$ direction of the molecule. The lens-shaped object in the center of each branched structure indicates the twofold sequence symmetry of the branched junction: Strand 1 has the same sequence as strand 3, and strand 2 has the same sequence as strand 4. In the central molecule, only the five nucleotides flanking the central branch on each side are shown. The central molecule is shown to branch migrate in each direction: To the right, the base pairs in the horizontal arms re-pair to join the vertical arms; to the left, the base pairs in the vertical arms re-pair to form the horizontal arms. The figures on the left and the right are free to migrate again, as they, too, are twofold symmetric. Repeated migration eventually results in the production of two linear molecules from each branched molecule.

During the three years I had been in Albany, my laboratory had been frustrated in its attempts to crystallize any of the molecules in which we were interested. The only scientific problem on which progress was being made was in a totally different field: As a postdoctoral research associate in Leonard Lerman's laboratory, Bruce Robinson (now Professor of Chemistry at the University of Washington) had done a study of DNA libration, and he wanted to apply his results to the branch migration of Holliday junctions, which are intermediates in genetic recombination. In a Holliday junction, four strands of DNA form four double helical arms about a central branch point [2]. In cells, the structure is formed by homologous strands of DNA, so it has a twofold axis of sequence symmetry. This symmetry permits the position of the branch point to relocate, through an isomerization called branch migration (Fig. 1). Bruce had asked me to build a model of a Holliday junction; when I did so in early 1979, we noted asymmetry in the molecule above and below the branch point, and we proceeded to model the effect that asymmetry might have on the kinetics of branch migration [3]. After we had been working on the kinetic model for a while, the idea struck me that it would be possible to build Holliday junction analogues with fixed branch points, if one used synthetic molecules whose sequences lacked sequence symmetry [4]. Not only are migratory junctions hard to characterize, because their branch points are all in different places, but they eventually resolve to become a pair of linear duplex molecules.

I was very excited by the prospect that synthetic Holliday junctions, with fixed branch points, might be assembled.

I mentioned this idea to a visitor, Malcolm Casadaban, who asked whether junctions with more than four double helices could be made. I didn't know at the time, but, upon thinking about it, I concluded that at least eight arms could be made to flank a branch point, without branch migration occurring [4]. Another influence at the time was the excitement of the early days of genetic engineering. Every week, in journal clubs and research conferences, I heard someone report the assembly of genetic constructs by the ligation of DNA molecules containing short single-stranded overhangs [5], called "sticky ends" (Fig. 2).

SCALES

The boundaries between the following scales are necessarily somewhat fuzzy.

MACROSCOPIC: The scale of phenomena with which we deal in everyday life, where the laws of classical physics appear to apply. Roughly describing phenomena and objects 10^{-6} or 10^{-7} m or larger.

MICROSCOPIC: Also called chemical, angstrom, or molecular. This scale is differentiated from the macroscopic scale in that quantum descriptions are necessary for most of the phenomena. Refers to phenomena and objects 10^{-9} m and smaller. Chemical bonds are 1–2 Å (10^{-10} m) in length.

NANOMETER: Also called *mesoscopic*. A scale just an order of magnitude larger than the microscopic scale, where classical physics appears to apply, but chemical phenomena, such as statistical-mechanical laws, are relevant. Refers to phenomena and objects 10^{-9} to 10^{-7} m or so.

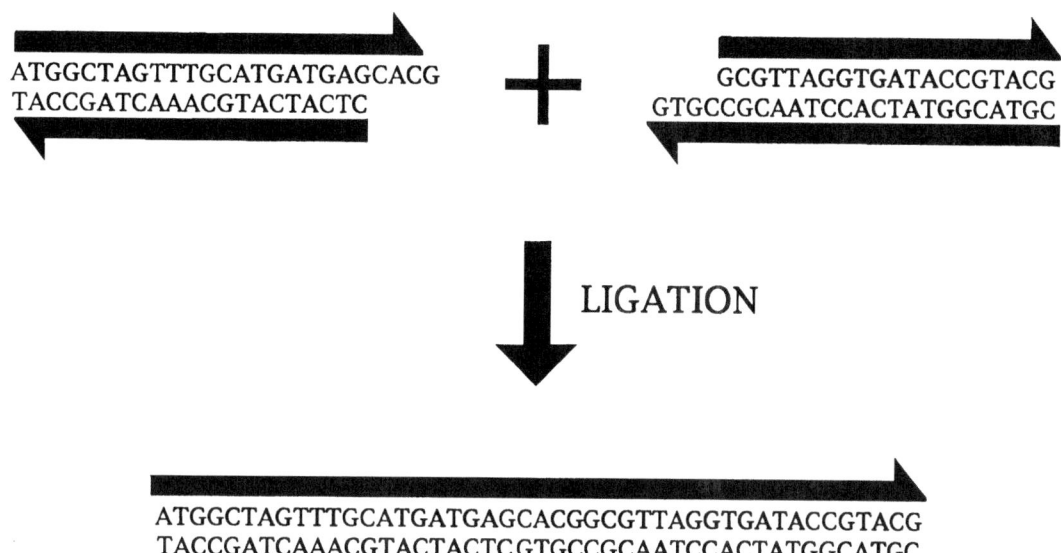

Fig. 2 Sticky-ended ligation. Two linear double-helical molecules of DNA are shown. The antiparallel backbones are indicated by the black lines terminating in half-arrows. The half-arrows indicate the $5' \rightarrow 3'$ directions of the backbones. The right end of the left molecule and the left end of the right molecule have single-stranded extensions ("sticky ends") that are complementary to each other. Under the proper conditions, these bind to each other and can be ligated to covalency by the proper enzymes and cofactors.

Fig. 3 Six-connected networks by M. C. Escher. Next page: (a) *Depth*, a woodcut; Above: (b) *Cubic Space Division*, a lithograph. *In Depth*, the connections between the fishlike creatures are implicit, whereas in *Cubic Space Division* the connections between vertices are explicit. Both drawings are reproduced courtesy of Vorpal Gallery, San Francisco and New York City.

In this context, I went over to the campus pub one afternoon to think about junctions that contain six double-helical arms. For some reason, perhaps my frustration in growing crystals, Escher's woodcut *Depth* (Fig. 3a) flashed through my mind as I was thinking of 6-arm junctions. The fishlike creatures in that picture are arranged parallel to each other in three dimensions, just like the molecules in a crystal. The six extremities of the fish (head, tail, right fin, top fin, left fin, bottom fin) are analogous to the six arms of a 6-arm junction, and the body center of each fish corresponds to its branch point. The branch points in the picture are the vertices of a network in which each vertex is connected to six other vertices. Escher's fish do not actually touch each other, but there is a complementary relationship along the lengths of their bodies, between the head of one fish and the tail of the fish in front of it. I imagined that complementary interactions could occur in three dimensions with 6-arm junctions, along the left–right axis and along the top–bottom axis, as well as along the front–back axis. The complementarity I envisioned would be furnished by DNA sticky ends. The arrangement in *Depth* is called a "six-connected network"; this arrangement is shown more explicitly in another Escher picture, called *Cubic Space Division* (Fig. 3b). Thus, it appeared possible to engineer a crystal, or a simple object, such as a polyhedron, out of DNA molecules containing branches at fixed points. Such polyhedra would be stick figures, with edges composed of double–helical DNA; their

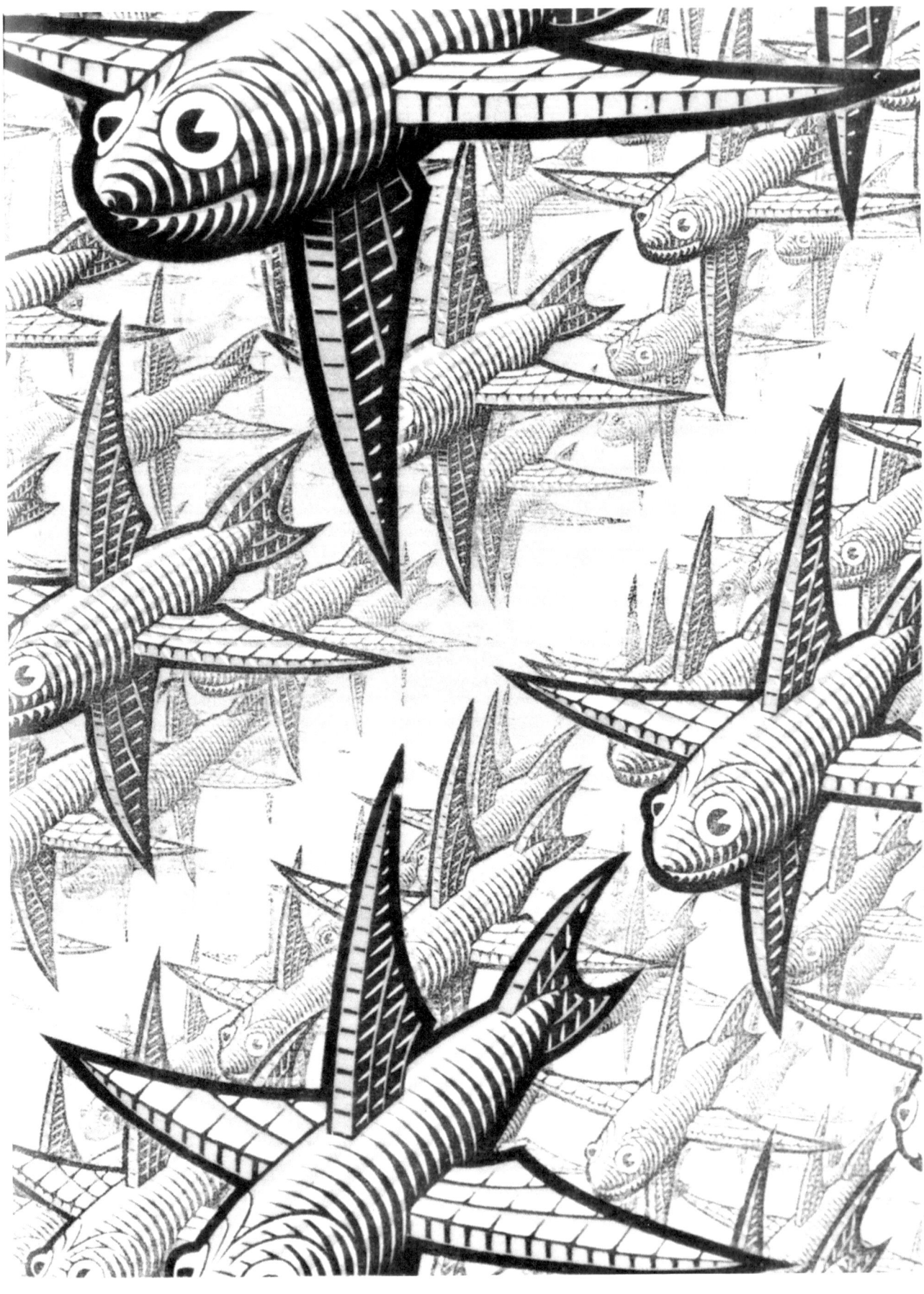

Fig. 4 A six-connected network acting as a host for macromolecular guests. The simplest conceptual network, the six-connected cubic lattice, is shown in this drawing. Macromolecular guests, represented as shaded kidney shaped objects, have been added to four edges that bound four unit cells. Note that it the guests are all aligned in the parallel fashion shown, the entire material will be a crystal, and it will be possible to determine the structure of the guests by crystallography.

Symmetry Is Antithetical to Control

The most favorable state of the DNA molecule appears to be the complementary linear double helix, when one considers molecules at adequate concentrations and temperatures, in neutral aqueous solutions that contain a sufficient supply of cations. I was fortunate to collaborate with Neville Kallenbach in the early years of the work with branched DNA; he was acutely aware that solution conditions are often key to obtaining the target DNA complex. Nucleic acid engineering is based on directing the tendency of DNA molecules to maximize base pairing. If one wishes to make a molecule containing branched helix axes or other unusual features, the target molecule will contain structures that do not correspond to the lowest energy form of DNA. For example, the formation of a 4-arm branched junction from two double helical molecules is disfavored by 1.1 kcal/mol at 18 °C [8]. Nevertheless, one can ensure the formation of the "excited state" branched molecule if one precludes the possibility of more favorable arrangements. Control in this system derives from minimizing DNA sequence symmetry [9]. The antithesis of control over chemical systems is symmetry, in its broadest sense, that is, multiple outcomes that are equivalent, or nearly so, from the standpoint of the free energy of the system. Thus, one must assign a sequence to make the target structure significantly more favorable than any other possible structure. This strategy will be familiar to readers of Sherlock Holmes mysteries: "…when you have eliminated the impossible, whatever remains, *however improbable*, must be the truth" [10]. The application of the sequence symmetry minimization procedure is an important step in the design of a molecular object from DNA. There are probably many other molecules with which one could work besides DNA, but success relies on the ability to predict interactions that represent free energy minima of the system. In general, these are difficult calculations to perform, but in the case of DNA the situation is simplified: The chemical calculations can be reduced to the rules of base pairing.

vertices would be at the branch points of junctions. A major goal of this work is to produce DNA lattices that can act as hosts for macromolecular guests, thereby enabling the determination of the guest structures by crystallography; a portion of a six–connected lattice acting as a host for a macromolecule is illustrated in Fig. 4.

In the time since that afternoon in the pub, our laboratory has built several complex geometric figures from DNA. When this research program was first envisioned, the synthesis of arbitrary sequences of DNA was extremely difficult, "Only a highly trained and skilled chemist could produce a single 12-unit [DNA] sequence in less than 3 months" [6]. Fortunately, this stumbling block was removed shortly thereafter, by automated solid support DNA synthesis [7]; in fact, the most reliable step in our constructions is the synthesis of defined sequences of DNA molecules, often containing more than 100 nucleotides.

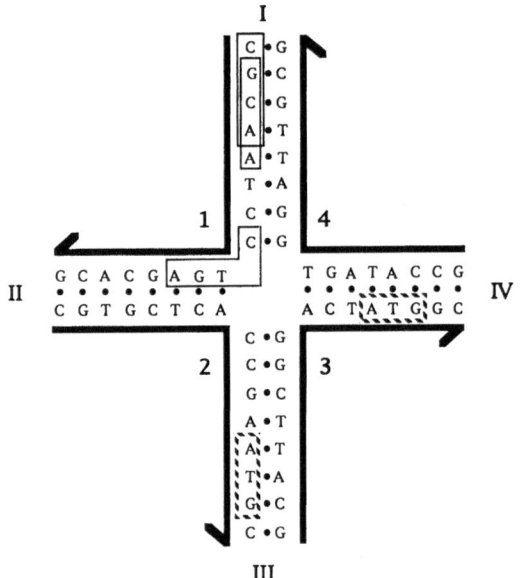

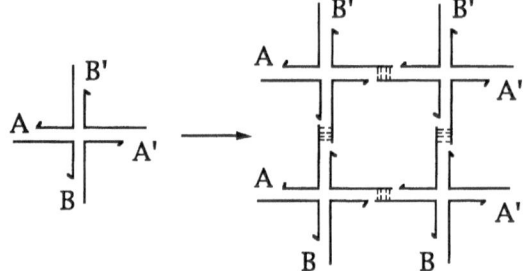

Fig. 6 Formation of a quadrilateral from an immobile junction with sticky ends. A is a sticky end, and A' is its complement. The same relationship exists between B and B'. Four of the monomeric junctions on the left are complexed in parallel orientation to yield the structure on the right. Note that A and B are different from each other, as indicated by the pairing in the complex. The enzyme DNA ligase can close the gaps left in the complex to form covalent bonds. Note that the complex has maintained open valences, so that it can be extended to form a two-dimensional lattice by the addition of more monomers.

Fig. 5 A stable DNA branched junction. The junction shown is composed of four strands of DNA, labeled with Arabic numerals. The 3' end of each strand is indicated by the half-arrowheads. Each strand is paired with two other strands to form double-helical arms; the arms are numbered with Roman numerals. The hydrogen-bonded base pairing that forms the double helices is indicated by the dots between the bases.

The sequence of this junction has been optimized to minimize symmetry and non-Watson–Crick base pairing. Because there is no homologous twofold sequence symmetry flanking the central branch point, this junction cannot undergo the branch migration isomerization reaction. At the upper part of arm I, two of the 52 unique tetrameric elements in this complex are boxed; these are *CGCA* and *GCAA*. At the corner of strand 1, the sequence *CTGA* is boxed. This is one of 12 sequences in the complex (3 on each strand) that span a junction. The complements to each of these 12 sequences are not present, whereas tetrameric elements have been used to assign the sequence of this molecule, there is redundancy in the molecule amongst trimers, such as the ATG sequences shown in boxes with dashed outlines.

The minimization of sequence symmetry is illustrated in Fig. 5, which shows a 4-arm branched junction containing 32 nucleotide pairs [11]. In principle, one would like to have 32 different nucleotide pairs with which to build this molecule, but only the conventional four pairs are readily available. It is fortunate that the pairing of DNA double helices is a cooperative phenomenon, because that lets us treat the sequence as a series of segments larger than individual nucleotide pairs. Each strand of the junction in Fig. 5 contains 16 nucleotides, which are broken up into 13 overlapping tetramers; two tetramers, CGCA and GCAA, are indicated by solid boxes. If we insist that each of the 52 tetramers in the entire molecule be unique, base pairing competitive with the designed molecule can come only from redundant trimers (e.g., the ATG sequences indicated by the boxes with broken outlines). As a further constraint, we insist that the molecule

contain no tetramer complementary to a bend; for example, there is no TCAG complementary to the boxed CTGA. Dyad symmetry around the branch point is also eliminated. The use of tetramers allows a "vocabulary" of 240 elements [$(4^4 = 256) - 16$ self-complementary sequences]. In order to design larger structures, one must divide the strands into longer elements of length n, and competition from redundant $(n-1)$-mers can assume greater importance [12]. When they are mixed together, the strands in Fig. 5 self-assemble into the well characterized branched complex illustrated [13].

The Construction of Polyhedra from DNA

Figure 6 illustrates the assembly of four 4-arm junctions into a quadrilateral. The opposite arms of the DNA molecules contain complementary sticky ends. The construction illustrated is unclosed, because it has sticky ends that could bind more junctions; in principle, the structure could be extended to generate an infinite two-dimensional lattice. When ligations like this are performed in practice, it turns out that branched DNA molecules do not have fixed "valence angles" between their helices in the same sense that carbon atoms have preferred angles between their bonds: The same branched molecule can cyclize to form a trimer, a tetramer, a pentamer, or higher products [14, 15]. The flexibility of DNA branches is a major problem that must be kept in mind when one is designing routes to specific structures [16].

Nevertheless, it is possible to construct particular target molecules, by directing the assembly of components with a series of unique sticky ends. Figure 7 illustrates a molecule that we constructed together with Junghuei Chen, using this

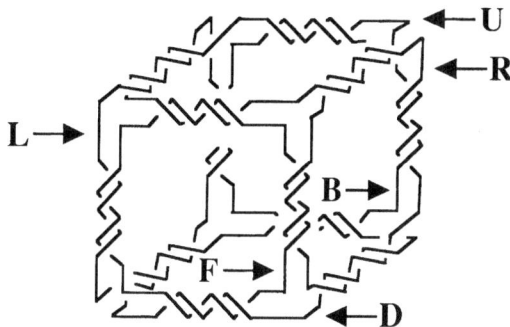

Fig. 7 A DNA molecule whose helix axes have the connectivity of a cube. The molecule shown consists of six cyclic strands that have been catenated together in this particular arrangement. They are labeled by the first letters of their positional designations, Up, Down, Front, Back, Left, and Right. Since each edge contains 20 nucleotide pairs of DNA, we expect that the edge lengths will be about 68 Å. From model building, the axis-to-axis distance across a square face appears to be about 100 Å, with a volume (in a cubic configuration) of approximately 1760 nm^3, when the cube is folded as shown.

strategy. This is a DNA molecule whose helix axes are connected like the edges of a cube [17]. Each edge of the object is designed to contain two double helical turns of DNA. Note that the graphical representation used in Fig. 7 places all the twisting in the middle of each edge for clarity, but the pairing goes all the way to the corners.

If a face of a polyhedron is flanked by edges containing an integral number of double-helical turns, a closed cyclic single strand of DNA will be part of each edge flanking that face. Inspection of Fig. 7 reveals each of the sis faces to be flanked by a single strand of DNA; these six strands are each doubly linked to their four neighbors, like the links of a chain. The molecule itself is thus a hexacatenane, and the six strands are bonded topologically to each other; although distinct molecules, the strands cannot be separated without breaking them. Each edge of the cube contains a unique site for scission by restriction endonucleases, enzymes that cut DNA at specific short sequences. Cleavage of individual edges allows one to establish that the molecule has indeed been synthesized properly [17]: An early stage in the synthesis of the molecule generates the linear triple catenane corresponding to the left (L), front (F), and right (R) sides of the molecule. Restriction of the complete cube at the L–F and F–R edges generates the linear triple catenane corresponding to the top, back, and bottom of the molecule (U–B–D in Fig. 7). By cleaving the edges connecting opposite faces of the cube, one can show that the structure is indeed a tetragonal prism, and not an octagonal or higher prism [17]. Proofs of synthesis are limited at this time to demonstrating topology, not structure.

The synthesis of the cube was performed in solution, and the purification of a key intermediate entailed its dissociation and reconstitution. From the standpoint of efficiency and generality, solution synthesis leaves a great deal to be desired. Consequently, a more effective means has been developed for constructing geometrical stick figures [18]. This synthetic methodology uses a solid support, allowing convenient removal of reagents and catalysts from the growing product. Each ligation cycle creates a robust intermediate object that is covalently closed and topologically bonded together. The method permits one to build a single edge of an object at a time. Control derives from the restriction endonuclease digestion of hairpin loops forming each side of the new edge. Intermolecular reactions are best done with asymmetric sticky ends, to generate specificity. Sequences are chosen in such a way that restriction sites are destroyed when the edge forms. For example, an *Alw*NI restriction site (CAGNNN | CTG, where | indicates the site of cleavage) on one arm could be combined with a *Dra*III site (CAGNNN | CTG) on another arm to generate an edge containing the sequence CAGNNN | GTG and its complement; the final sequence cannot be recognized by either enzyme. The NNN sequence corresponds to an arbitrary asymmetric sequence on the sticky end (e.g., ACG on one strand and CGT on the other), so that a unique product will form.

Together with Yuwen Zhang, we built a truncated octahedron by means of solid-support methodology (Fig. 8) [19]. This Archimedean solid contains six squares that flank its fourfold symmetry axes and eight hexagons that surround its three-fold symmetry axes. As with the cube, there are two turns of DNA per edge, so each of the 14 faces corresponds to a single cyclic strand. Thus, the final object is a complex 14-catenane. Each vertex of a truncated octahedron is bonded to three others (3-connected), but the molecule has been constructed from 4-arm junctions; consequently, each vertex is associated with another arm that could be used to join the polyhedra, although this has not been done. The extra arms are all hairpins extending from the strands that correspond to the square faces. The entire molecule contains 2550 nucleotides and has a molecular weight of ca. 790 kDa.

In this construction, the objects added to the support are squares and square groupings (Fig. 8). The object has been constructed by doing two intermolecular additions to a square attached to the support. In the first addition, a tetrasquare complex is added, and in the second addition, the final square is added. The structure at the lower left of Fig. 8 is a heptacatenane hexasquare complex. The square strands are already intact in this construct, and the hexagons are all formed from the outer strand. The hexagons result from successive intramolecular closures of the sticky

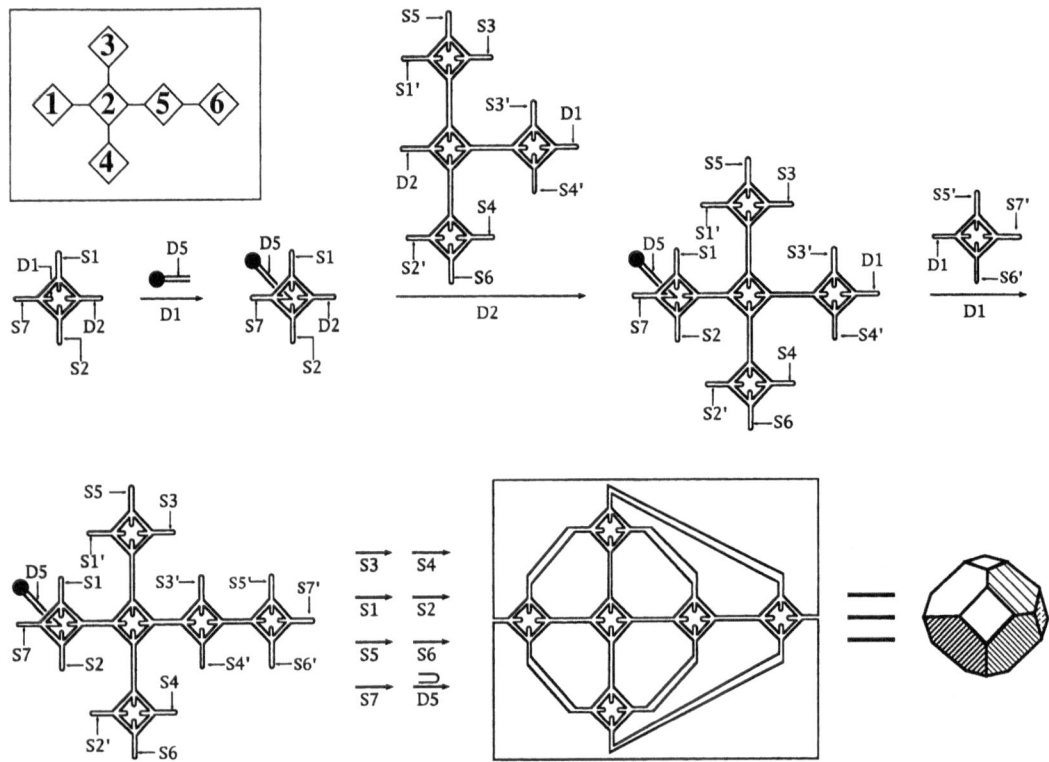

Fig. 8 The synthetic scheme used to synthesize a truncated octahedron on a solid *support*. The boxed diagram in the upper left corner indicates the numbering of the individual squares. Each square in the rest of the diagram is shown with its restriction sites indicated. Symmetrically cleaving restriction sites are labeled S, and are in pairs, with one of the pairs being primed; restriction sites that are cut distally are labeled D; restriction sites on the exocyclic arms are not indicated. The arms that will eventually be combined to form edges of the object are drawn on the outside of each square, and the exocyclic arms are drawn on the inside of the square. A reaction is indicated by a line above a restriction site name: This means that the restriction enzyme (or enzyme pair for those labeled S) is added, protecting hairpins are removed, and then the two sticky ends are ligated together. The product is shown in two forms. On the left, the S1–S6 closures are shown as triple edges, to emphasize their origins; the two strands of the edge formed by the S7 closure are separated to maintain the symmetry of the picture. On the right, a slightly rotated front view of a polyhedral representation of a truncated octahedron is shown without the exocyclic arms; the symmetry of the ideal object is evident from this view.

ends associated with the restriction enzyme site pairs, S1–S1′...S7–S7′. The molecules are isolated from each other on the support, so it is possible to expose this group of sticky ends by using symmetrically cleaving restriction enzyme pairs that recognize six nucleotide pairs each. Doing this in solution would lead to a mixture of products. Although the initial sites are destroyed after ligation, four-base cutting sites remain, for analytical purposes. For example, *Bgl*II (A | GATCT) is combined with *Bam*HI (G | GATCC) to generate an edge containing GGATCT, immune to both enzymes, but cleavable by *Dpn*II (GATC). The final step in the synthesis involves releasing the structure from the support and annealing it shut with a hairpin. The synthesis is demonstrated in two stages: first, by showing that all six cyclic strands corresponding to the square molecules are in the heptacatenane and, second, by digesting the final product to the tetracatenanes that flank the squares.

Control of DNA Topology: Single-Stranded DNA Knots

The cube and the truncated octahedron are catenanes, and an intimate relationship exists between catenanes and knots [20]: Removal of a node by switching strands, yet maintaining local strand polarity, converts a catenane to a knot, and a knot to a catenane (Fig. 9). This relationship is important, because it appears that cloning DNA structures in microorganisms will be achieved most readily by getting single-stranded DNA molecules to fold up into complex knots, whose restriction will lead to the desired stick figures [21]. It is not possible to clone branched structures directly, because a single round of replication will reduce a branch to linear double helices. Nevertheless, it is possible, in principle, to make an entire structure from a single strand, as illustrated for a pentagonal dodecahedron in Fig. 10. The key to this strategy is to add an extra external arm for every strand; for

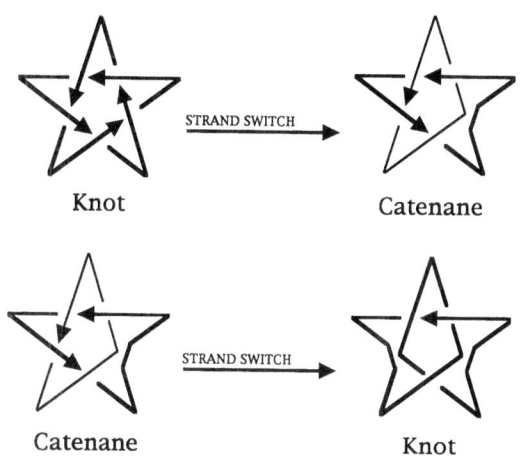

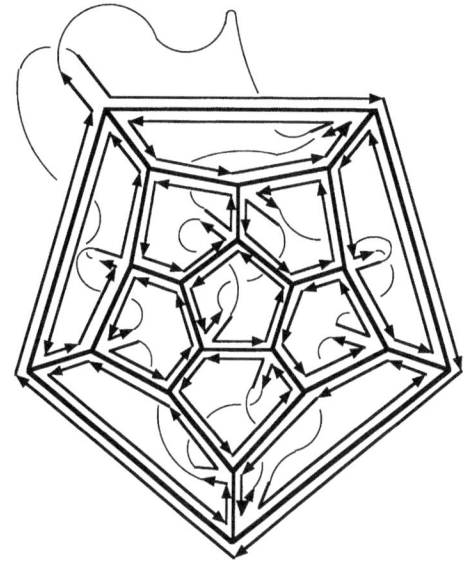

Fig. 9 Interconversions of knots and catenanes by switching strands at a node. The structure shown at the upper left is a 5_1 knot. The strand direction is indicated by the arrowheads appearing along the strand. When the two strands entering the lower node on the right exchange outgoing partners, the node disappears. This converts the knot to a catenane, shown at the upper right; the two linked cycles are drawn so as to retain their shapes, but they are drawn by pens of different thicknesses. The same catenane is redrawn at the lower left of the figure. The lower left node of the catenane undergoes the same strand switch, which converts it to a trefoil knot, shown at the lower right. The trefoil knot is a single strand, so it is drawn by only a single pen.

Fig. 10 A single-stranded representation of a pentagonal dodecahedron. A pentagonal dodecahedron is illustrated with 12 exocyclic arms, in a representation known as a Schlegel diagram. This is a two-dimensional representation of a three-dimensional object in which the central polygon is closest to the reader, the polygons removed from the center are distorted and further behind in the page, and the outer polygon is at the rear of the figure. The Schlegel diagram of the dodecahedron is shown in the thickest lines. Flanking these lines are short antiparallel arrows, drawn less heavily; these represent the double helical DNA corresponding to each edge of the dodecahedron, with an arrowhead indicating the $5 \rightarrow 3'$ polarity of the strand. Each of the 12 pentagons contains an exocyclic double-helical arm. In addition, each of the individual faces has been connected to a neighboring face via the exocyclic arms, so that the entire representation is a single long strand. The thin curved lines represent the connecting DNA that links the pentagons to each other. The structure is closed, so as to make a formal knot, but it would need to be cleaved in order to fold. Each exocyclic double-helical segment would be designed to contain a restriction site, so that it can be severed from the connecting DNA. This DNA will be cut away and removed upon formation of the structure. No attempt at topological representation is made here: All connecting DNA (thin curved lines) lies behind the polygonal DNA for purposes of clarity.

molecules whose edges contain only an integral number of helical turns, this corresponds to an extra arm per face. The external arms are connected to form the complex knotted structure shown. The sequence of such a single-stranded molecule could be cloned. External arms are needed on a polyhedral structure to form a lattice, so the target structure would have them anyway.

Together with Shouming Du, Hui Wang, and John Mueller, we have explored the possibilities for threading knots experimentally. A DNA molecule can be synthesized containing the sequence X-T-Y-T-X'-T-Y'-T, where X and Y correspond to one helical turn, X' and Y' are their Watson-Crick complements and T is a dT_n linker. When cyclized, the strand yields a trefoil knot with negative nodes $\left(3_1^-\right)$, because the nodes formed by ordinary right-handed DNA (B-DNA) correspond to negative topological nodes [22]. However, there is a left-handed form of DNA, Z-DNA [23], that is formed by special sequences, under the control of solution conditions [24]. One can choose for both X and Y sequences capable of forming Z-DNA under different conditions. Figure 11 shows that mild Z–promoting conditions will produce an amphicheiral figure-eight (4_1) knot, containing two positive nodes and two negative nodes, but strong Z–promoting conditions will produce a trefoil knot with positive nodes $\left(3_1^+\right)$ [25].

There is a general relationship between the nodes of DNA molecules and the nodes of single-stranded DNA knots: A half-turn of duplex DNA can be used to generate a node in a knot [26]. Figure 12 illustrates this point with a trefoil knot built from a branched junction. The three nodes of the knot shown are formed by perpendicular lines, whose polarity is indicated by arrowheads. The nodes act as the diagonals of a square, which they divide into four regions, two between antiparallel arrows and two between parallel arrows. The transition from topology to nucleic acid chemistry can be made by forming base pairs between strands in the antiparallel regions, to produce a half-turn of DNA. The axes of the helices are drawn perpendicular to the base pairs. A trefoil

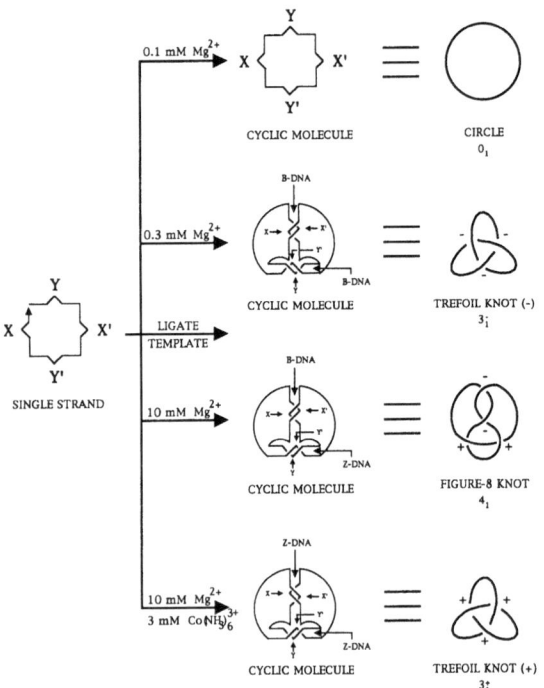

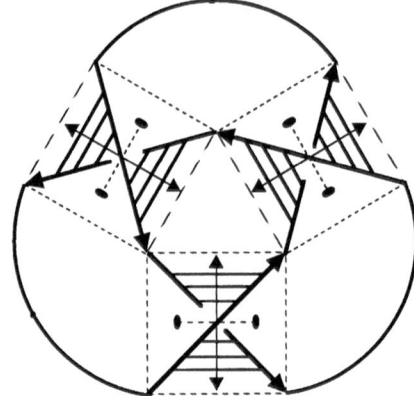

Fig. 12 The relationship between nodes and antiparallel B-DNA illustrated on a trefoil *knot*. A trefoil knot is drawn with negative nodes. The path is indicated by the arrows and the very thick curved lines connecting them. The nodes formed by the individual arrows are drawn at right angles to each other. Each pair of arrows forming a node defines a square in this figure, which is drawn in dashed lines. Each square is divided by the arrows into four domains, two between parallel arrows and two between antiparallel arrows. The domains between antiparallel arrows contain lines that correspond to base pairing between antiparallel DNA strands. Double-arrowheaded lines that represent helix axes are shown perpendicular to these lines. The amount of DNA shown corresponds to about half a helical turn. The helix axes are bisected by dashed lines terminating in ovals; these represent the local central dyad axes of each half-turn of DNA. It can be seen that three helical segments of this length could assemble to form a trefoil knot. The DNA shown could be in the form of a 3-arm DNA branched junction.

Fig. 11 The synthetic scheme used to produce three target knots and a circle from a single DNA strand. The left-hand side of this synthetic scheme indicates the molecule from which the target products are produced. The four pairing regions—X and its complement X′, Y and its complement Y′—are indicated by protrusions from the square; the dT_n linkers are represented by the corners of the square. The 3′ end of the molecule is denoted by the arrowhead. The 3′ end is between helical domains, and therefore a linking template complementary to the 3′ and 5′ ends of the strands is required in order to promote the ligation. The four independent solution conditions used to generate the target products are shown to the right of the basic structure. The pairing and helical handedness expected in each case is shown to the right of these conditions, and the molecular topology of the products is shown on the far right of the figure.

knot has been constructed recently from the branched junction motif, demonstrating the validity of equating a half-turn of DNA with a node in a knot projection [27].

Caveats and Prospects

It is important to realize that the constructions performed above have been conducted on the nanometer scale, not on the angstrom, or chemical, scale. From a structural engineering viewpoint, the difficulty of the chemical scale is that the laws of physics do not permit all conceivable arrangements of atomic nuclei to produce stable compounds; for example, carbon–carbon bonds 0.5 or 1.8 Å long are not available as structural components. By contrast, the nanometer scale appears to be the smallest macroscopic scale; there are no

evident principles that prevent the fabrication of any structural arrangements not forbidden by the impenetrability of matter. Thus, we have exploited a niche outside the usual chemical domain, in which elaborate structural assembly is much easier than usual; likewise, the helical nature of DNA has allowed us to generate specific catenanes and knots that are extremely difficult to build on the chemical scale [28, 29]. New arrangements of atoms on the chemical scale sometimes produce new chemical properties, owing subtle features of the charge density distribution in the product. It is unlikely that new chemical properties will arise on the nanometer scale, because the chemical features of the constituent residues are already fairly well fixed. However, the reorientation and juxtaposition of well-defined macromolecular elements can lead to new functionality. In much the same way that a piece of metal can be fashioned into a key, a screw, or a pipe without altering its internal structure, new shapes can lead to new functions.

An alternative to the "bottom up" assembly methods described above is top-down construction, which is exemplified by moving atoms with scanning tunneling microscopes (STM); the constructions of a corporate logo [30] and an atomic switch [31] are dramatic successes of this methodology. These top-down methods enjoy the advantage of lacking the undesirable byproducts that can arise between the large numbers of molecules found in a chemical reaction. Nevertheless, the efficiency of top-down methodology for interfacing with the nanometer scale is inherently low: A single STM is used by an individual investigator to construct one object. Chemists generate products in parallel on a scale that is almost unimaginably vast in the macroscopic world: A reaction involving a gram of a simple compound of molecular weight 60 generates roughly 10^{22} product molecules. The polyhedral constructions described above have been performed on an exploratory scale that is intermediate in yield between these two extremes: Typically, we produce about 10^{10} molecules, but larger synthetic scales appear possible.

It is important to remember that the nanometer scale does differ from the macroscopic scale. If we move a lever or turn a knob, the lever or knob goes where we force it to go. However, nothing is absolute on the molecular scale: Any transition will be characterized by a finite equilibrium constant Consider an intramolecular structural transition $(A \leftrightarrow B)$, and an open crystalline array (e.g., Fig. 4), 1 mm/edge, containing 10^{15} cubic unit cells, each having an edge length 100 Å; Imagine that it has been shifted from conditions favoring state A to conditions favoring state B by a free energy difference of 12 kcal/mol. At ambient temperatures, this transition is favored by a factor of about 5×10^{8}. Nevertheless, the crystal would retain about two million unit cells in state A. Furthermore, the particular unit cells containing A instead of B will vary with time.

As described above, the DNA array application that inspired this research program is the construction of macromolecular zeolites, to serve as hosts for globular macromolecular species, as an aid in crystallographic structure determination. The rate-determining step in macromolecular crystallography is the preparation of adequate crystals. The ability to assemble periodic arrays of cages that contain ordered guests would contribute to a solution of that problem. The assembly of periodic lattices is a difficult goal to achieve: We have shown that control over the synthesis of an individual object can be derived from minimization of sticky-end symmetry, but it is not possible to exploit symmetry minimization to build an entire crystalline array, since the lattice inherently contains translational symmetry. There are at least three key elements necessary for the control of three-dimensional structure in molecular construction involving high symmetry: (1) the predictable specificity of intermolecular interactions between components; (2) the structural predictability of intermolecular products; and (3) the structural rigidity of the components [32]. Branched DNA molecules meet the first two criteria (sticky ends are specific and they always form well-characterized DNA double helices

when ligated), but the search continues for a means to achieve the third one.

The medical and commercial importance of DNA has resulted in convenient technology for the modification [33] of DNA, permitting one to attach special functional groups, both on the bases and on the backbone. In addition, there are natural mechanisms by which drugs, particular proteins, or other DNA strands recognize and bind to specific sites on double-helical DNA. The self-assembly of the DNA molecules could in turn direct the assembly of attached species, such as molecular electronic components. Bruce Robinson and I have suggested that a crystalline array of such an assembly could lead to a biochip, in which the DNA is limited to a structural role [34]. Among other utilities envisioned for tethering molecules to DNA objects are the production or new catalysts, the solubilization of otherwise insoluble proteins and drugs, and the targeted delivery of such species [21]. As with any craft material, the structural applications of DNA are limited only by the imagination.

Acknowledgments The research described above has been supported by Grants N00014–89–J–3078 from the Office of Naval Research and GM–29554 from the National Institutes of Health. None of this work could have been done without the contributions of my colleagues, Junghuei Chen, Shou Ming Du, John Mueller, Yuwen Zhang, Tsu-Ju Fu, Yinli Wang, Hui Wang, Siwei Zhang, Bing Liu, and Jing Qi. The early work on branched DNA molecules was done in collaboration with Neville Kallenbach. Bruce Robinson's contributions have been invaluable at several stages. I would like to thank Muldoon Elder for helpful discussions about M. C. Escher.

References

1. Drexler, K. E. *Proc. Natl. Acad. Sci. U.S.A.* **1981**, *78*, 5275–5278.
2. Holliday, R. *Genet. Res.* **1964**, *5*, 282–304.
3. Robinson, B. H.; Seeman, N. C. *Biophys. J.* **1987**, *51*, 611–626.
4. Seeman, N. C. *J. Theor. Biol.* **1982**, *99*, 237–247.
5. Cohen, S. N.; Chang, A. C. Y.; Boyer, H. W.; Helling, R. B. *Proc. Natl. Acad. Sci.* U.S.A. **1973**, *70*, 3240–3244.
6. Alvarado-Urbina, G.; Sathe, G. M.; Liu, W.-C.; Gillen, M. F.; Duck, P. D.; Bender, R.; Ogilvie, K. K. *Science* **1981**, *214*, 270–274.
7. Caruthers, M. H. *Science* **1985**, *230*, 281–285.
8. Lu, M.; Guo, Q.; Marky, L. A.; Seeman, N. C.; Kallenbach, N. R. *J. Mol. Biol.* **1992**, *223*, 781–789.
9. Seeman, N. C. In *Concepts in Protein Engineering and* Design, Wrede, P., Schneider, G., Eds.; Walter de Gruyter: Berlin, 1994; pp. 319–343.
10. Doyle, A. C. "The Sign of Four" in *The Complete Sherlock Holmes;* The Literary Guild, New York, 1936; p. 118.
11. Seeman, N. C.; Kallenbach, N. R. *Biophys. J.* **1983**, *44*, 201–209.
12. Seeman, N. C. *J. Biomol. Struct. Dyn.* **1990**, *8*, 573–581.
13. Kallenbach, N. R.; Ma, R.-I.; Seeman, N. C. *Nature (London)* **1983**, *305*, 829–831.
14. Ma, R.-I.; Kallenbach, N. R.; Sheardy, R. D.; Petrillo, M. L.; Seeman, N. C. *Nucleic Acids Res.* **1986**, *14*, 9745–9753.

15. Petrillo. M. L.; Newton, C. J.; Cunningham, R. P.; Ma, R.-I.; Kallenbach, N. R.; Seeman, N. C. *Biopolymers* **1988**, *27*, 1337–1352.

16. Seeman, N. C.; Zhang, Y.; Chen, J. *J. Vac. Sci. Technol.* **1994**, *A12*, 1895–1903.

17. Chen, J.; Seeman, N. C. *Nature (London)* **1991**, *350*, 631–633.

18. Zhang, Y.; Seeman, N. C. *J. Am. Chem. Soc.* **1992**, *114*, 2656–2663.

19. Zhang, Y.; Seeman, N. C. *J. Am. Chem. Soc.* **1994**, *116*, 1661–1669.

20. White. J. H.; Millett, K. C.; Cozzarelli, N. R. *J. Mol. Biol.* **1987**, *197*, 585–603.

21. Seeman, N. C. *DNA Cell Biol* **1991**, *10*, 475–486.

22. Mueller, J. E.; Du, S. M.; Seeman. N. C. *J. Am. Chem. Soc.* **1991**, *113*, 6306–6308.

23. Wang, A. H.-J.; Quigley, G. J.; Kolpak, F. J.: Crawford, J. L.; van Boom, J. H.; van der Marel, G.; Rich, A. *Nature (London)* **1979**, *282*, 680–686.

24. Behe, M.; Felsenfeld. G. *Proc. Natl. Acad. Sci. U.S.A.* **1981**, *78*, 1619–1623.

25. Du, S. M.; Stollar, B. D.; Seeman, N. C. *J. Am. Chem. Soc.* **1995**, *117*, 1194–1200.

26. Seeman, N. C. *Mol. Eng.* **1992**, *2*, 297–307.

27. Du, S. M.; Seeman, N. C. *Biopolymers* **1994**, *34*, 31–37.

28. Dietrich-Buchecker, C. O.; Sauvage, J.-P. *Angew. Chem. Int. Ed. Engl.* **1989**, *28*, 189–192.

29. Stoddart, J. F.; Amabilino, D. B. *Mater. Res. Soc. Symp. Proc.* **1994**, *330*, 57–60.

30. Eigler, D. M.; Schweizer, E. K. *Nature (London)* **1990**, *344*, 524–526.

31. Eigler, D. M.; Lutz, C. P.; Rudge, W. E. *Nature (London)* **1991**, *352*, 600–603.

32. Liu, B.; Leontis, N. B.; Seeman, N. C. *Nanobiology* **1995**, *3*, 177–188.

33. Eckstein, F. *Oligonucleotides and Analogues;* IRL: Oxford, 1993, pp 49–313.

34. Robinson, B. H.; Seeman. N. C. *Protein. Eng.* **1987**, *1*, 295–300.

Nadrian C. Seeman was born in Chicago in 1945. He received his B.S. in biochemistry from the University of Chicago in 1966 and his Ph.D. from the Department of Crystallography at the University of Pittsburgh in 1970. He was a research associate in molecular graphics at Columbia University during 1970–1972 and was successively a Damon Runyon fellow and an NIH postdoctoral fellow in nucleic acid crystallography at MIT from 1972 to 1977. After 11 years on the faculty of the Department of Biological Sciences at SUNY/Albany, he joined the Department of Chemistry at New York University as Professor of Chemistry in 1988.

Dr. Seeman has received a Basil O'Connor Fellowship from the March of Dimes Birth Defects Foundation, and he has been the recipient of a Research Career Development Award from the National Institutes of Health. He was awarded the Sidhu Award for excellence in diffraction studies; this award recognized crystallography work on dinucleoside phosphates that demonstrated the first Watson-Crick A–U base pair in a single crystal. He is an author, with John Rosenberg and Alexander Rich, of a highly cited theoretical paper on the sequence-specific recognition of double-helical nucleic acids by proteins. Recently, he received a Science and Technology award from *Popular Science* magazine for the construction of a DNA cube. His research is funded by the U.S. Public Health Service–National Institutes of Health and the U.S. Office of Naval Research. The W. M. Keck Foundation has contributed to his work as well.

Water, Superwaters, and Polywater[a]

Irving M. Klotz[b]

Scientists are first of all human beings. Some are saints, a few are charlatans, and the overwhelming majority adhere to the unwritten codes of good scientific practice.

All human beings are subject to passions such as greed, vanity, ambition, generosity, compassion, integrity, and kindness. In addition, scientists are propelled by two other motivations: (a) an intense drive to discover a new phenomenon or a novel insight into nature and (b) a burning wish to be recognized by their peers, and by society.

Thus, from the very outset of their careers, scientists are exposed to various temptations. There are dozens of actions or inactions that are considered dishonorable within the scientific, community, ranging from egregious dishonesty to uncivil behavior. We rarely realize the extent to which our own actions are activated by our passions and motivations, but we can readily discern when *other* individuals act dishonorably, in serious or trivial ways. Einstein once said "Eigenen dreck stinkt nicht" (in polite English, "one is unaware of one's own malodors"). Certainly, one way to become more aware of the potential abysses into which we may stumble is to learn about the misperceptions, distortions, and deplorable actions of individuals in the past. Water provides an eminently suitable subject for this objective, for its history, from ancient times to the present, is replete with examples of missteps of a wide range of types and severities.

Water is the one chemical substance with which everyone is familiar. No matter how vague or primitive one's scientific conceptions, each of us recognizes from direct experience the vital role played by water in an individual's very existence. It is in us and around us. It comes as no surprise, therefore, to learn that Thales, the man Aristotle called the founder

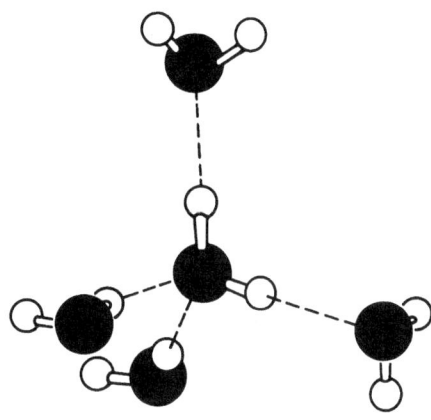

Fig. 1 Tetrahedral arrangement of the four water molecules around a central water molecule.

of Western philosophy of science, asserted that water is the source of all things: $\upsilon\delta\omega\rho$ $\pi\alpha\nu\tau\alpha$ (water is all).

Variations of this view prevailed for millennia. It has always been apparent to anyone watching an open container of water that the liquid disappears slowly; that the water is converted into air seems obvious. It has also been apparent for millennia that living organisms imbibe water and increase in size and weight; clearly, water can be converted into solids.

In fact, a "definitive" experiment that proved that water could be converted into solids was carried out in relatively recent times, the seventeenth century, by a distinguished savant, Joan Baptista van Helmont. His procedure was a quantitative, precise one, very modern in its conception and execution [1]:

> I took an earthen vessel, in which I put 200 pounds of earth that had been dried in a furnace, which I moistened with rainwater, and I implanted therein the trunk or stem of a willow tree, weighing five pounds. And at length, five years being finished, the tree sprung from thence did weigh 169 pounds and about three ounces. When there was need, I always moistened the earthen vessel with rainwater or distilled water, and the vessel was large and implanted in the earth. Lest the dust that flew about should be comingled with the earth, I covered the lip or mouth of the vessel with an iron plate covered with tin and easily passable with many holes. I computed not the weight of the leaves that fell off in the four autumns. At length, I again dried the earth of the vessel, and there was found the same 200 pounds, wanting about two ounces. Therefore 164 pounds of wood, bark and roots arose out of water only.

A brief biographical note about Professor Klotz has already appeared on page 126 of this book. Portions of this paper have been taken from his book, *Diamond Dealers and Feather Merchants, Tales from the Sciences;* Birkhäuser: Basel, 1966.

[a]*Chemical Intelligencer* 1995(4), 34–41.

[b]Department of Chemistry, Northwestern University, Evanston, IL 60208-3113, USA (deceased)

Copy of tile medallion, in the Great Hall, National Academy of Sciences, U.S.A.

Obviously, van Helmont had transmuted water into earthlike matter.

Van Helmont's experiment was later repeated by Boyle, who confirmed the former's observation. Boyle then went a step further, simplifying the design of the experiment: he grew some small plants in water alone. Again finding the plants had gained in weight and size, he concluded that water was transmuted into the plant and into the various substances that he was able to isolate from the plant. Furthermore, Boyle argued:

> the plants my trials afforded me, as they were like in so many other respects to the rest of the plants of the same denomination; so they would, in case I had reduced them to putrefaction, have likewise produced worms or other insects as well as the resembling vegetables are wont to do; *so that water may, by various seminal principles, be successively transmuted into both plants and animals.*

In other words, *water* is all.

Water

In modern science, water has lost its position as one of the fundamental elements. However, its importance and centrality in science has increased rather than diminished. It plays a preeminent role in physics, chemistry, geology, geophysics, and the life sciences. In physics it provides the reference standards for fundamental quantities. In chemistry it is the ubiquitous solvent for innumerable reactions. In geology and geophysics it is the most important terrestrial agent by which the surface of the earth is modified. In biology it is the matrix in which all living organisms maintain themselves and reproduce.

Under these circumstances it is no surprise that enormous efforts have been devoted, in the modern era of the past two centuries of chemistry, to explore exhaustively the properties

and behavior of water. Thus, more is known about water than about any single chemical entity. Nevertheless, as our knowledge increases, our understanding does not progress proportionately, because we ask increasingly more sophisticated and detailed questions. To say that water is H_2O tells us little beyond its chemical composition and the stoichiometry of its chemical reactions. To understand its remarkable properties and behavior, we need to know the distances between the constituent atoms and how H_2O molecules are arranged in solid and liquid phases.

Although spectroscopy has played an important role in establishing interatomic distances and angles in an isolated H_2O molecule, it is X-ray diffraction that has provided the crucial data for defining the spatial arrangements of water molecules in the solid and liquid states. The single most influential paper in this area was written in 1933 by J.D. Bernal, a brilliant, pioneering X-ray crystallographer and molecular structurist, and R. H. Fowler, a renowned theoretical physicist [2].[1]

The Bernal–Fowler paper set down the basic principles of the structure of water. The central point is that each water molecule, H—O—H, tends to be surrounded by and bonded to four other water molecules placed at the vertices of a tetrahedron (Fig. 1). Between each pair of oxygen atoms (large, solid circles in Fig. 1) is a hydrogen atom (small, open circles in Fig. 1). In normal ice the O—H···O distance is 2.76 Å, and in liquid water it is somewhat longer (2.88 Å at room temperature). Looking at a larger collection of water molecules in continuing tetrahedral arrangement, one finds hexagonal cages. In Fig. 2, a representation of the structure of ice, there is an O atom at every vertex, and every line between vertices has an O—H···O "hydrogen bond"[2] between oxygens. In normal ice, the boat and chair hexagons of Fig. 2 tend to be nearly perfect and fixed in geometry. In liquid water, defects, distortions, and deformations exist, and these disrupt the highly regular array of the solid state. The structure in any local region in the liquid also fluctuates with time.

Figure 2 illustrates the disposition of H_2O molecules in *one* form of solid ice, the one familiar to all of us, designated ice I. There are 11 solid ices. These ices are not familiar to

---

[1]This paper came about because weather prediction in the 1930s, as now, was not an exact science. Bernal and Fowler, concluding a visit to Moscow, had gone to the airport to board a plane to return to England. However, a dense fog set down on the airport and stayed for days. During this interval, Bernal and Fowler created their famous paper on the structure of aqueous solutions.

[2]The concept of the "hydrogen bond" as a molecular structural feature was discovered, or invented, by C. M. Huggins in 1919, when he was still a student of the University of California (see Ref. 3). However, Huggins named this interaction a "hydrogen bridge" and was unhappy with possible implications of the word "bond." The British chemist H. E. Armstrong, well known for his sardonic wit, coined the term "bigamous hydrogen." Armstrong, in a related context and in a derogatory manner, also referred to the great American thermodynamicist G. N. Lewis (after misspelling his name) as a "California thermodynamiter."

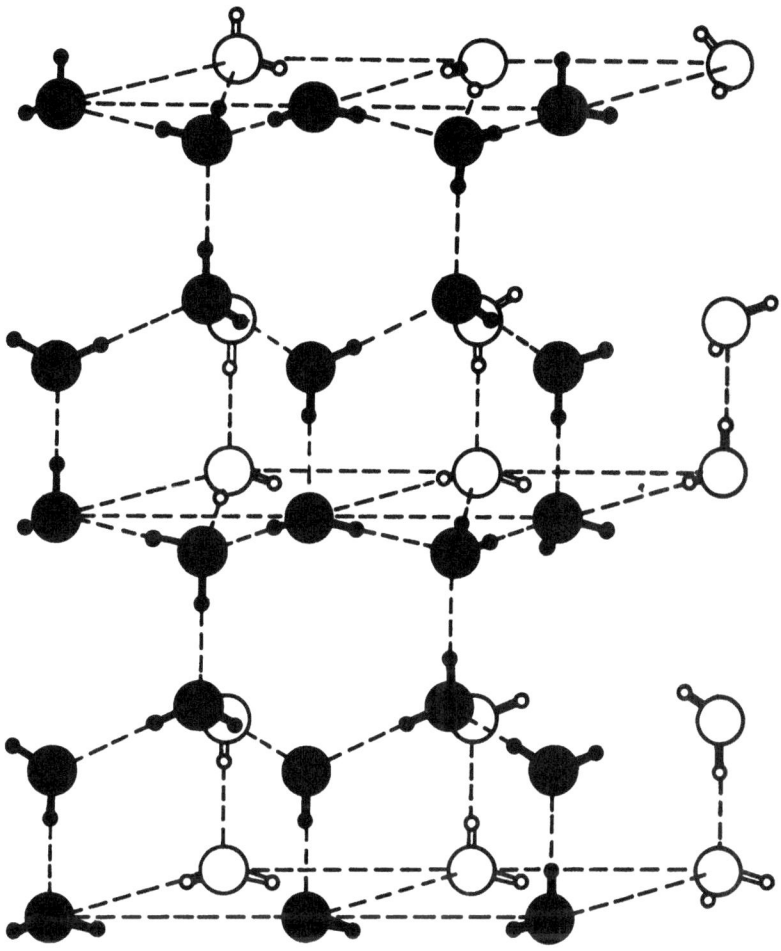

Fig. 2 Hexagonal arrays of H₂O molecules in ice, originally described in detail by J. D. Bernal and R.H. Fowler in 1933 [2]. With the inclusion of some distortions, defects, and bending, the model also can represent the arrangement of H₂O molecules in liquid water.

most people because they exist only at very high pressures. For example, ice VII can be obtained at pressures near 20,000 atmospheres, but in this extraordinary environment the crystal is stable even above 50 °C.

The high-pressure ices are much more dense than ice I. For the latter the density is 0.92 g/mL, for ice II it is 1.18, and for ice VII it is 1.56. In these, the arrangement of H₂O molecules must be more densely packed in order to put more mass in a given volume. Figure 2, the model for ice I, has much empty space in it. X-ray crystallography of the high-pressure ices has shown that the H₂O molecules are arranged in intertwining networks that fill in much of the open space of the tridymite diamondlike lattice (Fig. 2). For example, each H₂O molecule in ice VII is surrounded by eight others, at equal distances.

So far, I have mentioned only ices made from pure water. It is also possible to obtain even more novel "ices" from water to which a small amount of "impurity" has been added. These are called polyhedral "clathrate hydrates." The first one, chlorine hydrate, was discovered by Humphry Davy and by Michael Faraday at the beginning of the nineteenth

century. Subsequently, the inert gases argon, krypton, and xenon were found to form crystalline stoichiometric hydrates, and by now about a hundred polyhedral hydrates have been isolated. In composition, they are overwhelmingly water. X-ray crystallography has shown that the "impurity" or guest molecule is enclosed in a cage of water molecules. A few of the many different known cages are shown in Fig. 3. It is apparent that many different polyhedra can be formed, with pentagonal (and even quadrilateral) as well as hexagonal faces, and with cavities for the enclosure of a wide range of sizes of guest molecules.

Superwaters

Obviously, water is a remarkably versatile material, at the molecular as well as at the macroscopic level. It is not surprising, therefore, that many individuals are beguiled by it and lured into seeing features that have not been apparent to others who have exercised a more critical approach.

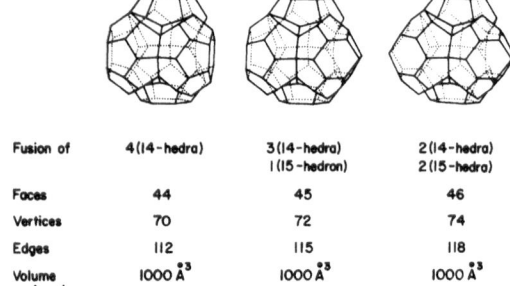

HYDRATE POLYHEDRA

	Dodecahedron	Tetrakai decahedron	Pentakai decahedron	Hexakai decahedron
Faces	12	14	15	16
Vertices	20	24	26	28
Edges	30	36	39	42
Volume enclosed	160 Å³	230 Å³	260 Å³	290 Å³

MULTIPLE FUSED POLYHEDRA

	4(14-hedra)	3(14-hedra) 1(15-hedron)	2(14-hedra) 2(15-hedra)
Fusion of			
Faces	44	45	46
Vertices	70	72	74
Edges	112	115	118
Volume enclosed	1000 Å³	1000 Å³	1000 Å³

Fig. 3 A selection of types of water polyhedra with enclosures of varying internal volume, from 160 Å³ to 1000 Å³. Other known types can enclose even larger volumes. At each vertex of a polyhedron there is an O atom. Between adjacent vertices, that is, along each edge, there is a hydrogen atom in an O—H⋯O bond. These clathrate hydrates form enclosures for guest molecules of a wide range of structure, for example, argon (Ar), chlorine (Cl_2), chloroform ($CHCl_3$), benzene (C_6H_6), and tetraisoamylammonium fluoride ($i\text{-}C_5H_{11}$)$_4$NF. Such polyhedral hydrates exist as beautiful crystals, often near 90% water in composition and yet not melting until the temperature is substantially above 0 °C.

For example, two brothers, Vadim and Igor Zelepukhin [4], from the Institute of Fruit Growing and Vine Growing in Kazakhstan described a form of water purported to be much more active biologically than ordinary water. These investigators reported that Soviet scientists have long recognized that water from freshly melted snow can stimulate some biological processes. For example, cut leaves absorb several times more meltwater than either tap water or boiled water. Mindful of the guidance of Michurin and Lysenko in Soviet genetics of the middle of the twentieth century, they also soaked cotton seeds in their "bioactive" water and found that the cotton plant arising from such seeds produced a greater yield, excelled in physiological characteristics, and could be spun into a superior fiber. Equally striking results were obtained with animals. As usual, a suitable theoretical explanation was offered: water from melted ice was said to retain in the liquid state some of the molecular order of the crystal;

such order was attributed also to water in living cells and was claimed to accentuate enzyme activity.

Such special insights into the properties of melted snow have not been restricted to scientists in the Soviet Union. A few years ago, three Frenchmen, J. R. Beaumont, L. Valageas Berger, and M.-M. F. E. Frey [5], obtained a patent to cover a new form of water, also prepared from melted snow, with very useful medical applications, particularly in burn therapy. A summary of this document follows:

PROCESS FOR OBTAINING A LIQUID FROM FRESH SNOW FOR BIOLOGICAL TREATMENTS AND PRODUCT COMPOSITIONS USING THIS LIQUID

This invention describes a process for preparing a biologically active liquid by melting fresh powdered snow at a temperature not exceeding 15 °C, and, after decantation, subjecting the liquid to an ageing process by storing it in inert containers containing an inert atmosphere (e.g., nitrogen). Care and precision are essential in the collection of the snow in order to preserve the special qualities of the fresh snow. The liquid may be sterilized by ultraviolet rays. After storing for several months it has an odor similar to fresh oysters. The liquid may be formulated into a practical composition by mixing with various constituents such as alcohols, perfumes or creams, and may be converted into an aerosol by using freon under pressure. The product has numerous applications in biological treatments. In superficial burns and sunburn, the product gives immediate relief from pain and leads to decongestion and rehydration of the skin, returning the epidermis to its initial physiological state without any blistering. The product may also be used as a lotion, eau de toilet, etc. It has many applications besides those described here.

In the United States also, there has been a succession of individuals during this century who have discovered forms of water with remarkable therapeutic properties. One of the most widely advertised of these was "Willard's Water" or "Catalyst Water," discovered by Professor J. W. Willard of the School of Mines in South Dakota. After treatment with a catalyst of secret composition followed by exposure to lignite coal, Professor Willard's water acquired potent curative powers [6]. A host of individuals testified that ailments such as skin ulcers, arthritis, cardiac arrhythmia, warts, sore throats, and cataract growth were cured or alleviated. Lesser achievements were the stimulation of hair growth on bald heads, relief from insomnia or from constipation, and the imparting of a more youthful facial appearance to women. Millions of bottles of this remarkable water were sold in the United States, at prices as high as $25 each. The explanation of its effectiveness, revealed by a chiropractor, was that the "catalyst" shrinks the size of the molecules of water so that they can more readily penetrate tissues.

Not to be outdone by Americans, many citizens of the former Soviet Union set a container of water on their television sets every morning so that a faith healer appearing on one of the TV channels can transmit curative powers that are then stored in the water for subsequent use [7].

Established scientists can also be carried away by deep-seated wishes and discover phenomena that fail to stand up to critical scrutiny. In the field of water, one of the most interesting current examples is that of the purported induction and maintenance of a unique molecular structure in liquid water after exposure to solute antibodies. The essence of the study is clearly described in the abstract of a paper in *Nature* [8] submitted by a French group:

When human polymorphonuclear basophils, a type of white blood cell with antibodies of the immunoglobulin E (IgE) type on its surface, are exposed to anti-IgE antibodies, they release histamine from their intracellular granules and change their staining properties. The latter can be demonstrated at dilutions of anti-IgE that range from 1×10^2 to 1×10^{120}; over that range, there are successive peaks of degranulation from 40 to 60% of the basophils, despite the calculated absence of any anti-IgE molecules at the highest dilutions. Since dilutions need to be accompanied by vigorous shaking for the effects to be observed, transmission of the biological information could be related to the molecular organization of water.

Let us look at the astonishing dilution of the active solute anti-IgE. If originally one has one liter of a 1 M solution of a solute, it contains 6×10^{23} molecules of solute. If that solution is successively diluted 10^{120}-fold, the original $\sim 10^{23}$ solute molecules are now distributed among 10^{120} liters; in other words, on the average one molecule would be present in 10^{97} liters. In essence, then, the probability of finding one IgE molecule in any 1 liter of the final dilution is 10^{-97}. Yet such a solution is purported to exhibit anti-IgE activity. The explanation offered by the investigators is that an anti-IgE molecule imprints a unique structural adaptation in the surrounding solvent water and this specific molecular arrangement of water molecules is retained even after dilutions far beyond that necessary to effectively remove all solute molecules. "Water could act as a 'template' for the [anti-IgE] molecule, for example, by an infinite hydrogen-bonded network, or electric and magnetic fields" [8].

Despite the incredulity of the referees, *Nature* published this "unbelievable" manuscript. A subsequent visit to the French laboratory by the editor of *Nature* accompanied by two other individuals led to an article [9] describing what they had seen and concluding that the "experiments [are] a delusion," providing "an unsubstantial basis for the claims made for them." The leader of the French team thereupon retorted with an angry criticism of the behavior and judgment of the visiting team, comparing the episode to "Salem witchhunts or McCarthy-like prosecutions [that] will kill science" [10].

A very recent careful reexamination of the degranulation experiments with anti-IgE antibodies by an English laboratory [11] failed to find any effects at dilutions of one part in 10^{12} or greater. "No aspect of the data is consistent with the previously published claims (of Benveniste [8, 10])." Nevertheless, Benveniste claims that he has repeatedly

reproduced his results. In science, the only high court that makes a decision in such controversial issues is time.

One does not even need experiments to generate heterodox insights into the structure and role of water. Enlightenment even of cosmic significance is obtained on occasion by revelation. One of the most interesting examples is the glacial cosmology picture of Hanns Hörbiger, an Austrian engineer who lived at the turn of this century [12]. In essence, his "world ice theory" (*Welteislehre*) claimed that the Milky Way, our galaxy, is constructed of ice crystals, encompassing also the solar system (Fig. 4). Furthermore, the planets have a casing of ice, and the moon has an ice mantle many kilometers thick. Also, he asserted, in the past, moons of ice had dropped onto the Earth and caused disastrous catastrophes. Even the Earth's climate today was said to be strongly affected by the falling in of galactic ice crystals.

During the Nazi period, Heinrich Himmler, in accord with his program to honor Nordic ancestors (*Ahnenerbe*), promoted *Welteislehre* as an appropriate area for scientific investigation and tried to recruit Werner Heisenberg into this field. Hitler also endorsed Hörbiger's *Welteislehre* [12]. The Nazi SS established a program to use *Welteislehre* for long-

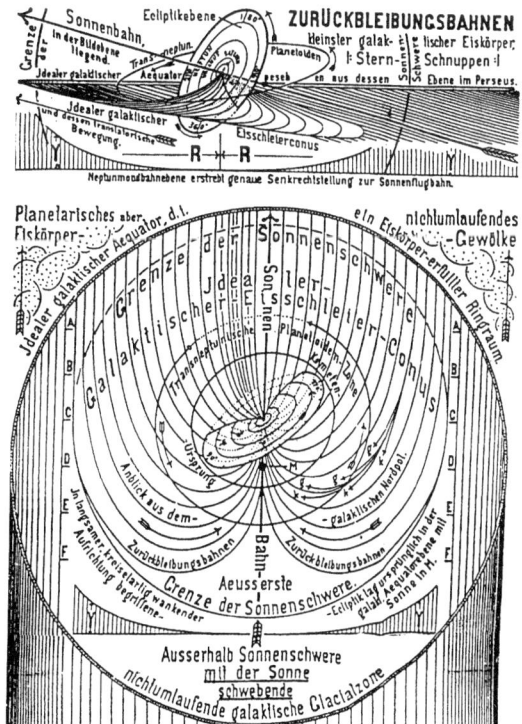

Fig. 4 Hanns Hörbiger's representation of the distribution of galactic ice in the vicinity of the sun and the planets. (See Philipp Faulth, *Hörbiger s Glacial-Kosmogenie*, 1913 and Ref. 12.).

range weather forecasting of military importance during World War II.

It is ironic that currently in the United States, the Afrocentric science movement, politically bitterly opposed to Nazi racial doctrines, is nevertheless fascinated by the concept that the water of the Nile River originated from the infall of extraterrestrial ice [13] onto the Earth.

Polywater

The claims for superwaters described above are all without experimental foundation or confirmation. In marked contrast is the anomalous water disclosed by B. V. Deryaguin at a Faraday Society meeting in Nottingham, England, in the mid-1960s [14].

Deryaguin, a Russian physical chemist of distinction, head of the Laboratory of Surface Phenomena at the Institute of Physical Chemistry of the Soviet Academy in Moscow, described some interesting experiments on the condensation of water into capillary tubes of tiny diameter (a few millionths of a meter) when the vapor pressure of water in the gas phase above the capillary is *below* saturation. That water in the gas phase at a pressure below the saturation vapor pressure of the liquid should condense into the liquid phase is astonishing. Taken at face value, this observation violates the laws of thermodynamics; if true, it means one could construct a perpetual motion machine. Such a conclusion is essentially unthinkable, even in the modern world of relativity and quantum mechanics. In fact, Einstein himself said [15] that Thermodynamics "is the only physical theory of universal content concerning which I am convinced that, within the framework of the applicability of its basic concepts, it *will never be overthrown*."

The alternative conclusion is that the "dew point," the saturation vapor pressure of the water in the capillary tube, is *below* that of ordinary liquid water. In that case, either the glass capillary is exerting some special effect on the water or the water in the capillary is different. A century ago, Lord Kelvin demonstrated on thermodynamic grounds that a very tiny bore capillary would affect the vapor pressure of a pure liquid within it and derived a mathematical equation to relate the vapor pressure to the capillary diameter. In principle, such effects should appear in tubes of less than one-millionth of a meter in diameter. Deryaguin demonstrated convincingly that the Kelvin effect could not be the basis of his observations by comparing a series of capillaries of different bores. "The absence of any perceptible dependence [of vapor pressure] on capillary radius indicates...there were bulk modifications of the ...liquid" [14].

Having convinced himself of the existence of a new form of liquid water, Deryaguin in his Nottingham paper proceeded to describe some of its unusual properties. It had a viscosity 15 times that of normal water, a thermal expansion (in the 20–40 °C range) one and a half times greater than that of normal water, a boiling point somewhere above 150 °C, and unusual freezing behavior in the range of −15 to −30 °C. Therefore, he concluded, "the usual state of water... is thermodynamically metastable." Since it was a new form of water, he christened it "orthowater." Alternative names were also used, but ultimately the term *polywater* was adopted generally.

In subsequent papers, Deryaguin and his colleagues described more unusual properties of polywater [16]. It had a density of 1.4 g/mL, a refractive index of 1.46, compared to 1.0 and 1.33, respectively, for normal water. Its viscosity was comparable to that of motor oil. It did *not* show a minimum in volume at 4 °C but continued to decrease in volume with decreasing temperature down to −12 °C. Its electrical conductance was higher than that of ordinary water. Further stability tests indicated that the new phase could be heated to over 400 °C without losing its anomalous properties.

In contrast to the other superwaters described, Deryaguin's experiments were confirmed in many laboratories throughout the world. In fact, in some universities capillary experiments were carried out even in introductory laboratory courses. Although polywater did not appear in every microtube, success rates of about 30 % were common.

Nevertheless, many felt uneasy about the interpretation of the observations, the claim that a new form of liquid water had been created. The grand old man of the physical chemistry of liquids, Joel Hildebrand (approaching his 100th birthday), expressed his doubts as follows [17]:

> Proponents of polywater in the pages of *Science* and elsewhere may be interested to learn why some of us find their product hard to swallow. One reason is that we are skeptical about the contents of a container whose label bears a novel name but no clear description of the contents. Another is that we are suspicious of the nature of an allegedly pure liquid that can be prepared only by certain persons in such a strange way. We choke on the explanation that glass can catalyze water into a more stable phase. Water and silica have been in intimate contact in vast amounts for millions of years; it is hard to understand why any ordinary water should be left.

On the other side were many who deeply hoped that Deryaguin was right. During a visit of Deryaguin to J.D. Bernal's laboratory at Birkbeck College (University of London), the host is recorded as saying, with regard to polywater [18], "In my opinion this is the most important physical-chemical discovery of the century." Deryaguin responded:

> I am glad to hear you say this: I would like to ask you something. Would it be possible for you to write something later about your opinion on the significance of this work.... It would be very important for me to get such an estimate.

Bernal said, "I will be glad to do this." An avowed Marxist and staunch supporter of the Soviet Union, he very much wanted a Soviet scientist to receive recognition for a great fundamental discovery.

A major boost to polywater proponents was provided by Ellis Lippincott and his co-workers in the United States [19] when they succeeded in obtaining an infrared spectrum of the tiny amount of liquid collected in a small capillary. What was striking about the published spectrum was the absence of the absorption peak near 3400 cm^{-1} that is so characteristic of normal liquid water. Instead, a strong peak was found much deeper in the infrared region, near 1600 cm^{-1}. Very properly, Lippincott interpreted this drastic shift as diagnostic of the formation of very strong O—H$\cdots$O bonds between water molecules. In fact, by analogy with other very strong hydrogen bonds, F—H$\cdots$F in alkali metal fluoride salts, Lippincott proposed that the oxygen-to-oxygen distance was 2.3 Å in polywater, in contrast to 2.8 Å in the normal liquid.

Shortly thereafter, additional evidence for the presence of strong hydrogen bonds in polywater was provided by observations in nuclear magnetic resonance experiments, in which again a peak was found that was very much displaced from that characteristic of normal water, in a direction consistent with stronger hydrogen bonding [18].

Such experiments, as well as a preliminary (later unsubstantiated) report of a distinctive pattern in X-ray diffraction, generated much enthusiasm among theoretical chemists. A variety of quantum-mechanical calculations were made, all leading to structures with stronger and shorter O—H$\cdots$O bonds. One of the more excitable and optimistic theoreticians was so elated with his results that he was prompted to say (to his later regret in his more reflective, critical moments): "We have presented arguments, supported by quantum mechanical calculations, which we believe *establish* [polywater's] existence and characterize its properties" [18]. In general, the quantum theoreticians favored a structure for polywater that was similar in atomic arrangement to the planar hexagons found for carbon in graphite. For normal water the puckered ring structure is similar to that for carbon in diamond.

In parallel with progress among the enthusiasts, reports of experimental work casting doubt on the purported nature of polywater began to appear more and more.

From the very beginning, there were questions about the purity of the new material. These lurked even in Deryaguin's mind. In his Faraday Society paper, he explicitly recognized the possibility of impurities from glass coming into the aqueous phase in the capillary, but he claimed that this factor could be dismissed on the basis of the observation of similar behavior in quartz capillaries. This claim was never convincing, however, as Hildebrand indicated [17]:

> There is another and, I think, much more plausible role for the necessary glass. Water and silica interact in wonderful variety, as may be read in a fascinating book by Ralph K. Iler, *The Colloid Chemistry of Silica and Silicates* (Cornell University Press, Ithaca, NY, 1955). It is easy to see why a spectroscopist might be excited by the term "polywater" to try to design new ways for water to polymerize which nature had overlooked, but I think that a chemist who feels curious about what is in those glass capillaries would have more success if he assumes that he is dealing with a system of two components.

In time, even Bernal moved into Hildebrand's camp, for Bernal's perceptive mind took control of his heart and he wrote: "One of the greatest difficulties in even accepting the existence of a more stable phase [of water] is its apparent absence in nature" [20].

Despite Herculean efforts, nobody succeeded in preparing macroscopic quantities of polywater. Critical investigators had to work, therefore, with microgram quantities, and they adapted a number of analytical techniques for their purposes. Analysis in the United States of polywater samples with electron microprobes showed the presence of sodium and boron. Analysis by electron spectroscopy disclosed the presence of sodium, potassium, sulfate, chloride, borate, and carbonate.

Deryaguin's rejoinder [18], when he was presented with these observations, was an undeviating one, quoted here from a 1971 paper delivered at a symposium of the American Chemical Society: "Unfortunately, many authors have obtained conflicting data for materials prepared without careful experimental considerations." In other words, other people's samples contained impurities because of sloppy techniques, but his material was pure.

Other damaging evidence also began to appear. Since deuterium (D) atoms are twice as heavy as H atoms, one would expect the atomic vibrations of O— D$\cdots$O in D_2O heavy water to be different from those of O—H$\cdots$O in H_2O water. If these vibrations were different, the infrared spectrum of polywater prepared from D_2O should be different from that prepared from H_2O. R. E. Davis, from Purdue University, reported, however, that these infrared spectra had the same features [21], a result incompatible with an assignment to hydrogen vibrations. At about the same time, S. W. Rabideau and A. E. Florin [22] from the Los Alamos Laboratory in New Mexico repeated the nuclear magnetic resonance experiments published by others and very sensibly carried out a necessary set of control scans—with an *empty* capillary. They reported: "Although an apparent broad absorption signal was observed [for polywater in a capillary] approximately 300 Hz downfield from the ordinary proton resonance, a *corresponding number of scans* at the same radio-frequency levels *with an empty capillary gave a hump in this same region.*" Such control scans had been omitted previously. Clearly, the anomalous signal attributed to polywater has its origin in material in the glass container. In fact, it had been shown by others previously that protons adsorbed on silica gel exhibit such a signal.

Simultaneously, several investigators reported that mixtures of silica (the major constituent of glass or quartz) and water showed properties very similar to those attributed to polywater, as in fact Hildebrand had expected. It became increasingly evident that the unusual properties of polywater were due to the impurities in it.

Suddenly and astonishingly, Deryaguin himself conceded that polywater was really normal water with impurities leached from the quartz in which it was prepared. He himself had carried out analyses of the contents of the capillaries and found dissolved silica. The evidence had become overwhelming. In 1973, in a short note to *Nature* [23], he wrote: "Consequently the anomalous properties of condensates may be explained, not by the formation of a new modification of water, as was previously supposed, but by the peculiar features of a reaction taking place between the vapour and solid surfaces in the process of condensation."

One must admire a man who had the intellectual honesty and emotional courage to triumph over a deep emotional and intellectual commitment and who was willing to face the public embarrassment of admitting that he had been proved wrong—an embarrassment that was all the more acute in the Soviet Union in that he had been proved wrong by Americans.

One can reasonably expect that new superwaters will continue to be discovered. It behooves us to be constantly aware of the potential pitfalls in this and other fields of science. There will always be individuals with a talent for tailoring observations to suit eccentric theories.

References

1. Nash, L. K. In *Harvard Case Histories in Experimental Science*; Conant, J. B., Ed.; Harvard University Press: Cambridge, Massachusetts. 1957; Vol. 2, Case 5.
2. Bernal, J. D.; Fowler, R. H. *J. Chem. Phys.* **1933**, *1*, 515–548.
3. Latimer, W. M.; Rodebush, W. H. *J. Am. Chem. Soc.* **1920**, *42*, 1419–1433.
4. Zelepukhln, V. D.; Zelepukhin, I. D. *Kluch kZhyvoivode*; Kainar: Alma Ata. 1987 [in Russian]. See also Maugh, T. H., II. *Science.* **1978**, *202*, 414; Aksyonov, S. I., Svintitskikh, V. A. *Stud. Biophys.* **1990**, *136*(2–3), 197–200.
5. Beaumont, J. R.; Valageas Berger, L.; Frey, M.-M. F. E. French Patent 2,033,769, 1970.
6. Gardner. M. *Skeptical Inquirer* 1993 (Fall), *18*, 15–20.
7. Randi, J. *Time*, April 13, 1992, p 80.
8. Davenas, E.; Beauvais, F.; Amara, J.; Oberbaum, M.; Robinson, B.; Miadonna, A.; Tedeschi, A.; Pomeranz, B.; Fortner, P.; Belon, P.; Sainte-Laudy, J.; Poitevin, B.; Benveniste, J. *Nature (London)* **1988**, *333*, 816–818.
9. Maddox, J.; Randi, J.; Stewart, W. W. *Nature (London)* **1988**, *334*, 287–290.
10. Benveniste, J. *Nature (London)* **1988**, *334*, 291.
11. Hirst, S. J.; Hayes, N. A.; Burridge, J.; Pearce, F. L.; Foreman. J. C. *Nature (London)* **1993**, *366*, 525–530.
12. Schröder. R. *Die Zeit* April **1991**, nos. 17–19, pp 41–42.
13. *Baseline Essays on Science*, Portland Public Schools, Portland, Oregon, 1990, p S-15. See also Klotz, I. M. *Phi Delta Kappan* **1993**, 75(3). 266–269.
14. Deryaguin, B. V. *Disc. Faraday Soc.* **1966**, *42*, 109–119.
15. Einstein, A. In *The Library of Living Philosophers*, Vol. VII, *Albert Einstein: Philosopher-Scientist*; Schilpp, P. A., Ed.; Open Court Publishing Co.: Peru, Illinois, **1973**; p 33.
16. Deryaguin, B. V. *Sci. Am.* **1970**, *223*(5), 52–71.
17. Hildebrand, J. H. *Science* **1970**, *168*, 1397.
18. Klotz, I. M. *Diamond Dealers and Feather Merchants, Tales from the Sciences*; Birkhäuser: Basel. **1966**; p 83.
19. Lippincott, E. R.; Stromberg, R. R.; Grant, W. H.; Cessac, G. W. *Science* 1969 *164*, 1482–1487.
20. Bernal, J. D.; Barnes, P.; Cherry, I. A.; Finney. J. L. *Nature (London)* **1969**, *224*, 393.
21. Davis, R. E. *Chem. Eng. News* **1970**, 48(41), 73, 79.
22. Rabideau, S. W.; Florin, A. E. *Science* **1970**, *169*, 48–52.
23. Deryaguin, B. V.; Churaev. N. V. *Nature(London)* **1973**, *244*, 430–431.

The Discovery of the Alpha Helix[a]

Linus Pauling

There are many of Dr. Linus Pauling's writings that remain unpublished—among them, for instance, one about Dutch elm disease, another about the development of the hen's egg, quite a few about vitamin C, not so many about quasicrystals—still, a few—and, of course, several about nuclear structure, rejected in his later years by the editors of Physical Review and Physical Review Letters.

Here at the Linus Pauling Institute, Dr. Pauling's research assistant, Dr. Zelek Herman, and I have for many years been compiling and updating Dr. Pauling's list of publications, and a few months after Dr. Pauling's death in August of 1994, I realized with sudden shock that I would no longer be able, happily and proudly (because I had helped in the preparation), to add another just-published paper to the list, which numbered at that time 1069 publications.

I remembered, however, one manuscript in the files on which I had at one time scribbled the notation "Was this ever published?" Its title was "The Discovery of the Alpha Helix," dictated in 1982 by Dr. Pauling on his ancient dictaphone and transcribed by me from the resulting now-ancient dictabelts. It had been written at the request of the editor-in-chief of W.H. Freeman and Company, as a chapter for a book. The paper had historical significance; had it ever been published? I decided to find out.

I communicated with publishers W.H. Freeman and Company and John Wiley & Sons, and finally with author Dr. Donald Voet, who told me by phone on April 25, 1995, that the essay had not been published by either Freeman or Wiley. Meanwhile, we had obtained a copy of Dr. Voet's book, for which the essay had been intended, and had determined that indeed it had not been included.

So I was free to find a publisher and get the paper published. At the suggestion of Dr. Robert Paradowski, Dr. Pauling's authorized biographer, I approached Dr. Hargittai, who kindly accepted the paper for publication in his lively new magazine, The Chemical Intelligencer.

Thus I shall once again, happily and proudly, be able to add a newly published paper to the list of publications of Professor Linus Pauling, and regain—for a little while—my lost sense of continuity.

DOROTHY MUNRO
Linus Pauling Institute of Science and Medicine
440 Page Mill Road, Palo Alto, CA 94306

Dorothy Munro, secretary/assistant to Linus Pauling at the Linus Pauling Institute from 1973 to his death in 1994. She continues there as Coordinator of Public Information.

I am still astonished to think that I have carried on research on proteins. I have never thought of myself as a biochemist. In 1922, when I began my career in research, the problem of understanding the simplest branches of chemistry, both inorganic and organic, seemed to me to be difficult and challenging, to such an extent that I was not able to envisage the progress that would be made during the next 15 years. The theory of the chemical bond was still in a very primitive stage. Gilbert Newton Lewis in Berkeley had in 1916 defined the chemical bond as a pair of electrons held jointly by two atoms, occupying the space between them, but how to make this compatible with the Bohr theory of the atom was a question that was not answered until 1927, when the quantum-mechanical theory of the chemical bond began to be developed.

My first work as a graduate student, under the tutelage of Roscoe Gilkey Dickinson, who had in 1920 been given the first Ph.D. degree ever awarded by the California Institute of Technology, was the study of the structure of minerals and other inorganic crystals by the X-ray diffraction technique. At that time this technique, which had been developed eight years earlier by W.L. Bragg and his father W.H. Bragg, had been applied in the determination of the structures of some hundreds of elements and inorganic compounds. I immediately

[a]*Chemical Intelligencer* 1996(1), 32–38.

became interested in the precise values of the interatomic distances—bond lengths—in the crystals. I tried to develop a system of atomic radii that would permit prediction of these bond lengths. The effort to understand interatomic distances was largely successful within a period of 10 years, but in fact I still continue to work on the problem of correlating interatomic distances and bond angles with the electronic structures of molecules and crystals.

Linus Pauling and Robert B. Corey. (Courtesy of California Institute of Technology).

By 1932 I felt reasonably well satisfied with my understanding of inorganic compounds, including such complicated ones as the silicate minerals. The possibility of getting a better understanding also of organic compounds then presented itself. There was as yet not any large amount of experimental information about bond lengths and bond angles in molecules of organic compounds. The first organic compound to have its structure determined, hexamethylene tetramine, had been investigated by Dickinson, together with an undergraduate student named Albert Raymond, in 1922. The carbon–nitrogen bond length had been found to be 1.47 Å, and the bond angles at both carbon and nitrogen were about 109.5°, the tetrahedral angle. By 1932, structure determinations had been made also of a few other crystals containing molecules of organic substances, but not a great many. In 1930, however, I had learned about a new method of determining the structure of molecules that had been invented by Dr. Herman Mark, in Germany. It was the electron diffraction method of studying gas molecules. The determination of the structure of a crystal of an organic compound, even with rather simple molecules, at that time was often difficult because the molecules tended to be packed together in the crystal in a complicated way. The method of electron diffraction of molecules had the advantage that a simple molecule always gave a simple electron diffraction pattern, so that one could be almost certain of success in determining the structure by this method. My student Lawrence Brockway began in 1930 to construct the first electron diffraction apparatus for studying gas molecules that had been built anywhere but in Mark's laboratory in Germany. Herman Mark had been good enough to say that he was not planning to continue work along this line and that he would be glad to see it done in the California Institute of Technology. He also gave me the drawings showing how the instrument could be constructed.

Within a few years we and other investigators had amassed a large amount of information about bond lengths and bond angles in organic compounds. This information had great value in permitting new ideas in structural chemistry, such as the theory of resonance, to be checked against experiment and even to be refined. For example, it was observed that in organic compounds many bonds between carbon atoms or a carbon atom and a nitrogen or oxygen atom were intermediate in length between a single bond and a double bond. This fact was interpreted as showing that the bonds were covalent single bonds with a certain amount of double-bond character. The observations were generally in accord with the results of quantum-mechanical calculations, and it became clear by 1935 that a far more extensive, precise, and detailed understanding of organic compounds had been developed than had been available to chemists in the earlier decades.

It was just at this time that I began to think about proteins. The first protein to attract my interest was hemoglobin. I had read that the equilibrium curve for hemoglobin, oxygen, and oxyhemoglobin was not represented by any simple theoretical expression of the sort that physical chemists had devised for chemical equilibria. I also knew that some eight years earlier it had been shown by Adair in Cambridge that the hemoglobin molecule contains four iron atoms—that is, four heme groups, each being a porphyrin with an iron atom linked to it—and that the molecule could combine with as many as four oxygen atoms. I formulated a theory, published in 1935, on the oxygen equilibrium of hemoglobin and its structural interpretation. The theory was that each iron atom can attach one oxygen molecule to itself, by forming a chemical bond with it. There is an interaction, however, between each heme group and the adjacent heme groups such that addition of the oxygen molecule to one iron atom changes the equilibrium constant for the combination of the other iron atoms with oxygen molecules. I had several ideas as to the nature of the heme–heme interaction, and somewhat later my student Robert C.C. St. George and I published a paper showing that the addition of a group such as the oxygen molecule to one of the iron atoms deforms the molecule through a steric hindrance effect in such a way as to make it easier for oxygen molecules to attach themselves to other iron atoms in the molecule (1951). While I was thinking about the oxygen equilibrium curve in 1935, it occurred to me that measurement of the magnetic properties of hemoglobin, carbon monoxyhemoglobin, and oxyhemoglobin should provide information about the nature of the bonds formed by the iron atoms with the surrounding groups (two distinct kinds of compounds of bipositive iron were known) and the electronic structure of the oxygen molecule in oxyhemoglobin. Charles Coryell and I carried out measurements of the magnetic properties of these compounds, showing that the iron atoms change their electronic structure when the oxygen molecule is attached, and also that the oxygen molecule changes from having two unpaired electron spins to having none. My first work on proteins accordingly dealt essentially with the physical chemistry and structural chemistry of the heme group and the attached ligand, rather than with the apoprotein, the globin.

The measurement of the magnetic susceptibility of solutions of hemoglobin and related substances turned out to be a valuable technique, and we immediately began applying it to determine equilibrium constants, rates of reaction, and other properties. A leading protein chemist, Dr. Alfred Mirsky, was sent to Pasadena by the Rockefeller Institute of Medical Research to work with us during the year 1935–36. He had been especially interested in the phenomenon of the denaturation of proteins by heat or chemical substances, such as hydrogen ion, hydroxide ion, urea, etc. After many discussions, he and I formulated a general theory of the denaturation of proteins. The theory involved the statement that a native protein consists of polypeptide chains that are folded in a regular way, with the type of folding determined and stabilized by the weaker interactions, especially hydrogen-bond formation. Denaturation, we said in our 1936 paper, is incomplete or complete unfolding of the polypeptide chains, producing molecules that can assume a large number of conformations, giving increased entropy and increased intermolecular interaction.

These considerations about the folding of the polypeptide chains in denatured protein molecules immediately raised the question, of course, as to the nature of the folding. It was a question to which I applied myself during the next 15 years.

Shortly after X-ray diffraction had been discovered, several investigators had made X-ray diffraction photographs of protein fibers. These photographs for the most part showed only rather diffuse diffraction maxima, insufficient to permit structure determinations to be deduced from them. There were two principal types, one shown by keratin fibers such as hair, horn, porcupine quill, and fingernail, and the other shown by silk. William T. Astbury and his collaborators in the early 1930s had reported that the diffraction pattern of a hair changes when the hair is stretched. He called the normal pattern alpha keratin and the stretched-hair pattern, which is somewhat like that of silk, beta keratin. In the early summer of 1937, when I was free of my teaching duties, I decided to try to determine the alpha-keratin structure. My plan was to use my knowledge of structural chemistry to predict the dimensions and other properties of a polypeptide chain and then to examine possible conformations of the chain, to find one that would agree with the X-ray diffraction data. The principal piece of information supplied by the rather fuzzy diffraction photographs of hair and other alpha-keratin pro-

teins came from a rather diffuse arc on the meridian, above and below (that is, in the direction of the axis of the hair). The measured position of this reflection indicated that the structural unit in the direction along the axis of the hair would repeat in 5.10 Å. This fact required that there be at least two amino acid residues for this apparent repeat distance of the alpha-keratin structure.

Because of the large amount of theoretical and experimental progress that had been made, I felt that I could predict the dimensions of the peptide group with reliability. This group is shown in Fig. 1 The alpha-carbon atom forms a single bond with a hydrogen atom, a single bond with the group R characteristic of the amino acid, a single bond to an adjacent main-chain carbon atom, and a single bond to the main-chain nitrogen atom. The single-bond lengths were known to within about 0.01 Å: 1.54 Å for C–C and 1.47 Å for C–N (as determined by Dickinson and Raymond as early as 1922 and verified in many compounds). However, for the other bond between carbon and nitrogen we have to consider the theory of resonance. According to this theory, there are two structures that can be written for a peptide group: in one the carbon–oxygen bond is a double bond, and the carbon–nitrogen bond is a single bond, and in the other the carbon–oxygen bond is a single bond (one of the electron pairs in the double bond having shifted out onto the oxygen atom, giving it a negative charge), and the carbon–nitrogen main-chain bond is a double bond (with the nitrogen atom having a positive charge). Because of the separation of charges, the second structure is less stable than the first, and the estimate that could be made is that it should contribute about 40 %, so

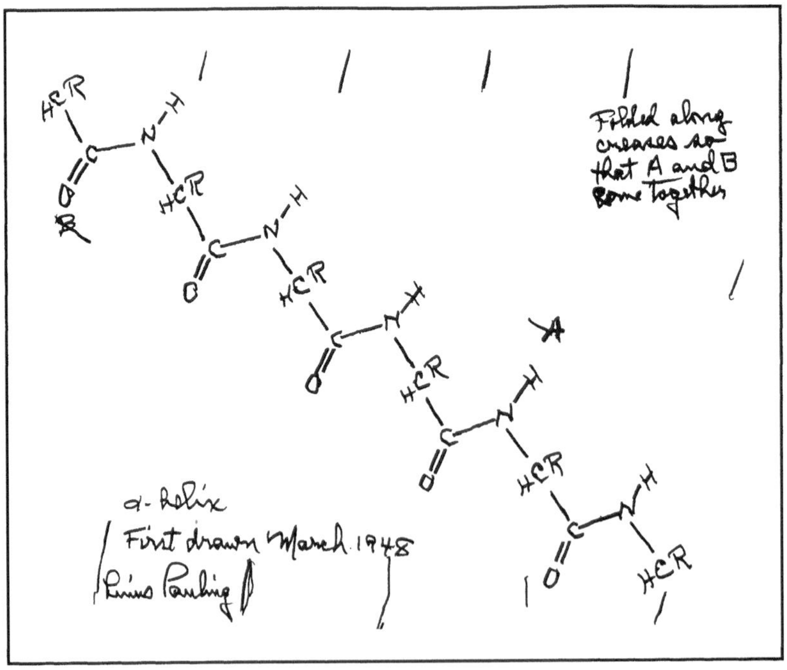

Fig. 1 α-Helix. First drawn in March 1948 by Linus Pauling.

that this bond has 40% double-bond character. The expected bond length is then 1.32 Å, rather than 1.47 Å. Moreover, because of the 40% double-bond character for this bond, these two atoms and the four adjacent atoms should all lie in the same plane, this quality of planarity being characteristic of compounds of molecules in which there are double bonds. In this way I reached the conclusion that these peptide groups in the molecule would have a well-defined rigid structure, with bond lengths and bond angles as shown in Fig. 1, and that there would be two degrees of freedom for the chain, rotation around the single bonds from carbon and nitrogen to the alpha carbon atom. Accordingly, the conclusion, on the basis of the theory of resonance, that the peptide group should be planar greatly restricts the possible structures.

Linus Pauling through the decades: 1960s, 1970s, and 1980s on the following pages. (Photographs courtesy of Linus Pauling Institute of Science and Medicine).

Despite this restriction, I was unable to find a way of folding the polypeptide chain to give a repeat in 5.10 Å along the fiber axis. After working for several weeks on this problem I stopped, having reached the conclusion that there probably was some aspect of structural chemistry characteristic of proteins and remaining to be discovered. This conclusion was, in fact, wrong, but it led to a large amount of experimental work.

Dr. Robert B. Corey was a chemist who, after getting his Ph.D. in chemistry in Cornell University and teaching analytical chemistry there for five years, had joined a leading X-ray crystallographer, Ralph W. G. Wyckoff, in the Rockefeller Institute for Medical Research. He worked with him on crystallographic problems for nine years and then came, in 1937, to spend a year as research fellow in the California Institute of Technology. He and Wyckoff had made some X-ray photographs of proteins, and he was interested in the problem of determining the structure of proteins. I told him about my failure to find the way of folding the polypeptide chains in alpha keratin and my conclusion that there might be some structural feature that we had ignored. I had assumed that the poly-peptide chain should be folded in such a way as to permit the NH group to form a hydrogen bond with the oxygen atom of the carbonyl group of an adjacent peptide group, with the N–H···O distance 2.90 Å, as indicated by structure measurements on compounds other than the amino acids. At that time there had been no correct structure determination made for any amino acid or any peptide. The state of X-ray crystallography was such that a year's work, at least, would be needed to make such a structure determination, even for such a simple compound as glycine, and the efforts of several investigators in other institutions to do such a job had resulted in failure. I suggested to Dr. Corey that he, together with graduate students, attack the problem of determining the structure of some simple amino acid crystals and simple peptides. He agreed, and within little more than a year he and two graduate students (Gustav Albrecht and Henri Levy) had succeeded in making completely satisfactory determinations of the structures of glycine, alanine, and diketopiperazine. This work was continued with vigor, with many students and postdoctoral fellows in chemistry in the California Institute of Technology involved in it, during the following years, interrupted to a considerable extent, however, by the Second World War.

In the spring of 1948 I was in Oxford, England, serving as George Eastman Professor for the year and as a fellow of Balliol College. I caught cold and was required to stay in bed for about three days. After two days I had got tired of reading detective stories and science fiction, and I began thinking about the problem of the structure of proteins. By this time Dr. Corey and the other workers back in Pasadena had determined with high reliability and accuracy the structures of a dozen amino acids and simple peptides, by X-ray diffraction. No other structure determinations of substances of this sort had been reported by any other investigators. I realized, on thinking about the structures, that there had been no surprises whatever: every structure conformed to the dimensions— bond lengths and bond angles and planarity of the peptide group—that I had already formulated in 1937. The N–H···O hydrogen bonds, present in many crystals, were all close to 2.90 Å in length. I thought that I would attack the alpha-keratin problem again. As I lay there in bed, I had an idea about a new way of attacking the problem. Back in 1937 I had been so impressed by the fact that the amino acid residues in any position in the polypeptide chain may be any of 20 different kinds that the idea that with respect to folding they might be nearly equivalent had not occurred to me.

I accordingly thought to myself, what would be the consequences of the assumption that all of the amino acid residues are structurally equivalent, with respect to the folding of the polypeptide chain? I remembered a theorem that had turned up in a course in mathematics that I had attended, with Professor Harry Bateman as the teacher, in Pasadena 25 years before. This theorem states that the most general operation that converts an asymmetric object into an equivalent asymmetric object (such as an L-amino acid into another molecule of the same L-amino acid) is a rotation–translation—that is, a rotation around an axis combined with a translation along the axis—and that repetition of this operation produces a helix. Accordingly, the problem became that of taking the polypeptide chain, rotating around the two single bonds to the alpha carbon atoms, with the amounts of rotation being the same from one peptide group to the next, and on and on, keeping the peptide groups planar and with the proper dimensions and searching for a structure in which each NH group performs a 2.90 Å hydrogen bond with a carbonyl group. I asked my wife to bring me pencil and paper and a ruler. By sketching a polypeptide chain on a piece of paper and folding it along parallel lines, I succeeded in finding two structures that satisfied the assumptions. One of these structures was the alpha helix, with 4.6 residues per turn, and the other was the gamma helix. The gamma helix has a hole down its center that is too small to be occupied by other molecules but is large enough to decrease the van der Waals stabilizing interactions, relative to those in the alpha helix. It seems to me to be a satisfactory structure in every respect but this one, but, so far as I am aware, it has not been observed in any of the protein structures that have been determined so far, and it has been generally forgotten.

I got my wife to bring me my slide rule, so that I could calculate the repeat distance along the fiber axis. The structure does not repeat until after 18 residues in 5 turns, the calculated repeat distance being 27.0 Å, which corresponds to 5.4 Å per turn. This value did not agree with the experimental value, given by the meridional arcs on the X-ray diffraction patterns, 5.10 Å. I tried to find some way of adjusting the bond lengths or bond angles so as to decrease the calculated distance from 5.4 Å to 5.1 Å, but I was not able to do so.

I was so pleased with the alpha helix that I felt sure that it was an acceptable way of folding polypeptide chains and that it would show up in the structures of some proteins when it finally became possible to determine them experimentally. I was disturbed, however, by the discrepancy with the experimental value 5.10 Å, and I decided that I should not publish an account of the alpha helix until I understood the reason for the discrepancy. I had been invited to give three lectures on molecular structure and biological specificity in Cambridge University, and while I was there I talked with Perutz about his experimental electron density distribution functions for the hemoglobin crystal that he had been studying. It seemed to me that I could see in his diagrams evidence for the presence of the alpha helix, but I was troubled so much by the 5.1 Å value that I did not say anything to him about the alpha helix.

On my return to Pasadena in the fall of 1948 I talked with Professor Corey about the alpha helix and the gamma helix, and also with Dr. Herman Branson, who had come for a year as a visiting professor. I asked Dr. Branson to go over my calculations and, in particular, to see if he could find any third helical structure. He reported that the calculations were all right and that he could not find a third structure. More than a year went by, and then a long paper on ways of folding the polypeptide chain, including helical structures, was published by W. Lawrence Bragg, John Kendrew, and Max Perutz, in *Proceedings of the Royal Society of London*. They described about 20 structures, and they reached the conclusion that none of them seemed to be satisfactory for alpha keratin. Moreover, none of them agreed with my assumptions, in particular, the assumption of planarity of the peptide group. Lord Todd has told the story of his having told Bragg, when they were just beginning their work, that the main-chain carbon–nitrogen bond has some double-bond character but that Bragg did not understand that that meant that the peptide group should be planar. My efforts during a year and a half to understand the 5.1-Å discrepancy had failed, but Dr. Corey and I decided that we should publish a description of the alpha helix and the gamma helix. It appeared in the *Journal of the American Chemical Society* in the fall of 1950. It was followed in 1951 by a more detailed paper, with Branson as coauthor, and a number of other papers on the folding of polypeptide chains. An important development had been the publication of X-ray photographs of fibers of synthetic polypeptides, in particular of poly-gamma-methyl-L-glutamate, by investigators at Courtaulds. These striking diffraction photographs showed clearly that the pseudo repeat distance along the fiber axis is 5.4 Å rather than 5.1 Å. There are strong reflections near the meridional line, corresponding to 5.1 Å, but they are not true meridional reflections. On the X-ray photographs of hair, the reflections overlap to produce the arc that seems to be a meridional reflection. It was this misinterpretation that had misled all of the investigators in this field. It was accordingly clear that the alpha helix is the way in which polypeptide chains are folded in the alpha-keratin proteins.

Moreover, we reached the conclusion, as did Crick, that in the alpha-keratin proteins the alpha helices are twisted together into ropes or cables. This idea essentially completed our understanding of the alpha-keratin diffraction patterns.

The apparent identity distance in the fiber X-ray diagrams of silk is somewhat smaller than corresponds to a completely extended polypeptide chain. We accordingly concluded that the polypeptide chains have a zigzag conformation in silk and the beta-keratin structure. We reported in detail three proposed sheet structures. The first one, which we called the rippled sheet, involves amino acid residues of two different kinds, one of which cannot be an L- amino acid residue, but can be a residue of glycine. It was known that Bombyx mori silk fibroin has glycine in 50 % of its positions, with L-alanine or some other L-amino acid residue (such as L-serine) in the alternate positions, so that the rippled sheet seemed to be a possibility for Bombyx mori silk fibroin. It turned out, however, that Bombyx mori silk fibroin has the structure of the antiparallel-chain pleated sheet. The third pleated sheet structure, the parallel-chain pleated sheet, is also an important one.

About 85 % of the amino acid residues in myoglobin and hemoglobin are in alpha-helix segments, with the others involved in the turns around the corners. In other globular proteins the alpha helix, the parallel-chain pleated sheet, and the antiparallel-chain pleated sheet all are important structural features. These three ways of folding polypeptide chains have turned out to constitute the most important secondary structures of all proteins. Dr. Corey, to some extent with my inspiration, designed molecular models of several

different kinds that were of much use in the later effort to study other methods of folding polypeptide chains. I used these units to make about 100 different possible structures for folding polypeptide chains. For example, if the hydrogen bonds are made alternately a little shorter and a little longer than 2.90 Å in a repeated sequence, an additional helical twist is imposed upon the alpha helix. Some of the models that I constructed related to ways of changing the direction of the axis of the alpha helix. I reported on all of this work at a protein conference in Pasadena in 1952, but then I became interested in other investigations and stopped working in this field.

It pleases me to think that our work in Pasadena in the Division of Chemistry and Chemical Engineering, first in collecting experimental information about the structure of molecules, then in developing structural principles, and then in applying these principles to discover the alpha helix and the pleated sheets, has shown how important structural chemistry can be in the field of molecular biology.

Quadruple Metal-Metal Bond: History and Outlook[a]

Author Petr Alexeevich Koz'min; Ada Stepanovna Kotel'nikova, 1927–1990.

Discovery of the Quadruple Re–Re Bond

By the mid-twentieth century, a dozen or so compounds had been found with direct bonding between transition–metal atoms via weak, relatively long single bonds. The only compound known to contain a short (2.41 Å) metal–metal bond was $K_3W_2Cl_9$ [1], in which the shortening of the W–W bond was related to the presence of triply-bonded metal atoms in the $[W_2Cl_9]^{3-}$ dimeric unit.

In the 1950s, the research group headed by Professor V. G. Tronev at the Institute of General and Inorganic Chemistry (IGIC) of the USSR Academy of Sciences focused on the synthesis of low-valent rhenium compounds. The first evidence for the existence of such compounds was reported in 1933 by Ida and Walter Noddack [2], who had discovered rhenium. They failed, however, to isolate individual compounds from solutions. Low-valent rhenium compounds are difficult to study because their synthesis requires a high degree of reduction of the metal, and also because they are unstable in aqueous media. The first compounds that were presumed to contain divalent rhenium were obtained by

Tronev and Bondin [3] through the reduction of ammonium perrhenate at about 300 °C.

By the late 1950s, Kotel'nikova and Tronev [4] had obtained samples of several low-valent rhenium chlorides. According to chemical analysis, the rhenium valence in these compounds was two. Surprisingly, these compounds appeared to be diamagnetic, contrary to expectations. Various assumptions regarding the mechanism of electron pairing were made, among them direct bonding between the rhenium atoms. Nobody, however, foresaw the existence of multiple metal–metal bonds.

In the mid-1960s, Professor V.G. Kuznetsov, Head of the Laboratory for X-Ray Diffraction Analysis, IGIC, provided me with a batch of $(pyH)HReCl_4$ (py = pyridine) crystals prepared by G.K. Babeshkina [5]. The presence of a "free" hydrogen atom was tentatively attributed by the synthetic chemists to the divalent state of rhenium. The majority of the crystals exhibited heavy twinning and were, hence, unsuitable for structure determination. After examination of hundreds of crystals, three single crystals were finally selected. Surprisingly, the X-ray diffraction analysis revealed that the structure comprised dimeric units with no bridging atoms, with the eight chlorine atoms in a square-prismatic arrangement and with the two rhenium atoms inside the prism bonded to each other (Fig. 1). The Re–Re bond was much shorter, 2.22 Å, than in metallic rhenium (2.74 Å). Although this extremely short distance between the rhenium atoms had raised much doubt among our colleagues, our report was submitted for publication in *Zhurnal Strulkturnoi Khimii* and was finally published in early 1963 [6]. Later in 1965, we published data on the structure of $(pyH)HReBr_4$ [7], which, again, contained the dimeric unit, with an even shorter Re-Re bond, 2.207 Å, than in the chlorine analogue. The first report, though, on the structure of the dimeric unit was presented already at a meeting in 1961 [8].

With our work on $(pyH)HReCl_4$, we had found the first example of a quadruply-bonded dirhenium compound, albeit unknown to us at that time. Given our unsophisticated experimental setup, the position of the hydrogen atom could not be determined, and thus we still thought that the metal–metal bond was formed by divalent rhenium atoms. It was only later established that there was no "free" hydrogen in this compound and that the rhenium was trivalent. Several months after our publication [6], the crystal structure of

[a]*Chemical Intelligencer* 1996(2), 32–36.

[b]Kurnakov Institute at General and Inorganic Chemistry, Russian Academy of Sciences, Leninskii pr. 31, Moscow 117907, Russia

B. Hargittai and I. Hargittai (eds.), *Culture of Chemistry: The Best Articles on the Human Side of 20th-Century Chemistry from the Archives of the Chemical Intelligencer*, DOI 10.1007/978-1-4899-7565-2_36, © Springer Science+Business Media New York 2015

169

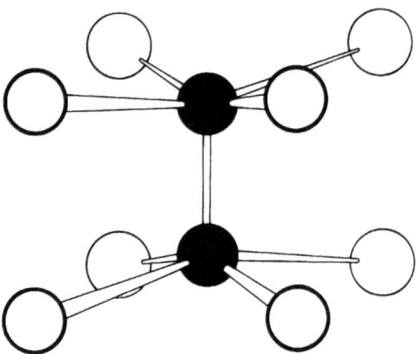

Fig. 1 Structure of a dirhenium complex.

$Cs_3Re_3Cl_{12}$ was reported independently by Robinson et al. [9] and by Bertrand et al. [10]. This compound was found to contain $[Re_3Cl_{12}]^{3-}$ trimeric anions, with rhenium–rhenium double bonds (2.48 Å) in the form of an equilateral triangle. Subsequently, the research group of F.A. Cotton, being aware of our results on the structure of the $[Re_2Cl_8]^{2-}$ unit [6], undertook an investigation of dirhenium molecules. Several compounds, including the pyridine derivative of rhenium tetrachloride, were shown to be identical to the molecule described by Tronev and co-workers [4, 5]. In 1964, Cotton et al. [11] reported preliminary X-ray diffraction data on the structure of $K_2Re_2Cl_8 \cdot 2H_2O$ and the $[Re_2Cl_8]^{2-}$ dimeric unit. Concerning the pyridine derivative, they noted the following: "The very short Re–Re distance reported by Kuznetsov and Koz'min initially led us to suspect that their structure determination was in error, especially when the pronounced tendency of this compound to give twin crystals (as reported by Kuznetsov and Koz'min and confirmed by us) and the trial and error method of refinement were taken into account... We have concluded, however, that the work of Kuznetsov and Koz'min is correct in all essentials..."

Based on chemical analysis of $KReCl_4 \cdot H_2O$, Cotton and his co-workers inferred that both the potassium and the pyridine derivatives of rhenium tetrachloride contain rhenium in the +3 oxidation state, the actual formula for the latter compound being $(pyH)_2Re_2Cl_8$, without "free" hydrogen. These findings suggested that the $[Re_2Cl_8]^{2-}$ ion contains a quadruple Re–Re bond formed by one σ, two π, and one δ bond, with eight electrons involved. In subsequent papers [12], Cotton has further developed his ideas on the nature of the quadruple metal–metal bond.

Other Quadruply-Bonded Dimetal Compounds

The discovery of the quadruple Re–Re bond had initiated the search for quadruply-bonded compounds of other transition metals, mostly in the vicinity of rhenium in the periodic

table. The first report on the structure of $[Tc_2Cl_8]^{3-}$ was published in 1965 by Cotton and Bratton [13]. Its structure is similar to that of the dirhenium analogue, with a very short (2.13 Å) metal-metal bond. This short Tc–Tc bond length and the eclipsed (prismatic) conformation of the complex were considered as evidence for the presence of a quadruple metal–metal bond. Subsequent work showed, however, that an "extra" electron occupies an antibonding orbital, and, hence, the order of the Tc–Tc bond is 3.5. Some years later, the dimeric $[Mo_2Cl_8]^{4-}$ anion, isoelectronic and isostructural with $[Re_2Cl_8]^{2-}$, was described by Brencic and Cotton [14] and identified as having an Mo–Mo quadruple bond (2.14 Å).

The quadruple chromium–chromium bond was first described in 1970 [15] based on a single-crystal X-ray diffraction study of $Cr_2(O_2CCH_3)_4 \cdot (H_2O)_2$. It is of interest to note that this compound was first prepared in the middle of the nineteenth century [16]. The original chemical formula was different since nobody could surmise the existence of the quadruple Cr–Cr bond at that time. The quadruple Cr–Cr bond length can vary over a wide range. The shortest quadruple metal–metal bond, 1.85 Å, was found in $Cr_2(2\text{-}CH_3O\text{-}5\text{-}CH_3C_6H_3)_4$ [17].

The history of research concerning quadruple metal–metal bonds, in particular, Re–Re, Mo–Mo, and Cr–Cr, clearly demonstrates that this kind of discovery can be made only on the basis of integrated studies involving, in addition to the necessary preparative work, structural characterization by X-ray crystallography and investigation of the electronic structure.

By the early 1970s, the investigation of quadruply-bonded dimetal compounds was essentially concentrated in two research centers. In the United States the large group of F.A. Cotton was engaged in the synthesis and crystallographic and spectroscopic characterization of dimeric complexes of rhenium, technetium, molybdenum, and chromium. At the IGIC a small group of synthetic chemists under A.S. Kotel'nikova cooperated with a group of X-ray crystallographers, P.A. Koz'min and M.D. Surazhskaya and, subsequently, T.B. Larina. They studied about 30 dirhenium compounds and a few dimolybdenum compounds. At about the same time, at the Institute of Physical Chemistry (IPC), USSR Academy of Sciences, Professor A.F. Kuzina and co-workers studied ditechnetium compounds whose structural characterization was carried out at the IGIC. By the mid-1980s, these two groups had prepared and characterized about 200 substances with quadruple metal–metal bonding.

A consequence of the strong M–M bonding is its maximum trans influence. In the dimeric complexes, the axial ligands are bonded to the metal atom much more weakly than the ligands in other positions and, hence, can be readily removed or replaced by other atoms or atomic groups. In 1972, two independent studies [18, 19] provided evidence for the inverse effect of axial ligands on the multiple metal–metal

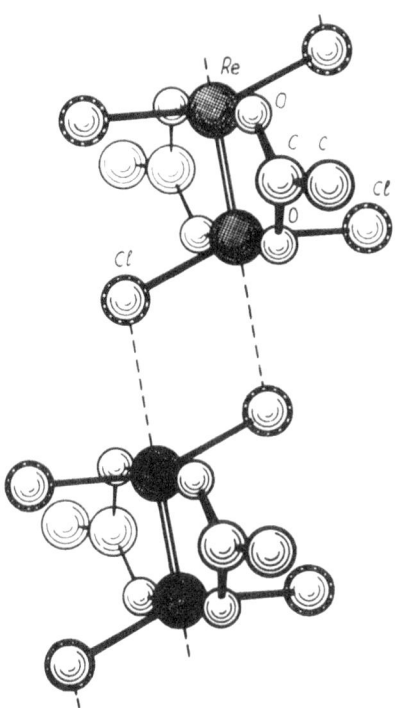

Fig. 2 Portion of the structure of a volatile dirhenium compound, $Re_2Cl_4(CH_3COO)_2$, with trans configuration.

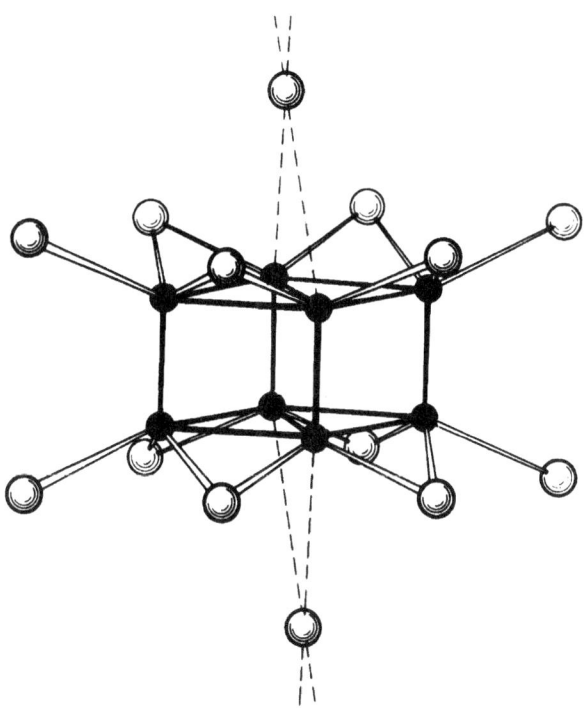

Fig. 3 Structure of the $[Tc_8Br_{12}]Br^-$ octameric unit.

bond; that is, the M–M quadruple bond is longer in the presence of axial ligands.

In addition to being of theoretical interest, the compounds with quadruple metal–metal bonds may also find practical application. Thus, for example, several rhenium compounds, including $Re_2Cl_4(CH_3COO)_2$ [20] (Fig. 2), sublime at 200–250 °C and can be deposited without structural change. When brought into contact with a surface heated to 350–450 °C, these dimeric complexes dissociate, and rhenium is deposited on the surface as a high-quality metallic coating. The relatively low decomposition temperature of these compounds allows the use of glass substrates, porcelain, mica, quartz, steel, brass, and other nonmetallic and metallic materials.

By the early 1980s, the interest in quadruple metal–metal bonds had waned somewhat, and it seemed that studies would be limited to dimeric complexes only. However, radically new findings emerged, with unexpected results.

Twenty Years Later

In mid-1981, S.V. Kryuchkov and co-workers (IPC) synthesized a compound of technetium and bromine [21], which was expected to contain dimeric units with a multiple metal–metal bond. Our first attempts to solve its structure were unsuccessful, although the data indicated the presence of short Tc–Tc bonds. We thought that either twinning or a disordered arrangement of atomic groups hindered our work,

and the investigation was suspended. After a while, we selected a crystal that had no apparent sign of twinning, and the investigation was resumed, supplemented at this stage by "direct" methods of structure determination. Interestingly, it turned out that the crystal consists of fairly intricate clusters, rather than of dimeric units, with the formula $[Tc_8Br_{12}]Br[H_2O)_2H$ [22], in which eight technetium atoms are arranged in the form of a rectangular prism with a rhombus at the base (Fig. 3). Pairs of technetium atoms (lateral edges of the prism) form short (2.16 Å) multiple bonds. There are also weaker Tc–Tc bonds. The complex contains eight bridging and four terminal bromine atoms. The technetium atoms sit at the vertices and are connected by the short diagonals of the rhombus. Bridging Br^- anions link adjacent complexes. In view of the eclipsed configuration of this octameric unit, the four short Tc–Tc bonds can formally be regarded as quadruple bonds. Actually, however, the complex contains a multicenter system of Tc–Tc bonds, and partitioning of this system into two-center bonds is purely formal.

The first hexameric technetium complex, $[(CH_3)_4N]_3[Tc_6Cl_{14}]$, was synthesized at the IPC [21] and then structurally characterized at the IGIC [23]. The crystal contains $[Tc_6Cl_{14}]^{3-}$ hexameric anions (Fig. 4). The complex is made up of six technetium atoms arranged in the form of a triangular prism, six bridging and six terminal chlorine atoms, and, in addition, two chlorine atoms each weakly bound to three technetium atoms. The technetium atoms form three multiple bonds (2.16 Å) and six weak bonds (2.7 Å). An analogous structure was found for a rhenium complex [24],

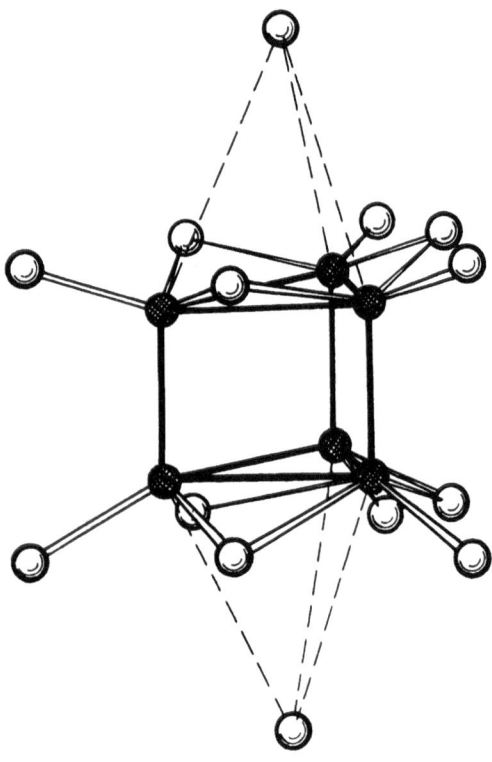

Fig. 4 Structure of the $[Tc_6Cl_{14}]^{3-}$ hexameric unit.

with three multiple Re–Re bonds (2.26 Å) along the lateral edges of a triangular prism and six weaker metal–metal bonds (2.65 Å). To date, the IPC and IGIC groups have reported eight polynuclear complexes of technetium and rhenium, viz., $[Tc_8Br_{12}]$, $[Tc_8Br_{12}]^{1+}$, $[Tc_6Cl_{12}]^{1-}$, $[Tc_6Cl_{12}]^{2-}$, $[Tc_6Br_{12}]$, $[Tc_6Br_{12}]^{1-}$, $[Re_6Br_{12}]^{1+}$, and $[Re_6Br_{12}]$, The formulas do not include weakly bonded halogen atoms. These complexes exhibit structural features that are absent in the quadruply-bonded dimetal compounds. First of all, both octameric and hexameric units include pairs of homologous complexes differing by one electronic charge. The existence of these pairs is likely to result from the fact that in a cluster one electronic charge is shared by several metal atoms and, hence, cannot cause substantial changes in their nonformal charges. In the pairs of complexes under consideration, an extra electron may produce different effects. In the octameric technetium complexes, the difference in charge leads to a very small change in both multiple and weak metal–metal bond lengths. In the $[Tc_6Cl_{12}]^{2-}$ complex, the multiple Tc–Tc bonds are longer by 0.06 Å and the weak bonds are shorter by 0.12 Å than in the $[Tc_6Cl_{12}]^-$ complex, presumably as a result of a substantial rearrangement of the bonding system. By contrast, on going from $[Tc_6Br_{12}]$ to $[Tc_6Br_{12}]^-$, the multiple Tc–Tc bonds are shortened by 0.03 Å and the weak bonds are lengthened by 0.04 Å. The change in the charge on the rhenium hexameric unit is not accompanied by any changes in the metal–metal bond lengths, and the structural

parameters of the complex remain unchanged within experimental error. Mentioning a practical point, one can envision the use of the hexameric rhenium complex as a smallest element in memory devices.

Meanwhile, radically new experimental findings continued to emerge. In 1986, Kryuchkov et al. reported the preparation and crystal structure of $K_2[Tc_2Cl_6]$ [25]. The dimeric units are linked by shared chlorine atoms to form chains. Surprisingly, it was found that the $[Tc_2,Cl_8]$ groups in the chain have a staggered, rather than an eclipsed, conformation. Prior to this work, it was believed that an eclipsed conformation was a necessary condition for the existence of quadruple metal–metal bonding. The authors considered the technetium–technetium bond in $K_2[Tc_2Cl_6]$ to be quintuple, involving electrons from the Tc–Cl bonds, supposing that the metal–metal and metal–ligand bonds compete for electrons.

What More Is To Come?

In order to predict further advances in studies of compounds with M–M bonds of orders above three, it is necessary to find out which transition elements are capable of forming such bonds. So far, quadruple bonds have been reported only for Group VI and VII transition metals, which seem to have an optimal number of outer electrons (6 or 7) for their formation. With fewer outer electrons, the transition element, after forming metal–ligand bonds, lacks electrons for a quadruple metal–metal bond. Conversely, with more outer electrons, a fraction of them remain paired and cannot be involved in metal–metal bonds. Thus, it is unlikely that quadruple metal–metal bonding will be found for other transition elements, although the possibility cannot be ruled out for elements such as manganese, osmium, ruthenium, niobium, and vanadium. It seems more likely though that quadruple bonds will be found to form between atoms of different transition elements with a total of 14 to 16 outer electrons.

I expect the range of ligands to be extended, and new compounds containing chains and sheets of dimeric units linked by bridging groups to be found. I also hope that new polynuclear complexes of more complicated structures with multiple metal–metal bonds will be prepared.

Modern inorganic chemistry is, in large part, the chemistry of the conditions at the surface of the Earth, characterized by a narrow temperature range, low pressure, and certain natural abundances of the elements. Given that nonmetals such as oxygen, nitrogen, and carbon constitute the great bulk of the atoms, most natural compounds contain few or no metal atoms. In conventional complexes, the metal:ligand ratio is relatively small (1:4 to 1:8). I expect that under relatively high temperatures and pressures and, what is more important, at high concentrations of heavy transition metals, conditions that are likely to occur inside the Earth, there exist compounds differing in composition, structure, and properties from the coordination compounds studied to date. Therefore, it is of interest to prepare compounds with a larger metal:ligand ratio by further reducing the metal at high temperatures and pressures. Investigations of such compounds may, however, involve substantial difficulties. In a private communication, the late A. S. Kotel'nikova mentioned to me the following intriguing fact. In some preparations of quadruply-bonded dirhenium compounds, excessive reduction of the metal occurred, yielding a viscous black mass, which transformed into powder upon slow drying. This mass was unsuitable for preparing crystals, and upon reoxidation it transformed into a quadruply-bonded dirhenium compound. Nevertheless, I trust that it will be possible to prepare compounds with high contents of metals in very low oxidation states and, in particular, layered and chain compounds with "infinite" systems of multiple metal–metal bonds.

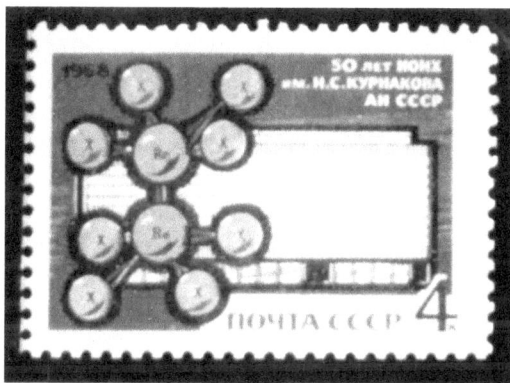

The square prismatic structure of the [Re₂Cl₈]²⁻ ion on a Soviet stamp (1968). The stamp was issued to commemorate the fiftieth anniversary of the N. S. Kurnakov Institute of General and Inorganic Chemistry of the Academy of Sciences of the USSR. The then new building of the Institute is shown in the background.

References

1. Watson, W. H.; J. *Acta Crystallogr.* **1958,** *11(6),* 689.
2. Noddack, I.; Noddack W. *Z. Anorg. Allg. Chem.* **1933,** *215,* 129.
3. Tronev, V.G.; Bondin, S. M. *Dokl. Akad. Nauk SSSR* **1952,** *86,* 87.
4. Kotel'nikova. A. S.; Tronev. V. G. *Zh. Neorg. Khim.* **1958,** *3,* 1008.
5. Babeshkina, G. K.; Tronev, V. G. *Zh. Neorg. Khim.* **1962,** *7.* 215.
6. Kuznetsov, V. G.; Koz'min. P. A. *Zh. Strutt Khim.* **1963,** *4,* 55.
7. Kuznetsov, V G.; Koz'min, P. A. *Acta Crystallogr.* **1963,** *16,* A41.
8. Koz'min, P. A.; Kuznetsov, V. G. *Abstracts of the 4th All-Union Conference on Crystal Chemistry.* Shtiints: Kishinev, **1961;** pp 74–75.
9. Robinson, W. T.; Fergusson, J. E.; Penfold, B. R. *Proc. Chem. Soc.* **1963,** 116.
10. Bertrand, J. A.; Cotton. F. A.; Dollase, W. A. *Inorg. Chem.* **1963,** *2,* 1166.
11. Cotton, F, A.; Curtis, N. F.; Harris, C. B.; Johnson, B. F. G.; Lippard, S. J.; Mague, J. T.; Robinson, W. R.; Wood, J. S. *Science* **1964,** *145,* 1305.
12. Cotton, F. A. *Inorg. Chem.* **1965,** *4,* 334; *Quart. Rev* **1966,** *20,* 389.
13. Cotton. F. A.; Bratton, W. K. *J. Am. Chem. Soc.* **1965,** *87,* 921.
14. Brencic, Ju. V.; Colton, F. A. *Inorg. Chem.* **1969,** *8,* 7.
15. Cotton, F. A.; DeBoer, B. G.; La Prade, M. D.; Pipal, J. R.; Ucko, D. *A. J. Am. Chem. Soc.* **1970,** *92,* 2926.
16. Peligot. E. *C. R. Acad. Sci.* **1844,** *19,* 734.
17. Cotton, F. A; Koch, S. A.; Millar, M. *Inorg. Chem.* **1978,** *17,* 2084.
18. Koz'min, P. A. *Dokl. Akad. Nauk SSSR* **1972,** *206,* 1384.
19. Cotton, F. A.; Norman, J G., Jr. *J. Am. Chem. Soc.* **1972,** *94,* 5697,
20. Koz'min, P. A.; Surazhskaya. M. D.; Larina. T. B. *Koord. Khim.* **1979,** *5,* 1542.
21. Kryuchkov. S. V.; Kuzina, A. F.; Spitsin, V. I. *Dokl. Akad. Nauk SSSR* **1982,** *266,* 127
22. Koz'min. P. A.; Surazhskaya, M. D.; Larina. T. B. *Dokl. Akad. Nauk SSSR* **1982,** *265,* 1420.
23. Koz'min, P. A.; Larina, T. B.; Surazhskaya, M. D. *Dokl. Akad. Nauk. SSSR* **1983,** *271,* 1157; Koz'min, P, A.; Surazhskaya, M. D.; Larina, T. B. *Koord Khim.* **1985,** *11,* 1559.
24. Koz'min, P. A.; Kotel'nikova, A. S.; Larina. T. B.; Mekhtiev, M. M.; Surazhskaya, M. D.; Bagirov, S. A.; Osmanov, N. S. *Dokl. Akad. Nauk SSSR* **1987,** *295,* 647.
25. Kryuchkov, S. V.; Grigor'ev, M. S.; Kuzina, A. F.; Gulev, B. F.; Spitsin, V. I. *Dokl. Akad. Nauk SSSR* **1986,** *288,* 389.

The Quadruple Bond in Historical Perspective[a]

F.A. Cotton[b]

This article by Professor Koz'min gives an interesting idiosyncratic perspective on Russian work in this field. His observations provide insight into how the early developments were, and to some extent still are, viewed in Moscow. His account supplements in a few details the account that appears in the book [1] written by R.A. Walton and me. I would, however, like to add a few further notes of my own.

It is interesting that the $[W_2Cl_9]^{3-}$ ion, whose structure was reported *very* long ago by a fine Swedish crystallographer, Cyrill Brosset [2, 3] (later work cited by Koz'min merely improved the accuracy), would retrospectively be regarded as the first example of a metal–metal triple bond, though not recognized as such until after other triple bonds (and the quadruple bond) had been described. In 1947, Pauling [4] used his bond number formalism to assign a W–W bond number of 1.70.

With regard to the question of analyses and "free" hydrogen atoms, another perspective can be found in Ref. 1. It is also not the case that we "undertook" our structure analysis of $K_2Re_2Cl_8 \cdot 2H_2O$ after "being aware of" the Russian results, although we did not complete it until after we were aware of them. We also refined the $(pyH)_2Re_2Cl_8$ structure [5] and

showed that the Re–Re distance is actually 2.24 Å (not 2.22 Å) and thus in full agreement with that in our potassium compound.

One final point concerns Kryuchkov's $K_{2\infty}[Tc_2Cl_6]_\infty$. In a 1991 paper [6] with Kryuchkov as a co-author, it was shown that the Te–Te distance is *not* anomalous and is in fact quite consistent with the (expected) presence of a triple bond and not with the orbitally impossible pentuple bond once suggested.

References

1. Cotton, F. A., Walton, R. A. *Multiple Bonds between Metal Atoms*, 2nd ed.; Oxford University Press, 1993.
2. Brosset, C. *Nature* **1935**, *135*, 874.
3. For another account of the history of metal–metal bonds, with special attention to the pioneering work of Brosset, see Cotton, F. A. *J. Chem. Educ.* **1983**, *60*, 713.
4. Pauling, L. *Chem. Eng. News* **1947**, *25 (41)*, 2970.
5. See footnote 19 in Bratton, W. K.; Cotton, F. A. *Inorg. Chem.* **1969**, *8*, 1299.
6. Cotton, F. A.; Daniels, L. M.; Falvello, L. R.; Gregoriev, M. S.; Kryuchkov, S. V. *Inorg. Chim. Acta* **1991**, *189*, 53.

[a]*Chemical Intelligencer* 1996(2), 36.

[b]Department of Chemistry, Texas A&M University, College Station, TX 77843, USA (deceased)

Opprobrium: Occurrence, Preparation, Properties, and Uses[a]

Derek A. Davenport[b]

In January 1979 I was called by Joseph F. Bunnett asking if, as a former student of Hughes and Ingold, I would be willing to give the banquet address at the IUPAC Conference on Physical Organic Chemistry due to be held on the campus of the University of California at Santa Cruz in August 1980. At first I demurred, pointing out that I no longer had any claims as a physical organic chemist and that, while I was familiar with the classical literature, I had not kept up with the non-classical variety in spite of occasional ear-bendings by my colleague H.C. Brown. However, as is his wont, Joe Bunnett persisted, my memories of a sabbatical leave spent at Santa Cruz grew fonder by the minute, and with his promise of an all-expenses-paid trip my uncharacteristic reluctance to speak was overcome. Besides August 1980 was a long way off—or so I thought.

A week or two later Joe was back on the line: IUPAC needed a title—immediately. Why is it that international meetings always have such early deadlines and yet so frequently fail to get their programs out on time? Recalling the title of a 1953 paper by Ingold in which he had put H. C. Brown to the verbal sword, I suggested "On the Comparative Unimportance of the Invective Effect in Physical Organic Chemistry." After assuring Joe that I did indeed mean "invective" and not "inductive" (what a numbingly boring talk that might have made) I once again put the matter out of my mind.

As August 1980 rapidly approached I cobbled together a slide–lecture gleaned from Wohler and Kolbe, Peep and Poachy [1], Lewis and Langmuir, and particularly from the published writings of Christopher K. Ingold and the private files of Herbert C. Brown. The general atmosphere of the conference proved to be somewhat somber and self-doubting since traditional physical organic chemistry was now no longer the darling of the funding agencies, and only the more adventurous had so far had the courage to take their tools, techniques and terminologies elsewhere. I was not sure that such an audience, well wined and dined though they were,

would be receptive to a recusant tweaking their fragile egos. The lecture started inauspiciously with the projector seemingly locked in automatic advance. I felt like Charlie Chaplin in *Modern Times*. Fortunately that problem was quickly corrected and the lecture hit its natural stride. The audience proved surprisingly receptive and with a sense of having successfully run the gauntlet I arrived at my, or rather Milton's, concluding moral:

> Where there is much desire to learn, there of necessity will be much argument, much writing, many opinions; for opinions in good men is but knowledge in the making.

Subsequently I have given modified versions of that talk on many occasions and it was later published in a somewhat bowdlerized [2] and in a shortened form. [3] One such occasion was the post-crustacean address at an Inorganic Gordon Conference. Not surprisingly that particular audience was more than usually receptive to the skewering of their organic colleagues. When, a couple of years later, they invited me back for a second lobster quadrille I decided that it was now the inorganic chemists' turn to be served *en brochette*. My title "Opprobrium: Occurrence, Preparation, Properties and Uses" had a simple origin. I consider myself a product of the Prince Albert Chemical Curriculum (PACC) in which the occurrence, preparation, properties, and uses of each of the elements were systematically, indeed remorselessly, elaborated. In some ways I think PACC may have been more successful than many of its acronymic successors.

There was one problem: I felt I could no longer rely solely on the published literature and the rich resources of the Herbert C. Brown Archives [4]. My thoughts turned first to F.A. Cotton, an accepted master of inorganic invective. Al amiably allowed that though he might have had an adverse referee's report once in a while they had all vanished from his memory—and his files. They had done nothing of the kind, of course, but what could I do? Next I wrote to 25 or so of the leading inorganic chemists in the United States, saying in part:

> In the case of modern inorganic chemistry I lack the rich resource of invective found in the "Herbert C. Brown Archives" so I am writing to you in the hope that you might help supply the deficiency. Would you be willing to supply one or two examples of choice referee invective from your files? I hasten to add that

[a]*Chemical Intelligencer* 1996(2), 45–50.

[b]Department of Chemistry, Purdue University, West Lafayette, IN 47907, USA

I mean invective directed *at* you not *by* you. For the latter I will trust to other suppliers. Although entries are not limited to 25 words or less they should be pithy, quotable, and not necessarily fair or true in order to be useful for my purpose. The name and citation of the target will be identified in the talk while the name of the attacker (even if known) will not. I might even offer a door prize for the maximum of correct guesses as to the identity of the latter. I trust you don't find the whole idea too frivolous.

Perhaps they did for, with the exception of Roald Hoffmann's, the responses were disappointing. One would think that only Roald had ever heard a discouraging word from a referee. Harry Gray vaguely recalled someone writing about "a manuscript that wouldn't receive a passing grade in a high school science course," but he put it down to the fact that in one of his references the spelling of the referee's name had been confused with that of a popular cognac. With characteristic hubris Harry claimed that the following was typical of his referee reports:

> Both of these communications are clearly written, concise accounts of some initial spectroscopic and electron transfer studies of an interesting Ir(I) dimer. Each is loaded with information derived directly from well-designed experiments which clearly support the authors' conclusion. Clearly written, pertinent communications on interesting work deserve brief, to-the-point reviews. Compliment the authors, publish their contributions quickly without change, and be thankful they chose your journal.

As a consequence of the dearth of new material I was forced to go back and comb the literature for examples. Fortunately there is no shortage of historical, if not historic, chemical opprobrium, many of which follow this preamble. Since the Gordon Conference audience was even better wined and dined than the one in Santa Cruz the lecture was a boisterous success.

Not long afterwards I was asked to contribute to a poster session on chemical trivia scheduled for an American Chemical Society meeting in New Orleans. Though I was not planning to attend the meeting, I agreed to submit an abstract.

I was not present to "explain" my poster but all 200 xerox copies placed alongside disappeared.

OPPROBRIUM: OCCURRENCE, PREPARATION, PROPERTIES AND USES.

Derek A. Davenport, Department of Chemistry, Purdue University, West Lafayette, Indiana 47907.

Compared with philosophers, poets, and politicians, chemists are generally speaking an amiable lot. On occasion, however, the presumed malevolence, crassness, deviousness, presumption, pomposity, intemperance or plain stupidity of a colleague has caused the chemist's mask of amiability to slip. A few of the more memorable occasions are recalled.

One of these must have fallen into the hands of the managing editor of the ACS publication *Today's CHEMIST* for on June 28, 1988 she wrote:

> For the August 1988 issue of *Today's* CHEMIST, the newest publication of the American Chemical Society, we would very much like to reproduce your paper entitled "Opprobrium: Occurrence, Preparation, Properties, and Uses." I am enclosing a typeset version of the article, and we propose to place it in the book as the very first article...
> We look forward to the presentation of your article to our readers, and thank you in advance for your consideration of our request.

I was somewhat startled to receive a proof since I had not submitted anything. Bowing to a *fait accompli,* however, I corrected the proof and sent it in. Shortly thereafter I received a phone call from someone at the ACS whose name I did not catch but whose tone of voice and measured pace betrayed the lawyer. Our conversation ran more or less as follows.

–You are Derek *A.* Davenport?

–Yes, I am.

–The author of "Opprobrium: Occurrence,...?"

–Yes.

–Some of this is strong stuff.

–I certainly hope so.

–This Berzelius and Engestrom, are they alive?

–No, long dead. In no position to sue.

–Well what about F. A. Cotton? He's not yet dead is he? (Did I detect a hint of regret in his voice?)

–No, he's very much alive.

–What about Harry Gray?

–Him too.

–But Cotton's pretty rough on Gray isn't he?

–Not particularly. Besides it's already been published.

–It has. Where?

–JACS.

–Oh.

Convinced that my legal hurdles had been overcome I sought out the August 1988 issue of *Today's CHEMIST.* Like Macavity, the article was not there. When it failed to surface in the next issue I wrote to enquire the reason. "Your piece," I was told, "does not convey the kind of image that *Today's CHEMIST* is trying to project and was consequently withdrawn." Apparently an article not officially submitted does not merit the courtesy of an official rejection.

No doubt it would be wise to accept my fate— "If at first you don't succeed, quit"—but I am resubmitting the "piece" to what I hope may prove a more broad-minded and less pusillanimous editor and to a readership less touchy, or so I hope, about its *amour propre.*

Since feuilletons *such as this are not capable of bearing the weight of too much scholarly apparatus I have left the quotations that follow unreferenced though almost all can be readily tracked down in a good library.*

A Crestomathy of Chemical Opprobrium

"…me thinks the Chymists, in their searches after truth, are not unlike the Navigators of Solomons Tarshish Fleet, who brought home from their long and tedious Voyages, not only Gold and Silver, and Ivory, but Apes and Peacocks too; For so the Writings of several (for I say not, all) of your Hermetick Philosophers present us, together with divers Substantial and noble Experiments, Theories, which either like Peacocks feathers make a great shew, but are neither solid nor useful; or else like Apes, if they have some appearance of being rational, are blemish'd with some absurdity or other, that when they are Attentively consider'd, makes them appear Ridiculous."

—ROBERT BOYLE ON VULGAR SPAGYRISTS

"I have searched the feeble lucubrations of this author without success for some trace of ingenuity, acuteness or learning that might compensate for his obvious deficiency in the powers of solid thinking or of calm and careful investigation…This manuscript uncovers no new truth, reconciles no contradictions, arranges no anomalous facts, suggests no new experiments and leads to no new inquiries…As this paper contains nothing which deserves the name either of experiment or discovery, and as it is in fact destitute of any species of merit, it should certainly be admitted to your Proceedings, to join the company of that multitude of other paltry and unsubstantial papers which are being published in your journal every month.

–HENRY BROUGHAM ON THOMAS YOUNG

"The peculiarity of the style and tendency of this attack led me at once to suspect, that it must have been suggested by some other motive than the love of truth; and I have both internal and external evidence for believing that the articles in question are either wholly, or in great measure, the productions of an individual, upon whose mathematical works I had formerly thought it necessary to make some remarks, which, though not favourable, were far from being severe; and whose optical speculations, partly confuted before, and already forgotten, appeared, to their fond parent, to be in danger of a still more complete rejection from the establishment of my opinions."

—THOMAS YOUNG ON HENRY BROUGHAM

"Mr. Dalton's aspect and manner were repulsive. There was no gracefulness belonging to him. His voice was harsh and brawling; his gait stiff and awkward; his style of writing and conversation dry and almost crabbed."

—HUMPHRY DAVY ON JOHN DALTON

"He was a very coarse experimenter, and almost always found the results he required, trusting to his head rather than his hands."

—HUMPHRY DAVY ON JOHN DALTON

"The most devoted sacrifice of experiment to hypothesis which the modern history of chemistry presents. Berzelius' theory is a measure of his facts, and like the Grecian robber who extended his victim on a bed, and cut them shorter, if they were too long, or stretched them, if too short, this chemical Procrustes screws his results to the preconceived limits of an empirical axiom."

—ANONYMOUS REVIEWER

"It is impossible not to admire the ingenuity and talent with which Mr. Dalton has arranged, combined, weighed, measured, and figured his atoms; but it is not, I conceive, on any speculations upon the ultimate particles of matter, that the true theory of definite proportions must ultimately rest."

—HUMPHRY DAVY ON JOHN DALTON

"You have sought to introduce the doctrine of definite proportions, and that is good; but I do not at all like the way you have done it. I ask you to immediately banish from your writings, especially when they treat this doctrine, a word which you use all too often. This word is 'about'. To the extent that this word exists in chemistry, the doctrine in question would be an aborted foetus, sick and ready to die. Moreover, the use of this word makes all your numerical determinations, without exception, erroneous."

—JÖNS JAKOB BERZELIUS ON HUMPHRY DAVY

"The second difficulty which Dr. Berzelius states is so obscurely expressed that it requires an acute atomist to perceive the force of it…He has discovered a law (which for the sake of argument I shall take to be true)…"

—JOHN DALTON ON JÖNS JAKOB BERZELIUS

"There is very essential difference between the researches of Mr. Dalton and myself. Mr. Dalton has chosen the method of the inventor, by setting out a principle from which he endeavours to deduce the experimental results. For my part, I have been obliged to take the road of the ordinary man…I have endeavoured to mount from experiment towards the first principle; while Mr. Dalton descends from that principle to experiment. It is certainly a great homage to the speculations of Dalton if we meet each other on the road."

—JÖNS JAKOB BERZELIUS ON JOHN DALTON

"Well this is a very interesting paper for those that take any interest in it."

—JOHN DALTON

(continued)

"The reason for this intentionally barbaric and dissipated appearance is the philosophical school which is known in Sweden as phosphorism and in Germany as Naturphilosophie. Its basis is ignorance of everything real, love of poetry and the fine arts, and a trusting uncritical devotion to the views of such persons who through incomprehensibility have acquired a reputation for profundity, especially when their foolishness goes so far that the government find it necessary in the name of sound reason to remove the idiots to where they can do no more damage."

—JÖNS JAKOB BERZELIUS ON NATURPHILOSOPHIE

"It is becoming increasingly unfashionable to undertake [scientific] investigation any more; none of my friends or former students does anything worthwhile...Mosander is so preoccupied with making money from his health resort that little issues from his own hands. Walmstedt diligently makes his own cardboard filing boxes to save a few pennies.

Lynchnell has become such a drunkard that he has lost his teaching position at Upsala...Wallquist does nothing but come in once a week to get his salary...Professor Engestrom has always been a dolt and has now become a boozer."

—JÖNS JAKOB BERZELIUS ON HIS COLLEAGUES

"According to some...the ultimate elements of matter are atoms, of which it is proved by certain reasonings, that they are each one sixth of one of the motes that float in the sunbeam."

—WILLIAM WHEWELL

"...Your sickness appears to be a specific illness of chemists. One could call it Hysteria Chemikorum, which originates from the combined damaging influence of mental exertion, ambition, and the vapors and fumes. Davy suffered from it, Mitscherlich, I—on the whole probably all great chemists."

—FRIEDRICH WÖHLER TO JUSTUS VON LIEBIG

"A consequence of this is the spread of the weed of the apparently scholarly and clever, but actually trivial and stupid, Naturphilosophie, which was displaced fifty years ago by exact natural science, but which is now brought forth again...by pseudo-scientists who try to smuggle it, like a freshly dressed and freshly rouged prostitute, into good society, where it does not belong...It is not possible to criticize this work even halfway thoroughly because the play of phantasy therein dispenses completely and entirely with factual basis and is absolutely unintelligible to the sober scientist..."

"A Dr. J. H. van 't Hoff, of the Veterinary School at Utrecht, has no liking, it seems, for exact chemical investigation. He has considered it more convenient to mount Pegasus (apparently borrowed from the Veterinary School) and to proclaim in his *La chimie dans l'espace* how the atoms appear to him to be arranged in space, when he is on the chemical Mt. Parnassus which he has reached by bold flight."

—HERMANN KOLBE ON JACOBUS VAN 'T HOFF

"We have been led to believe that not only have we atoms, but that these atoms possess imaginary prongs, and that there is an imaginary clasping between them...in a sort of hermaphroditism which it is scarcely possible to refer to."

—WILLIAM ODLING

"There is scarcely a page without a footnote, and some of the pages are practically little else than footnotes...The author, indeed, recommends that these should be read only by the advanced student...but we are afraid that no intelligent reader will follow this advice when once he has begun to dip into them. They are, in fact, like the postscripts of ladies' letters—often more important, more instructive, more suggestive, and more characteristic, than the main body of the text."

—THOMAS THORPE ON DMITRII I. MENDELEEV

"On p. 164, in the description of the experiment of burning phosphorus in oxygen, it is recommended that "the cork closing the vessel should not fit tightly, otherwise it may fly off with the spoon". That the cork should fly off with the spoon is contrary to a well-established precedent: if anything is to fly away with the spoon, it should, of course, be the dish on which the bell jar is represented as resting."

—THOMAS THORPE ON DMITRII I. MENDELEEV

"He frequently compares molecules to planetary systems. His inclination to speculate and generalize is everywhere evident, and while it often leads to valuable and suggestive results, it is occasionally carried to an extreme...there is a possible mysterious connection between the eight groups of the Periodic System, the eight major planets, and the eight satellites of Saturn, a view which may be pardoned (only) in a mind which perceived the Periodic Law."

—H. N. STOKES ON DMITRII I. MENDELEEV

"...but to be perfectly candid I think there is a chance that the casual reader may make a mistake which I am sure you would be the last to encourage. He might think you were proposing a theory which in some essential respects differed from my own, or one which was based upon some vague suggestions of mine which had not been carefully thought out."

—GILBERT N. LEWIS ON IRVING LANGMUIR

"In the first place we must realize that no one can or should have a proprietary right in a theory for all time... Strictly speaking if the originators of the theory must be mentioned...it should be called the Thomson–Stark–Rutherford–Bohr–Parson–Kossel–Lewis–Langmuir theory."

—IRVING LANGMUIR ON GILBERT N. LEWIS

"...a certain appearance of originality is gained by giving new names to old functions...It is perhaps not sufficiently emphasized that the use of activity can be only a passing phase in the study of thermodynamics."

—J. R. PARTINGTON ON GILBERT N. LEWIS

(continued)

"G.N. Lewis introduced the activity concept to meet the difficulty (of nondilute electrolytes). We can always make the experimental data agree with the theoretical values by multiplying the data by the ratios of the theoretical values to the experimental data…"

—WILDER BANCROFT ON GILBERT N. LEWIS

"We might consider Mrs. Eddy and G.N. Lewis as the Gold Dust Twins of Christian and Physical Science. Mrs. Eddy eliminates sickness but admits error. Lewis admits sickness but eliminates error."

—WILDER BANCROFT ON GILBERT N. LEWIS

"We have seen 'cyclical processes' limping about eccentric and not quite completed cycles, we have seen the exact laws of thermodynamics uncritically joined to assumptions comprising half-truths or no truth at all, and worst of all we have seen ill-begotten equations supported by bad data."

—GILBERT N. LEWIS AND MERLE RANDALL

"I have returned from a short vacation for which the only books I took were half a dozen detective stories and your Nature of the Chemical Bond. I found yours the most exciting of the lot."

—GILBERT N. LEWIS ON LINUS PAULING

"Dr. Pauling has been so successful…that his advocacy of the doctrine of the infallibility of Pasadenean research and the somewhat pontifical style in which the book is written are understandable and should not be taken amiss."

—GEORGE KISTIAKOWSKY ON LINUS PAULING

"The style of the book is poor, even by the lowest chemical standards…One can only hope that the original German version of the book was equally illiterate and that the translator has merely carried out his task too conscientiously…Many books are in parts like the curate's egg and many are uniformly mediocre; this book however, is uniformly bad from cover to cover."

—MICHAEL DEWAR ON A BOOK BY KARAGOUNIS

"There have been those who have wondered, often to themselves, and occasionally out loud, just where these various attempts to be precise and quantitative about the nature of acidity and basicity become too quixotic to be valuable. It is not impossible that the venerable sport of jousting at windmills is being practiced by the more zealous defenders of the various acid and base 'religions'."

—F. ALBERT COTTON (AND GEOFFREY WILKINSON) ON RUSSELL DRAGO

"In spite of extensive further investigations of these processes by F.A. Cotton and his students…they have not been able to demonstrate that their findings have any consequences on chemical reactivity. There have been those who wondered to themselves, and occasionally out loud,

just when these extensions of the original discovery of Piper and Wilkinson become too quixotic to be valuable. It is not impossible that the venerable sport of jousting at windmills is now being practiced…"

—RUSSELL DRAGO ON F. ALBERT COTTON AND GEOFFREY WILKINSON

"I hope other inorganic chemists know the meaning of "Opprobrium." I had to look it up. A few years ago Drago took out after us but that's no real distinction…"

—FRED BASOLO

"When reviewing a book written by two friends…it is most agreeable to be able to express sustained enthusiasm for their work…I am very sorry to have to say that…I find myself somewhat dissatisfied with the part which, though small, represents in my judgment the heart of the book…The mere outline of the 'rules of the game' as it is played by these particular authors adds something, though nothing of great importance, to the descriptions already in the literature…It is just plain bad scholarship to make absolutely no mention of this… and instead to write Fanny Farmer's *Cookbook* for the devotees of the lickety-split school of quantum mechanics."

—F. ALBERT COTTON ON HARRY GRAY AND CARL BALLHAUSEN

"No more fiction for us: we calculate. But that we may calculate we have to make fiction first."

—FRIEDRICH NIETZSCHE

"It isn't even wrong."

—WOLFGANG PAULI

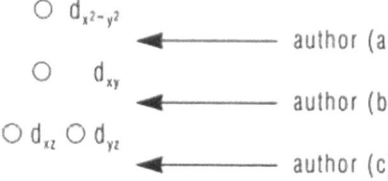

"Clearly the level of Od_{z^2} has been found. It only remains to determine which of the three authors has found it."

—HARRY GRAY

"La théorie c'est bon mais ça n'empêche pas d'exister." [Theory is fine but it doesn't prevent things from happening.]

—JEAN MARTIN CHARCOT

"I have attempted the kind of book which 'one can read in bed without a pencil'…I have made no concessions to the reader who refuses to inspect steric models in conjunction with study of the text."

—F. ALBERT COTTON IN PREFACE TO *CHEMICAL APPLICATIONS OF GROUP THEORY*

(continued)

"In the preface the author claims that the book is of the type which can be read in bed without a pencil. I tested this out by retiring with the book on three successive evenings…I was able to follow the discussion in this chapter even though my wife would not allow me to take any models to bed to aid my study."

—HARRY GRAY IN REVIEW OF *CHEMICAL APPLICATIONS OF GROUP THEORY*

"This paper would not be acceptable for publication in *Physical Review.* The authors should calculate the binding energy of this structure and compare it with graphite, not just propose it as a possible structure. The extended Hückel method contains errors of order 3 eV; it is absolutely useless except for publishing papers in chemistry journals. You chemists should raise your standards."

—REFEREE II OF PAPER BY ROALD HOFFMANN

"This paper is worthless and totally unsuitable for publication. I agree entirely with the comments by Referee II—which apply equally to Part V. My comments on Part V apply even more forcibly to this paper; it is bad enough to use Hückel theory for ground states of molecules, but to apply it to excited states is very much worse. There is some hope that ground states may be representable, at any rate approximately, by simple determinant wave functions, since SCF functions for ground states do not mix with those for singly excited states. Singly excited functions constructed from ground state orbitals do, however, mix with one another; any treatment of excited states in terms of single determinant functions of this type is therefore totally unsound. Would this article be suitable for publication:

1. with change? NO 2. after minor revision? NO 3. after major revision? NO Is JACS the best medium for publishing this article? NO. If not, what other Journal would you suggest? NOWHERE."

—REFEREE III OF ANOTHER ROALD HOFFMANN PAPER

"Most people, on finding that calculations, carried out by a method of demonstrated unreliability, had led to results in direct conflict with experiment, would quietly forget them; but not Dr. Hoffmann! If the facts disagree with theory, so much the worse for the facts. Frankly I am amazed that this paper should have been submitted for publication; it is certainly quite unsuitable for any reputable scientific journals, let alone JACS."

—PRESUMABLY A DIFFERENT REFEREE III

"The speculations in this paper are the sort of thing that one expects to hear at research seminars, or in social chemical gatherings over a glass of beer; certainly many of them have been made at my own seminar by bright young students. No one else, however, has had the conceit or effrontery to think them worth publishing, let alone in a communication written in the first person. This paper seems to me entirely unsuitable for publication in any reputable scientific journal, let alone JACS."

—YET ANOTHER REFEREE III

"Like the Irish, theoretical chemists are fair people—they never speak well of one another."

—APOCRYPHAL

"In a fight between you and the world, back the world."

—FRANZ KAFKA

"Your manuscript is both good and original; but the part that is good is not original, and the part that is original is not good."

—SAMUEL JOHNSON

"Ich sitze in dem kleinsten Zimmer in meinem Hause. Ich habe Ihre Kritik vor mir. Im nächsten Augenblick wird sie *hinter* mir sein." ["I am sitting in the smallest room of my house. I have your review before me. In a moment it will be *behind* me."]

—MAX REGER

"This paper should either be reduced by 50 % or oxidized by 100 %."

—COMMUNICATED BY RALPH PEARSON

References

1. Davenport, D. A. "Sulfur, Stereochemistry and the Reverend William Archibald Spooner", *J. Chem. Educ.* **1981,** *58,* 682–683.
2. Davenport, D. A. "On the Comparative Unimportance of the Invective Effect in Physical Organic Chemistry", *CHEM TECH* **1987,** *17,* 526–531.
3. Davenport, D. A. "On Opinion in Good Men: An Oblique Tribute to H. C. Brown", *Aldrichimica Acta* **1987,** *20*(1), 25–27.
4. Davenport, D. A. "The Herbert C. Brown Archives", *CHOC News* **1988,** *5*(1), 6–8.

The Surprising Periodic Table: Ten Remarkable Facts[a]

Dennis H. Rouvray[b]

The periodic table is now universally recognized as one of the crowning achievements of nineteenth century physical science. As a fundamental classificatory scheme for the chemical elements, it stands on a par with other great classifications in the sciences, such as the classification of plant species by Linnaeus in the 1750s or the classification of subnuclear particles by the physicists Gell-Mann and Ne'eman in the 1950s. The periodic table, now a major part of our chemical heritage, has never been held in higher esteem by the chemical community at large. For the vast majority of chemists, its authority is unassailable, its usefulness is undoubted, and its capacity to stimulate new thinking is unarguable. The periodic table is also deeply reassuring in that it accounts for and assigns a specific position to every element; even elements still waiting to be synthesized have their rightful place awaiting them. It would appear to be one of the very few theoretical constructs in chemistry that comes close to being treated as sacrosanct. This enthusiastic acceptance of the periodic table raises an important question: Is the periodic table now in danger of acquiring the aura of a scientific icon? Of being viewed as a 125-year-old masterpiece so sublime in its conception that it commands something akin to veneration?

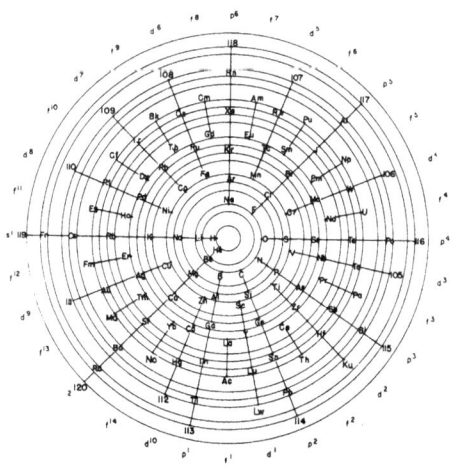

[a]*Chemical Intelligencer* 1996(3), 39–47.

[b]Department of Chemistry, University of Georgia, Athens, Georgia 30602

Definitely not! There is absolutely no danger of this happening. Although certain modern chemistry textbooks may attempt to lull us into believing this, the reality is quite different. The periodic table has never been nor is it ever likely to become the object of veneration by chemists, even though it enjoys considerable esteem and is widely thought of as being extremely valuable. From the earliest days of its inception, the periodic table has been fought over: it has been attacked, contested, and disputed. At times the wrangling became so heated that the imminent demise of the periodic table was foreseen. On one occasion it was even suggested that the periodic table had no scientific basis whatsoever and that it would make just as much sense to list the elements alphabetically [1]. Did all of these battles take place in the dim and distant past? By no means! Controversy has continued to surround the periodic table and still does so today. For instance, there is a lack of unanimity on the precise designations to be used for the groups and subgroups in the periodic table [2] and no general agreement on the shape the periodic table should assume [3]. In the last few years alone, two new controversies have arisen that we shall discuss more fully later on: one concerning whether the periodic table is "complete," and the other relating to the naming of postactinide elements. It should thus be evident that the periodic table never really had the possibility of becoming a scientific icon—nothing about it has ever been settled for that long!

Our aim here will be to delve into a number of these controversies. We shall attempt to understand why they erupted, in what ways they affected the overall development of the periodic table, and how they are generally interpreted today. In all, we shall consider 10 such controversies that range in time from the earliest history of the periodic table down to the present day. Our analysis may come as something of a surprise to some, for it will reveal that many cherished beliefs about the periodic table are little more than comforting myths. We shall be pointing out that much of what currently masquerades as knowledge about the periodic table is actually comprised of erroneous perceptions that have unfortunately been continually reinforced over the years by frequent repetition in articles and books. In an attempt to set the record straight, we shall examine the periodic table from 10 different perspectives. This will illuminate 10 different aspects of the history of the periodic table spanning its entire development. Each of the 10 aspects we cover will highlight one particular fact—a fact that

B. Hargittai and I. Hargittai (eds.), *Culture of Chemistry: The Best Articles on the Human Side of 20th-Century Chemistry from the Archives of the Chemical Intelligencer,* DOI 10.1007/978-1-4899-7565-2_39, © Springer Science+Business Media New York 2015

we consider to be remarkable. We use the adjective *remarkable* because we expect that the facts we present will indeed be surprising or little known to most readers or will fly in the face of conventional wisdom on the subject. Let us now move to a discussion of the first remarkable fact.

Fact 1

The first remarkable fact we consider takes us back to the very beginning of the periodic table. Precisely when was this? We cannot say with any certainty because different observers have differing ideas on the subject. Our first remarkable fact is thus that the periodic table has no clear-cut beginning; its precise origins remain fuzzy and contentious. Some commentators see in the work of Johann Döbereiner the origins of the periodic table. It was his paper [4] of 1829 that pointed out the existence of triads of elements, that is, groups of three elements in which the middle element displays properties that are the mean of those of the other two. Examples of Döbereiner's triads are calcium–strontium–barium and chlorine–bromine–iodine. However, in a paper dating from 1817, Döbereiner's work on the first of these triads is already mentioned [5]. So, the origin of the periodic table would date back to 1817. This date, however, could be pushed back even further. For instance, it may be argued that the table of the elements given by Antoine Lavoisier in his famous book [6] *Traité élémentaire de chimie* was really the first periodic table. In his table, published in 1789, Lavoisier listed the then known elements according to their physical nature (see Fig. 1) but admitted that his work

TABLEAU DES SUBSTANCES SIMPLES.

	NOMS NOUVEAUX.	NOMS ANCIENS CORRESPONDANTS.
Substances simples qui appartiennent aux trois règnes, et qu'on peut regarder comme les élémens des corps.	Lumière	Lumière.
	Calorique	Chaleur.
		Principe de la chaleur.
		Fluide igné.
		Feu.
		Matière du feu et de la chaleur.
	Oxygène	Air déphlogistiqué.
		Air empiréal.
		Air vital.
		Base de l'air vital.
	Azote	Gaz phlogistiqué.
		Mofette.
		Base de la mofette.
	Hydrogène	Gaz inflammable.
		Base du gaz inflammable.
Substances simples, non métalliques, oxydables et acidifiables.	Soufre	Soufre.
	Phosphore	Phosphore.
	Carbone	Charbon pur.
	Radical muriatique	Inconnu.
	Radical fluorique	Inconnu.
	Radical boracique	Inconnu.
Substances simples, métalliques, oxydables et acidifiables.	Antimoine	Antimoine.
	Argent	Argent.
	Arsenic	Arsenic.
	Bismuth	Bismuth.
	Cobalt	Cobalt.
	Cuivre	Cuivre.
	Étain	Étain.
	Fer	Fer.
	Manganèse	Manganèse.
	Mercure	Mercure.
	Molybdène	Molybdène.
	Nickel	Nickel.
	Or	Or.
	Platine	Platine.
	Plomb	Plomb.
	Tungstène	Tungstène.
	Zinc	Zinc.
Substances simples, salifiables, terreuses.	Chaux	Terre calcaire, chaux.
	Magnésie	Magnésie, base de sel d'Epsom.
	Baryte	Barote, terre pesante.
	Alumine	Argile, terre de l'alun, base de l'alun.
	Silice	Terre siliceuse, terre vitrifiable.

Fig. 1 Lavoisier's table of simple substances or elements, dating from 1789. As the elements are classified according to their nature, this table may be seen as a forerunner of the periodic table.

drew heavily on the results of another French chemist, Guyton de Morveau, who published [7] an earlier table in 1782.

It should be evident by now that the process of seeking for the true origins of the periodic table could be extended almost indefinitely and could certainly be continued back as far as the ancient Greeks, with their supposed four elements of air, earth, fire, and water [8]. Any date that we may wish to consider as the starting point for the periodic table would thus be arbitrary. The chemist whose name is commonly associated with the introduction of the periodic table, the Russian Dmitri Mendeleev, opted for an origin of very recent vintage. He wrote [9] that "the decisive moment in the development of my theory of the periodic law was in 1860, at the conference of chemists in Karlsruhe…at which I heard the views of the Italian chemist S. Cannizzaro. I regard him as my immediate predecessor, because it was the atomic weights we found which gave me the necessary reference material for my work." This quotation brings us to the second remarkable fact involving the periodic table.

Fact 2

The second remarkable fact concerns the precise date on which the periodic table may be said to have appeared in its final form. Once again, it is not possible to give a precise date. Although the periodic table evolved into something approaching its final form during the 1860s, it has not ceased to evolve since then. Indeed, the development of the periodic table is probably best characterized as a virtual continuum of small evolutionary advances spread over a considerable time span rather than as a giant leap forward that resulted from one unique act of creation. This, in turn, means that no single individual can be said to be responsible for the introduction of the periodic table. Essentially, it involved a joint venture in which many contributors participated in different ways. This fact probably explains why no award of the Nobel Prize was ever made for the periodic table. In Fig. 2 a time line is shown that extends from the year 1860 to the year 1870. Along this line we indicate the major contributions to development of the periodic table within this time frame. Figure 3 presents four of the periodic tables put forward during this decade, and each has an insert showing a picture of its author. Both of these figures demonstrate graphically that progress on the periodic table was evolutionary rather than revolutionary and support the contention that no one person can claim to have single-handedly invented the periodic table.

What then are we to make of the oft-repeated claims that Dmitri Mendeleev was the true founder of the periodic table? Mendeleev's claim to fame rests very largely on his own insistence on priority in this matter and on the boldness with which he made predictions about the properties of ele-

1860	
1861	
1862	B. de Chancourtois: Helical Periodic System of Elements
1863	J. Newlands: Classification of Elements into Groups
1864	L. Meyer: Classification of Elements into Groups J. Newlands: Listing of Elements by Ordinal Numbers W. Odling: Periodic Classification of Elements, with Subgroups
1865	J. Newlands: Law of Octaves or Periodic Law for Elements
1866	G. Hinrichs: Classification of Elements into Series
1867	G. Hinrichs: Periodic Classification of Elements
1868	L. Meyer: Periodic Classification of Elements W. Odling: Another Periodic Classification of Elements
1869	G. Hinrichs: Another Periodic Classification of Elements D. Mendeleev: Periodic Classification of the Elements

Fig. 2 A time line extending from the beginning of the year 1860 through the beginning of the year 1870. For each year the significant contributions to development of the periodic table are listed. The time line shows clearly that Mendeleev was a comparative latecomer on the scene.

ments that were completely unknown at the time. He even left gaps for these elements in his version of the periodic table. In all, Mendeleev predicted the existence of 11 new elements, and of these he got 8 right. His predictions concerning the elements now numbered 21 (scandium), 31 (gallium), and 32 (germanium) proved to be uncannily accurate. After all of these elements had been discovered (within 20 years of his prediction) and had been shown to possess almost exactly the properties he had foretold, his fame soared. Other contributors to the periodic table were all but forgotten [10], even though they had produced similar tables to his several years before him (see Fig. 3). To the end of his life in 1907, however, Mendeleev remained firmly convinced of his priority as originator of the periodic table. On different occasions, he stated [10] that "the question of the periodicity of the elements owes nothing at all to Messrs Newlands and Meyer" and that "if Odling felt his table had any theoretical significance, he would certainly have mentioned it"!

Fact 3

Our third remarkable fact pertains to the contested usage of the expressions "periodic table" and "periodic law." Strictly speaking, the former is a misnomer, for what we commonly

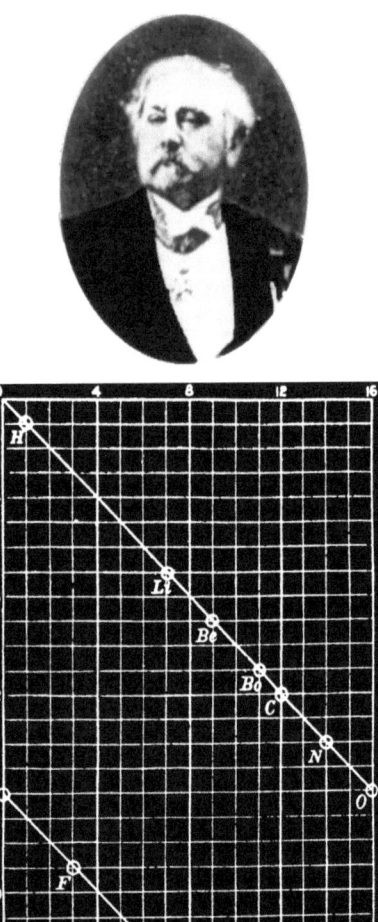

Fig. 3 This figure consists of four parts, four of the periodic tables put forward during the 1860s, with pictures of their originators. The tables shown are those of de Chancourtois (1862, above), Newlands (1864, top next page), Mendeleev (1869, bottom next page), and Odling (1864, bottom following page).

elements each, and two long periods of 32 elements each. The table is perhaps best characterized group-theoretically in terms of the direct product group $SU(2) \times SO(4,2)$, though we shall not explore this here [11]. In order to be a table, the periodic table has to be depicted in one of its several tabular formats. This, however, is not the only possible way of depicting it, as we discuss below. A better designation for the periodic table might therefore be the chart of periods or the period system of the elements [12]. Whereas a table is a highly structured entity with two or more clearly defined axes, a chart or system is any graphic representation that presents information in an orderly fashion. On the question of whether we are justified in referring to a periodic law, as Mendeleev was wont to do, there has been much controversy. As pointed out by Shapere [13], everything ultimately turns on what is meant by the word *law*. If this word refers to no more than some observed empirical relationship that enables us to make predictions about elements, its use here is justified. If, on the other hand, we are to understand a mathematical relationship that is capable of providing a rationale for the structure of the periodic table, the word law would be inappropriate. Clearly, if we opt for the latter interpretation, it cannot be argued that the periodic tables developed during the 1860s are based on a law. Mendeleev indicated some awareness of this when he stated [14] that "although greatly enlarging our vision, even now the periodic law needs further improvements in order that it may become a trustworthy instrument in further discoveries."

Fact 4

It was not long before the so-called law of Mendeleev was put to the test and found to have serious shortcomings. From his writings, it is evident that Mendeleev regarded his periodic table as little more than a rough-and-ready manifestation of some precise mathematical law that was yet to be discovered. This view was certainly borne out when the noble gases began to be discovered in the 1890s by William Ramsey and his co-workers. The periodic table as it then was had been neither capable of predicting the existence of the noble gases nor of accounting for their properties. Even as late as 1900, the inadequacies of the periodic table were still being remarked upon. Ramsey and Travers [15] complained that they had "not been able to predict accurately any one of the properties of these [noble] gases from a knowledge of those of the others; an approximate guess is all that can be made. The conundrum of the periodic table has yet to be solved." Our fourth remarkable fact is thus

refer to as the periodic table is neither periodic nor is it necessarily a table. In order to be strictly periodic, the table would need to have all of its periods of the same length. Actually, the table consists of periods of different length, namely, a preperiod of two elements, two short periods of eight elements each, two medium-length periods of 18

that the periodic table was of no use at all in predicting either the existence or the properties of the noble gases. In fact, the discovery of these gases proved to be an embarrassment for proponents of the periodic table. Eventually, Mendeleev suggested that a new zeroth group be added to his table to accommodate the new discovery [16]. Other discoveries made subsequently, such as the discovery of the lanthanides, were equally controversial in that it was difficult to assign appropriate positions for them within the periodic table. One particular discovery, which we now address, proved to be so highly contentious that it almost resulted in the complete demise of the periodic system of the elements.

Fact 5

We refer here to the discovery of the existence of chemical isotopes early in the twentieth century. Isotopes are atoms with the same number of protons but with differing numbers of neutrons. It was their discovery that brought home to chemists the fact that the position of an element in the periodic table is determined not by its atomic mass but rather by its atomic number, that is to say, by the number of protons present in the atomic nucleus. With this new insight, it immediately became evident why it had been necessary to invert the order of certain element pairs in the periodic table. In particular, it had proved necessary to invert the pairs argon and potassium (atomic masses 39.95 and 39.10, respectively); tellurium (127.60) and iodine (126.91); and cobalt (58.93) and nickel (58.69). The discovery of isotopes, however, did more than account for certain apparently anomalous placings of elements in the periodic table. It unleashed a vigorous debate on whether the periodic table was adequate to accommodate all the isotopes that were being discovered. There was a widespread feeling that the periodic table needed to be replaced by some much more cumbersome structure that would incorporate all the isotopes of every element [17]. Such a new structure would, of course, have meant the end of the periodic table as we know it today. Fortunately, work by Paneth and von Hevesy [18] demonstrated that the presence of isotopes in an element had no effect on its chemical reactivity. The only exception to this general rule was hydrogen, the isotopes of which did differ markedly in their physical and chemical properties. Apart from this single exception, however, the replacement of one isotope of an element by another was of no chemical significance. The periodic table succeeded in coming through its worst crisis unscathed. Our fifth remarkable fact is thus that the appearance of isotopes on the scene put the future of the periodic table in jeopardy and very nearly led to its complete abandonment.

```
                              Triad.
                    ┌──────────┬──────┬──────────┐
                    │ Lowest   │ Mean │ Highest  │
                    │ term.    │      │ term.    │
    I.    Li  7 +17 = Mg 24    Zn 65   Cd 112
    II.   B  11
    III.  C  12 +16 = Si 28            Sn 118            Au 196
    IV.   N  14 +17 = P  31    As 75   Sb 122 +88 = Bi 210
    V.    O  16 +16 = S  32    Se 79·5 Te 129 +70 = Os 199
    VI.   F  19 +16·5 = Cl 35·5 Br 80  I  127
    VII. Li 7 +16 = Na 23 +16 = K 39   Rb 85  Cs 133 +70 = Tl 203
    VIII.Li 7 +17 = Mg 24 +16 = Ca 40  Sr 87·5 Ba 137 +70 = Pb 207
    IX.                        Mo 96   V 137   W 184
    X.                         Pd 106·5         Pt 197
```

```
                              Ti=50     Zr=90      ?=180.
                              V=51      Nb=94      Ta=182.
                              Cr=52     Mo=96      W=186.
                              Mn=55     Rh=104,4   Pt=197,4
                              Fe=56     Ru=104,4   Ir=198.
                     Ni=Co=59           Pl=106,6   Os=199.
    H=1                       Cu=63,4   Ag=108     Hg=200.
          Be=9,4    Mg=24     Zn=65,2   Cd=112
          B=11      Al=27,4   ?=68      Ur=116     Au=197?
          C=12      Si=28     ?=70      Sn=118
          N=14      P=31      As=75     Sb=122     Bi=210
          O=16      S=32      Se=79,4   Te=128?
          F=19      Cl=35,5   Br=80     I=127
    Li=7  Na=23     K=39      Rb=85,4   Cs=133     Tl=204
                    Ca=40     Sr=57,6   Ba=137     Pb=207.
                    ?=45      Ce=92
                    ?Er=56    La=94
                    ?Yt=60    Di=95
                    ?In=75,6  Th=118?
```

Fact 6

Even after this crisis had been weathered, disputes about the precise shape and format of the periodic table continued to erupt. It seems that chemists have always had the greatest difficulty in agreeing on what the periodic table should look like. This brings us to our sixth remarkable fact: there has never been full and general acceptance on the shape that the periodic table should assume. Endless debate on the subject has still not resolved the issue, though the tabular block form is now the one commonly presented in most chemistry textbooks. A book that was published to celebrate the centenary of Mendeleev's table in 1969 [19] mentioned that no fewer than 700 different periodic tables had been generated in the century that had elapsed since 1869. These the author was able to classify into 146 basic types—a classification which revealed a gradual evolution to ever more graphic presentation of the data, with the best developed tables being those that displayed detailed information about the atoms, including their electronic configuration. All of the tables were ultimately reducible to the form of some curve in space, and the tables themselves could be depicted either as a top view of the curve, an elevation of the curve, or simply as the curve itself. Basically, these space curves assume the shape of a helix on a cone or cylinder, a set of squares, concentric circles, or figures of eight. The latter curves are designated mathematically by the rather exotic name of lemniscates. In Fig. 4 we present one example of each of these four basic types. Is it possible to say which of these tables is best? No, for the simple reason that each table was designed with a specific purpose in mind, for which its author(s) felt it was the best. The block-form tables in common use today for teaching and other purposes are actually elevations of helical space curves. It should come as no surprise that Mazurs [19] has stated that only those tables that include all the elements—the representational, the transition, and the inner transition elements—can be regarded as genuine periodic tables.

Fact 7

At this point we consider what role quantum theory has in prescribing the overall structure of the periodic table. Perhaps surprisingly, the periodic table lacks a completely sound theoretical basis at present, in spite of the prodigious efforts of quantum theoreticians and (to a lesser extent) group theoreticians to provide one [11]. Our seventh remarkable fact concerning the periodic table is thus that even today the periodic table cannot be said to rest on an entirely firm foundation. Let us now take a closer look at its foundation. The periodic table developed originally from experimentally observed analogies that occur in the properties of the elements. Although analogies still form its foundation, the role of the analogies has been refined in recent years. By their very nature, periodic tables summarize a very large number of property–atomic number mappings for the elements. It is these mappings that are ultimately responsible for the symmetry features of the table, which arise from the placing of similar elements into groups. However, individual mappings never exactly reproduce this symmetry, for each such mapping will exhibit periodicity to a differing degree. To allow for these differences, Jensen [20] proposed that elemental properties be classified according to their ability to reflect this symmetry. Properties such as the electronic configuration appeared at the top of the list while properties such as electrical conductivity were at the bottom. This listing suggested to him the idea of primary, secondary, and tertiary analogies between elements, with primary analogies being those that come closest to producing the group symmetry. For primary analogies to occur, the valence electrons in the atoms of the elements concerned need to be in precisely the same configuration.

				Ti＝50	Zr＝90	?＝180.
				V＝51	Nb＝94	Ta＝182.
				Cr＝52	Mo＝96	W＝186.
				Mn＝55	Rh＝104,4	Pt＝197,4
				Fe＝56	Ru＝104,4	Ir＝198.
			Ni＝Co＝59		Pl＝106,6,	Os＝199.
H＝1				Cu＝63,4	Ag＝108	Hg＝200.
	Be＝9,4	Mg＝24	Zn＝65,2		Cd＝112	
	B＝11	Al＝27,4	?＝68		Ur＝116	Au＝197?
	C＝12	Si＝28	?＝70		Sn＝118	
	N＝14	P＝31	As＝75		Sb＝122	Bi＝210
	O＝16	S＝32	Se＝79,4		Te＝128?	
	F＝19	Cl＝35,5	Br＝80		I＝127	
Li＝7	Na＝23	K＝39	Rb＝85,4		Cs＝133	Tl＝204
		Ca＝40	Sr＝87,6		Ba＝137	Pb＝207.
		?＝45	Ce＝92			
		?Er＝56	La＝94			
		?Yt＝60	Di＝95			
		?In＝75,6	Th＝118?			

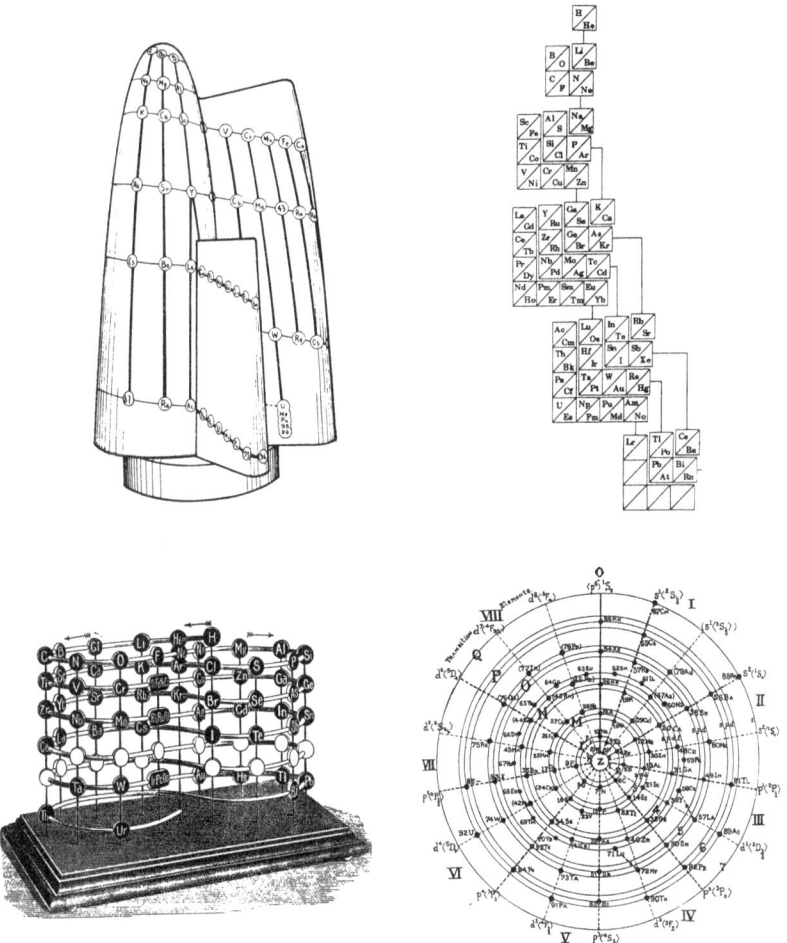

Fig. 4 The four basic geometrical forms that have been used to construct periodic tables. From top left and moving clockwise, the forms depicted are those of the helix, the set of squares, concentric circles, and figures of eight (lemniscates). (Reproduced from Refs. [38–41], respectively).

Since no one property ever yields perfect periodicity and periodic tables actually reflect an averaged symmetry structure for the elements over the entire range of elemental properties, we may ask whether quantum theory is able to provide a more substantial basis for the periodic table. We have already noted that the electronic configuration of an atom heads the list in reflecting periodicity. Even this descriptor, however, does not yield perfect periodicity, and a number of anomalous situations result from its use [21]. For instance, the ground-state electronic configuration of the nickel atom is [Ar]$3d^8 4s^2$ even though the configuration [Ar]$3d^9 4s^1$ is of lower energy. In general, there exists no straightforward relationship between the ground-state configuration of an atom and the chemical behavior of the corresponding element [21]. The nature of this relationship has been explored by Löwdin [22] and others [23, 24] who have pointed out that for atoms in which the electrons experience a coulombic field, for example, the hydrogen atom, the electronic energy will depend only on the principal quantum number n. For all other atoms, the electrons will not be in a coulombic field because there is shielding of the valence electrons by the inner shell of electrons. Since this shielding effect increases with the azimuthal quantum number, l, the chemical properties of the elements depend not only on the number but also on the disposition of the valence electrons. Moreover, arguments of this type assume that each electron in a many-electron atom can be assigned a well-defined set of quantum numbers. This, however, is not the case because electronic motion is influenced by the repulsions of other electrons. It would be true only within the framework of approximate one-electron theories of atoms [24]. Quantum mechanics—even with the use of modern ab initio procedures—is thus not able to predict electronic configuration from theory and so is unable to provide a firm foundation for the periodic table. In fact, the best that can be done with current methods is to predict which ground-state configuration an atom will assume from a number of possible alternatives [17].

Fact 8

Since both the overall shape and the theoretical basis of the periodic table are such contentious themes, it may come as no surprise that the designation of the periods and groups within the table is equally controversial. Our eighth remarkable fact is that, in spite of numerous attempts to settle these matters, there is still much debate on the best way to designate both the periods and the groups in the periodic table. Let us start at the beginning of the periodic table, with hydrogen and helium. Both of these elements, which together form the first "short period" of the periodic table, are somewhat anomalous in their behavior. Hydrogen with its $1s^1$ electron configuration has been placed at the head of both alkali metal and halogen groups. Helium with its $1s^2$ configuration would appear to belong at the head of the alkaline earth group but, in terms of its properties, clearly should be put at the head of the noble gas group. For these reasons, a number of authors do not regard hydrogen and helium as forming a period at all and consider the first period as that beginning with lithium and ending with neon [25]. Moving on now to the lanthanide and actinide elements, much debate has focused on where these begin and end. Jensen has persuasively argued that the lanthanide series should start with lanthanum and end with ytterbium and that the actinide series should start with actinium and end with nobelium [26]. This has initiated a trend away from the earlier groupings of cerium through lutetium for the lanthanides and thorium through lawrencium for the actinides. Jensen marshaled many arguments in favor of the switch, perhaps the most convincing being spectroscopic evidence which shows that the ideal ground-state configuration for f-block elements is [noble gas]$(N-2)fxNs^2$ rather than the previously thought [noble gas]$(N-2)fx^{-1}(N-1)d^1Ns^2$, where N is the period number and x is an integer lying in the range $1 \leq x \leq 14$.

The optimal designation to be used for each group of elements within the periodic table has been the subject of considerable dispute ever since the earliest periodic tables were put forward in the mid-1860s by pioneers such as Newlands and Odling [27]. Over the years, a plethora of competing naming schemes has been advanced. Many of these pertain to the 18 groups in the so-called "long" periodic table. Strictly speaking, what is often called the long periodic table is more correctly referred to as the medium-length periodic table; the genuinely long periodic table spans the groups 1 through 32. For 18-group periodic tables, designations of the groups have included IA through VIIIA, M1 through M8, 1 through 8, and 1L through O,8R, with the transition metals generally being denoted by some suffix such as d or T [28]. One of the most recent and one of the more popular schemes has been that of the American Chemical Society [29] which starts with groups 1 and 2, continues with groups 3d through 12d, and finishes with groups 13 through 18. In an attempt to get some kind of international agreement, the British scientific periodical *New Scientist* asked its readers to comment on the proposed revision of the nomenclature suggested by the American Chemical Society in 1984. Although the response was overwhelming, the opinions expressed were surprisingly divergent [2]. In spite of this, the periodical went ahead to produce its own version of the medium-length periodic table, in which the groups are numbered from 1 through 18 without any suffixes. This may be the best that can be done at present. It is all too evident that the last word has not yet been said on this subject.

Fact 9

When it comes to naming certain of the elements in the periodic table, passions also run high. There have been rumblings over the naming of the elements numbered 11, 19, and 74 for many years. These are still referred to in English as sodium, potassium, and tungsten even though their respective symbols are Na, K, and W. Repeated exhortations to change the names to natrium, kalium, and wolfram have fallen on deaf ears. However, this minor ripple pales into insignificance compared to the turbulence generated over the naming of the post-actinide elements. The underlying problem was that American and Russian scientists could not agree who had first synthesized a number of these elements. Since the honor of naming a new element is usually bestowed upon its discoverer, more than priority disputes were at stake here. In the case of element 104, the Russians went ahead and called it kurchatovium after the Russian nuclear physicist Igor Kurchatov. The Americans, however, preferred the name rutherfordium after the New Zealand-born physicist Ernest Rutherford. In order to calm matters as far as possible, the International Union of Pure and Applied Chemistry (IUPAC) suggested that the newer elements be given Latin names derived from their atomic number, with 0 being *nil*, 1 being *un*, 2 being *bi*, and so on. This resulted in the names shown on the left-hand side of our Fig. 5. Perhaps not surprisingly, these names were universally criticized as being clumsy and inelegant. In a surprise move in 1994, IUPAC proposed that the elements be given specific names after all; these are listed on the right-hand side of Fig. 5. This new move caused consternation among American chemists who had already given the name seaborgium to element 106 in honor of Glenn Seaborg, the discoverer of several heavy elements. Seaborg himself said [30] that he anticipated "widespread argument around the world over this action" and believed "it will not stand in the long run." Our ninth remarkable fact is thus that the business of naming the elements in the periodic table is by no means finally settled and is still capable of arousing a great deal of passion.

Interim IUPAC Latin Name	Element Number	Official IUPAC Element Name	Offical Symbol
Unnilunium	101	Mendelevium	Md
Unnilbiium	102	Nobelium	No
Unniltriium	103	Lawrencium	Lr
Unnilquadium	104	Dubnium	Db
Unnilpentium	105	Joliotium	Jl
Unnilhexium	106	Rutherfordium	Rf
Unnilseptium	107	Bohrium	Bh
Unniloctium	108	Hahnium	Hn
Unnilennium	109	Meitnerium	Mt

Fig. 5 The nomenenclature for the elements with atomic numbers ranging from 101 through 109. Listed are the interim Latin names, the new IUPAC official names, and the new symbols.

Fact 10

Perhaps one of the most astonishing controversies concerning the periodic table was initiated in 1989 when Leland Allen of Princeton University in New Jersey began to suggest that the periodic table was incomplete [31]. In a subsequent paper entitled "The Periodic Table Is Misleading, Incomplete, and Unduly Neglected" [32], he went further and expressed his surprise "that some manifestation of atomic energy is not a part of it." The addition of an energy parameter would, he claimed, solve a number of puzzling and infrequently addressed features of the periodic table [33]. Such a new addition to the periodic table could be expressed as a new dimension—one that is presently lacking only because of historical accident. What should the new dimension be? Allen announced that the average one-electron valence shell energy of the ground-state free atom was the "missing third dimension." He depicted his three-dimensional version of the periodic table as shown in Fig. 6, with the height of each element representing its energy. The premises on which this periodic table is based, however, are less than robust. In the first place, there has been no shortage of commentators who wished to reform the periodic table during its long history [27, p 257]. Secondly, three-dimensional periodic tables are not a novelty; one very similar to Allen's table was published in 1964 [34]. Thirdly, the choice of the one-electron valence shell energy is unfortunate because, as we pointed out in discussing our seventh fact, approximate quantum theory is not able to predict the ground-state electronic configuration of atoms and so cannot provide a reliable foundation for the periodic table.

Allen's notions can also be criticized on other grounds. Scerri [35], for instance, has argued that the success of the periodic table in no way depends on the quantum-chemical description of matter. Modern physics has not fundamentally altered the periodic table. Scerri further maintained [35] that Allen "is not seriously proposing a 3D representation of the periodic system but is merely punning on the word dimension." Our 10th and final remarkable fact concerning the periodic table is thus that it is complete, apart from the eventual addition of a number of super-heavy elements when these are synthesized. Although it is quite within the realms of possibility that other ingenious schemes will be devised in the future to "complete" the periodic table—either by the addition of new dimensions or by other devices—none is likely to have a lasting impact. In fact, it will take no less than a radical upheaval in our current understanding of atomic phenomena, or perhaps the introduction of profoundly new ideas on the nature and structure of matter, to bring about any significant changes in the periodic table.

Concluding Remarks

In spite of numerous efforts over the years to reform, to radically alter, or even to abandon the periodic table entirely, it has survived. It has stood the test of time and has succeeded in coming through virtually unscathed. Is it now on the way to becoming a scientific icon? No, and it is unlikely that it will, for there is at present little unity about such things as its precise shape and structure. Even its theoretical foundations are still disputed. While such rumblings will doubtless

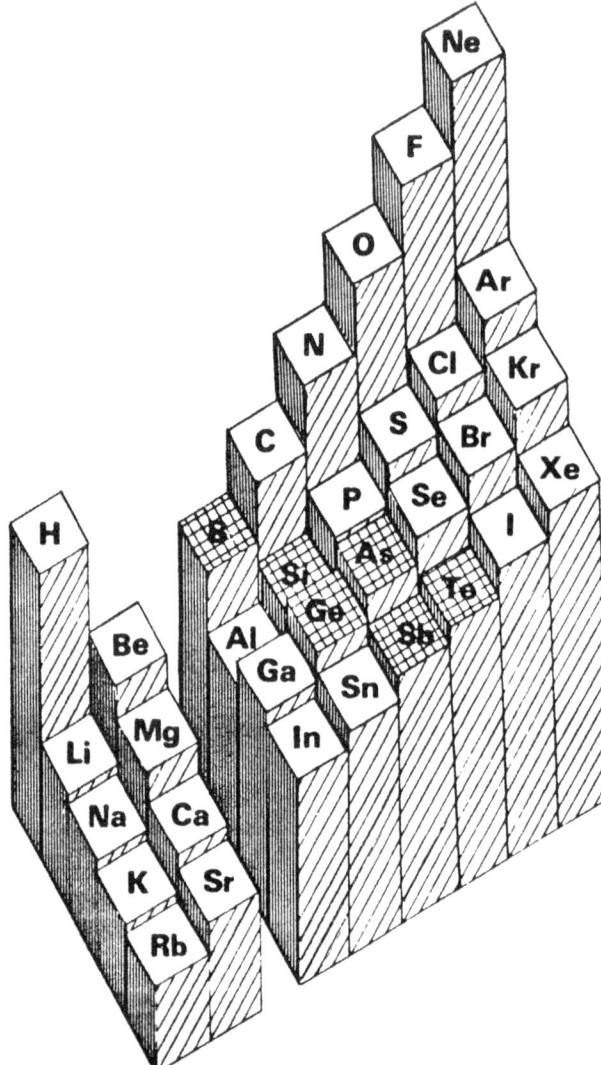

Fig. 6 A representation of the periodic table showing a third dimension, as advocated by Leland Allen. The third dimension is basically one of energy and derived from the averaged ionization potentials of the s and p electrons in the valence shell of the atoms. (Reproduced from *J. Am. Chem. Soc.* **1989**, *111*, 9004)

The periodic table was of even greater significance for the distinguished Italian author Primo Levi. He was sufficiently moved by it to write an entire book, called simply *The Periodic Table* [37], in which each element formed the basis of a short story. The stories were drawn from his own extraordinary life. The element cerium, for instance, caused him to recount his harrowing experiences as a prisoner in the Auschwitz concentration camp during the Second World War. His very survival was at stake, and it was only his training as a chemist and knowledge of the periodic table that saved him. If ever the existence of the periodic table was justified, it was here.

References

1. Newlands, J. A. R. *Chem. News (London)* **1866**, *13*, 113.
2. Emsley, J. *New Sci.* **1985**, *105* (March 7), 32.
3. Fernelius, W. C.; Powell, W. H. *J. Chem. Educ.* **1982**, *59*, 504.
4. Döbereiner, J. W. *Ann. Phys. Chem.* **1829**, *15*, 301.
5. Wurzer, F. *Ann. Phys. (Gilbert)* **1817**, *56*, 331.
6. Lavoisier, A. L. *Traité Elémentaire de Chimie*; Cuchet: Paris, 1789.
7. Guyton de Morveau, L. B. In *Encyclopédie Méthodique*; Pankoucke: Paris. 1782.
8. Rouvray, D. H. *Endeavour (New Series)* **1977**, *1*, 23.
9. Quoted in Hartley, H. *Studies in the History of Chemistry*; Clarendon Press: Oxford, **1971**; p 85.
10. Rouvray, D. H. *Chem. Br.* **1994**, *30*, 373.
11. Kibler, M. *J. Mol. Struct. (Theochem)* **1989**, *187*, 83.
12. Brock, W. H. *The Fontana History of Chemistry*; Fontana Press: London, **1992**; p 321.
13. Shapere. D. In *The Structure of Scientific Theories*; Suppe, F., Ed.; Illinois University Press: Chicago, **1974**; p 535–537.
14. Mendeléeff, D. I. *J. Chem. Soc. (London)* **1889**, *55*, 634.
15. Quoted in Freund, I. *The Study of Chemical Composition*; Dover Publications: New York, 1968; p 500.
16. Grant, G. M. R. *Chem. Br.* **1994**, *30*, 388.
17. Scerri, E. R. *Chem. Br.* **1994**, *30*, 379.
18. Paneth, F. A.; von Hevesy, G. *Sitzung. Kaisl. Akad. Wiss. (Vienna)*, *Abt. 2A* **1914**, 122, 1037.
19. Mazurs, E. G. *Graphic Representations of the Periodic System during One Hundred Years;* University of Alabama Press: Tuscaloosa, Alabama, 1974.
20. Jensen, W. B. In *Symmetry: Unifying Human Understanding;* Hargittai, I., Ed.; Pergamon Press: New York, 1986; p 487.
21. Jørgensen, C. K. *Angew. Chem. Int. Ed. Engl.* **1973**, *12*, 12.
22. Löwdin, P.–O. *Int. J. Quant. Chem.* **1969**, *3*, 331.
23. Longuet-Higgins, H. C. *J. Chem. Educ.* **1957**, *34*, 30.
24. Vanquickenborne, L. G.; Pierloot, K.; Devoghel, D. *J. Chem. Educ.* **1994**, *71*, 469.
25. Cooper, D.G. *The Periodic Table*, 4th ed.; Butteworths: London, 1968; p 10.
26. Jensen, W. B. *J. Chem. Educ.* **1982**, *59*, 634.
27. van Spronsen, J. W. *The Periodic System of Chemical Elements: A History of the First Hundred Years*; Elsevier: Amsterdam, 1969; p 107.
28. Fluck, E. *Pure Appl. Chem.* **1988**, *60*, 431.
29. Loening, K. L. *J. Chem. Educ.* **1984**, *61*, 136.
30. News Report, *Sci. News* **1994**, *146*, 271.
31. Allen, L. C. *J. Am. Chem. Soc.* **1989**, *111*, 9003
32. Allen, L. C. *Croat. Chem. Acta* **1991**, *64*, 389.
33. Allen, L. C. *J. Am. Chem. Soc.* **1992**, *114*, 1510.
34. Strong, L. E., Ed. *Chemical Systems*; McGraw-Hill: St. Louis, Missouri, 1964; back cover.

continue well into the future, it seems probable that they will be confined mainly to specialist protagonists. For the rest of us, indeed for the chemical community at large, the periodic table will stand as a magnificent summary of the myriad disparate facts of chemistry. It will remain an indispensable tool to scientists and a powerful source of inspiration to those gifted with creative poetic vision. On first having the periodic table explained to him, the British author and novelist C.P. Snow responded [36] in terms of rapture:

> For the first time I saw a medley of haphazard facts fall into line and order. All the jumbles and recipes and hotchpotch of the inorganic chemistry of my boyhood seemed to fit themselves into the scheme before my eyes—as though one were standing beside a jungle and it suddenly transformed itself into a Dutch garden.

35. Scerri, E. R. *J. Phys. Chem.* **1993,** *97,* 5786.
36. Snow, C. P. *The Search*; Scribner: New York, 1958; p 27.
37. Levi, P. *The Periodic Table.* Translated by R. Rosenthal, Abacus Books: London, 1985.
38. Stedman, D. F. *Can. J. Res.* **1947,** *25B*, 199.
39. Treptow, R. S. *J. Chem. Educ.* **1994,** *71,* 1007.
40. Sibaiya, L. *Am. J. Phys.* **1941,** *9,* 122.
41. Crookes, W. *Chem. News* **1898,** *78, 25.*

The Afterlife of George C. Pimentel[a]

Jeanne (Mrs. George) Pimentel[b]

Biographical sketches refer to George Pimentel's 40 years of teaching at the University of California, Berkeley; his cutting edge research—the first chemical laser, the matrix isolation technique; his three years' service as Deputy Director of the National Science Foundation, and many more years of service on numerous committees; his visionary and effective leadership of projects of such long-lasting influence as the ChemStudy high school course and *Opportunities in Chemistry* ("The Pimentel Report"); and his many awards and honors. People who came in contact with him knew more than all that. We remember his energy and humor, confidence and humility, generosity and stubbornness; his demanding standards and his sentimentality; above all, his enormous enthusiasm for life. And we have been changed by it.

Pimentel's office always looked chaotic, but he could (almost) always find what he wanted. The champagne bottles testify to celebrations, mostly for his students' Ph.D.'s. (Photographer unknown; all photographs are from the Pimentel Family Archives).

"He went to the ball park every day, and he let them know he came to play" was his chosen epitaph. It wasn't written down anywhere but had stayed in my mind since he had tossed it out light-heartedly many years before his death. I imagined it was a quote from someone like Lou Gehrig or Babe Ruth, but a search turned up no source. I now think he made it up himself; it was certainly apt. George was poet and performer, as well as researcher and educator, athlete and family man, public servant and adventurous romantic. He was a passionate player in everything he attempted.

George had no doctrinal religious beliefs, though he was tolerant and respectful of other people's (with one hippocratic condition, that they do no harm). So he didn't believe in an afterlife. Philosophy had always interested him, and he had his own personal credo, which changed relatively little from college days to the end of his life—except for some moderation in his idealism brought on by experience. George used to say he believed he would die young, and in fact when we met he seemed mildly surprised to have reached the age of 47; to me, he seemed to regard life as an unexpected gift. So he faced an early death from cancer relatively serenely, feeling gratitude for a full and exciting life. And he died knowing that he had made some difference in other people's lives, and that was all he asked of eternity. He couldn't help but know it was true, in spite of his modesty, because—fortunately—through the years some of the people whom he had affected expressed their appreciation to him personally. One student wrote immediately after taking his Freshman Chemistry course:

> Thank you not only for the chemistry learned, but also for sharing your approach to science. I shall long remember the idea which you stressed from the first lecture onwards: that there are other objectives, besides the vocational, to be gained through the study of chemistry, such as an understanding of man's place in the universe and an appreciation for man's accomplishments.

Another wrote after being in medical practice for seven years:

> Chemistry One was by far the most challenging and exciting course of my academic years…and at times I had an almost spiritual feeling that you had led me closer to an understanding of our world…I responded very intently to your inspirational and demanding teaching.

One fine arts major showed her appreciation of Chem 1A by sending George a poem she had written about molecules; she later married one of his postdocs, and during George's illness they each wrote to him:

> *She:* George, the fact that you valued my opinion—as uninformed and sophomoric as it must have been—made me feel that I might count for something, after all…I realize what I might have missed in life had you not been there.

[a]*Chemical Intelligencer* 1996(3), 53–58.

[b]754 Coventry Road, Kensington, CA 94707-1404, USA
e-mail: Jpim@aol.com

B. Hargittai and I. Hargittai (eds.), *Culture of Chemistry: The Best Articles on the Human Side of 20th-Century Chemistry from the Archives of the Chemical Intelligencer*, DOI 10.1007/978-1-4899-7565-2_40, © Springer Science+Business Media New York 2015

He: You showed concern, gave me support, and encouraged me. Your support of my early efforts at teaching led me to a career which I thoroughly enjoy. You also seemed to sense that C—and I were made for each other.

A student in Pimentel's popular freshman chemistry course recalls: "He never put us to sleep with slides—he covered all the blackboards in PSL and you had to keep up with him." (Credit:© Dennis Galloway).

George realized that these letters represented the appreciation of scores more who never expressed it. He regretted the lack of time and the weight of numbers that prevented him from giving personal attention to more students, but he would always respond to a call for help, spoken or not. I remember a student coming to his office to discuss her grade and ending up crying on his shoulder about the death of a parent; I remember him going to the apartment of a grad student at 7:30 a.m. to rouse a talented young man who had a phobia about taking finals.

George was opportunistic about helping other people besides his own students. When he discovered his auto mechanic had been a disillusioned physics major, he somehow got him to become a graduate student in chemistry, a field in which he now excels. He helped my daughter's high-school-dropout boyfriend to master his night-school calculus homework and become a qualified computer technician.

Friends and colleagues have related other typical incidents.

George and I found ourselves flying home together across the Atlantic, and Pan Am had put us in a three-seat row together with a personable young American teacher who had been enjoying her summer holiday in Europe. Naturally, as a matter of symmetry—group theory was the key to molecular spectroscopy after all—we shifted seat assignments so that she was seated between us. And what had we been over in Denmark for? For a meeting on molecular spectroscopy. And what was molecular specstrop—what was that word? At which

George and I eyed each other across the young lady, and George began: "Well, you've seen a rainbow of course?"—What better way to spend the time of the interminable Atlantic crossing than to correct an area of ignorance in the mind of a young teacher? Our trip became possibly the most high-level private tutorial in the history of one-on-one, or rather two-on-one, instruction. I believe she enjoyed it, and learned something; I know I enjoyed it, and I learned something about presenting spectroscopy "from scratch," and also something about George Pimentel.

[When George was visiting our university,] after a full schedule of visits with appropriate groups, and after George had given a good seminar and we had enjoyed good discussions, the afternoon drew toward the dinner hour; and he and I went down tired but happy to my lab area—good spectroscopic labs are always in the basement—to pick up our papers and our gear. And there we encountered one of my graduate students; naturally I presented him to our distinguished visitor. And as they met, I saw the reinflation of a late-afternoon bone-weary Pimentel into a curious scholar, a colleague to my student, asking with simulated eagerness what he was working on, and what was developing, and could he hold the cell and look into it (George, who had designed and constructed and fine-tuned a dozen far more sophisticated cells!)—and the eagerness wasn't simulated; as teachers from Socrates on have known, the best way to teach is to lead the student into explaining matters out of his own knowledge: the three of us shared the joy of science, and the student learned. And the insatiable and uncontrollable eagerness to teach was, then and always, an integral part of George Pimentel.

Pimentel always had time for a student of any age who wanted to learn. (Credit: Kenneth Hodges/Saturday Science Academy, Clark Atlanta University).

Through George, my own life changed and improved in a very practical way that sustains me beyond his death. He supported me in achieving, in middle age, the college education denied me in my youth. I believe he held the devastating cancer at bay just long enough to see me graduate with the degree that gave me a new and fulfilling career; he died three weeks later. Furthermore, he unconsciously instilled in me, an English major, some of his own passion for science: something that I know now will never leave me.

I came to know well both his impulsive generosity to strangers and his deep loyalty to friends: often when I met him at the airport on his return from a grueling trip, he would have offered a ride far out of our way to a lonely traveler; once when an unfortunate old friend was released from a long jail sentence, George flew across the country and drove a hundred miles to be there when he got out.

As he achieved recognition, he cared less for the honors heaped upon him than for the opportunities it gave him to reach more people and teach them about science—and to enjoy new adventures. As Deputy Director of the NSF, he jumped at the opportunity to go to Antarctica in an uncomfortable military transport plane and get out on the ice with fellow scientists. He had a veritable lust for action. In 1967 he applied to become an astronaut, mostly because he felt scientist-astronauts, not "hot-shot Charlies," would provide the most benefit to posterity, but partly for the sheer physical challenge and risk. In spite of being overage, he excelled in the rigorous trials and was accepted in the program but was told he would never fly because of a minor imperfection on his retina. He declined such passive participation, preferring to return to his own research. (In fact, he did get to go to Mars, vicariously, when the infrared spectrometer designed in his lab was part of the payload of Mariner 6 & 7.) His enthusiasm for science prevailed over adventure early in his career. Having just missed the action of World War II, he made himself useful at the Office of Naval Research in Washington, D.C., urging peaceful uses of nuclear energy. He was invited to attend the nuclear tests at Bikini Island in the South Pacific, but the schedule conflicted with his plans to enter graduate school and science won out; he returned to Berkeley.

George's success in life following a rather underprivileged youth gave him the joyful experience of succeeding against odds, and he wanted others to have that satisfaction: always in sports, and often in life, he rooted for the underdog. In addresses to graduating students, he sometimes gave "inspirational advice on how best to succeed" that reflected his attitude:

Pimentel became a father while in graduate school. His three daughters by his first wife, Betty, were a constant joy to him. (Credit: Lorraine Reid, courtesy Betty Pimentel).

My formula consists of three elements: Work hard—be smart—and be lucky. Now I realize that it isn't easy to plan to be lucky, so you have to take that element as it comes. My own experience is that it isn't so easy, either, to plan to be smart. But the third element you can do something about. You can find something you think is worth doing, something that you enjoy doing, and then you can give it your best effort. I guarantee that it will work for you as well as it has for me. I find that whatever rewards came to me from working hard, I treasured above all other things. And while I was working hard at things I liked to do, every now and then I would inadvertently do something smart, enhancing still more the pleasure of the outcome. And finally, inevitably, every now and then, something lucky is bound to happen.

Athletics were always part of Pimentel's life. He played squash with his colleagues, football and softball with his students, volleyball at picnics. (Credit: James Hansen, courtesy LOOK Magazine).

He concluded by predicting an eventual benefit from their college years: "You will see how this experience permitted you to measure yourself and to reach down deep inside for your very best." Which is what George did, always, whether grappling with a problem in science, fielding a baseball, or coping with the disaster of terminal illness.

Although he valued greatly the serious respect and admiration of his peers, he never sought adulation and he hated pomp and ceremony. (Though I'm sure he would have enjoyed the recognition of a Nobel Prize, he was honestly relieved at his near miss, knowing how much that honor takes over the recipient's life.) In Washington, D.C., he was known for his "California" attire, complete with cowboy boots and bolo tie. When an appointment on Capitol Hill necessitated a business suit, he would unroll his tie from his pocket on the way up in the elevator. Even in Washington our frequent parties at our home encouraged casual dress, and once when an embassy attaché arrived in a tuxedo George put a napkin over his arm and asked him to serve champagne. One exception to his informality: at Berkeley, he would regularly don cap and gown and march in the faculty procession at Commencement or on Charter Day to show his respect for tradition and his esteem for his beloved university—and by example impart that sense to students and the public.

Such adherence to principle was apparent throughout his life. He quit working on the Manhattan Project—his first job after graduating from UCLA—when he realized he was playing a sedentary role in the development of a weapon of mass destruction: that was all his adventurous spirit needed to decide to follow his pilot brother; he joined the Navy to take an active role in trying to end the war before the atom bomb was needed. An early opponent of the Vietnam War (he used to join his grad students in leafletting campaigns), he participated in a massive rally at the UCB Greek Theater sparked by the Cambodian invasion and the Kent State shootings but courageously disagreed with the majority of speakers—popular radicals who incited students to attack the university as representing the establishment. (George, considered a conservative on campus, advocated a more patient approach: impeaching the President of the United States!) Toward the end of his life, he crusaded to improve the image of chemistry and counteract the growing phobia against its toxic effects, so that humanity could better enjoy the beneficial effects of chemistry; though some saw his stance as antienvironmental, it was more truly antiextremist. Many years before, he had joined an appeal by distinguished scientists to the government to halt the development of the SST aircraft because of its damage to the ozone layer, and he was a member of the National Academy of Sciences committee that warned early on about the effects of chlorofluorocarbons. Though he welcomed the travel opportunities that science gave him, he always questioned their purpose: invited to South Africa by its government in 1971, he accepted only because he could give equal time to all racial groups. After many years of advocating changes in science

policy at the national level, he felt obliged to "put up or shut up" and accepted a presidential appointment as Deputy Director of the National Science Foundation under Jimmy Carter. (He had previously refused a position under Nixon, and he left when Reagan was elected—not because of their political party but because he was opposed to their policies.) He resigned from an advisory panel on "Star Wars," which made use of his brainchild, the chemical laser.

Part of his enjoyment of science was its collaborative nature, on both a large and a small scale. Interaction with international colleagues meant the excitement of sharing data that would advance the common quest for knowledge. He loved working with graduate students, encouraging them to come up with their own imaginative ideas, giving freely of his experience and perception, and willingly sharing the credit for success. He lamented the trend toward cutthroat competition for glory and would have been saddened by the current apparent increase in ethical lapses in the realm of scientific research.

However, George was certainly no saint—he had his share of human faults and failings. And, in spite of his popularity, he did have his detractors, and he himself disliked and disagreed with some individuals. But I never knew him to behave vindictively: if he couldn't just avoid antagonists, he would take them on in fair fight. One distinguished colleague with whom he worked long and hard on a blue-ribbon committee wrote:

> We didn't just admire him, we were all terribly fond of him as well. George has left an indelible mark on the physical sciences, on his students, collaborators and colleagues, and on national science policy. He amazed us all by his dedication and boundless energy and enthusiasm. What made him so special is that he managed to accomplish all this in a gentle way, without ever stepping on anyone's toes.

In his personal life, perhaps his three daughters came off second in the competition for his time—as happens to many such children—but they were always first in his heart, and they are quick to acknowledge his rich legacy to them, so well expressed by his daughter Jan.

> He cared so much for people that he very easily embraced them and included them in his own enormous family. At times this made life difficult for those of us who had to share him with all of you. And though we might have lost some things, we reaped countless other benefits from this arrangement. The door to my father's house was always open, and there was a continuous stream of stimulating and exciting company going through the door. . . . My father had an almost childlike zest for life….He got intense pleasure out of a Smokehouse hot dog, a root beer float, a tag at home plate, or seeing a big dog. He saw himself as an incredibly wealthy man….Perhaps this is the greatest lesson we can learn from George and perhaps it has nothing to do with chemistry.

His concern for his children was practical too—his version of a chicken in every pot was a Macintosh for every child. Although his brilliant scientific and mathematical mind easily comprehended the most complex mainframe computer, he delighted in the friendly home version and shared his enthusiasm with us all; he helped each one acquire a "Mac" as it became appropriate. His eldest daughter and my son currently earn their living as Mac experts. Though he inspired countless students to make science a career, none of the five children of our combined family has done so, though one enjoyed teaching science for a while, and another doesn't discount the eventual possibility: my son Vincent relinquished his original ambition to be on the cutting edge of research as he observed its demands on George; growing up in the same house gave him a certain perception regarding his stepfather. When I spoke of writing (or encouraging) some kind of biography of George, he felt that, for appeal beyond the family, it would need a theme—and he suggested one:

> It concerns his role as a now-rare kind of academic who felt immense privilege in being able to do his work and to be immersed in academe.

The "Pimental Report," commissioned by the National Academy of Sciences, was later revised for high school students and the general public. (Credit:© Andrée Abecassis).

Moreover, he had a deep sense of the importance of academic industry for society and championed this view quietly, relentlessly, and eloquently for many years, even when his opinions became frequently in the minority.

His views were interesting in part because his life straddled the great historical divide of World War II. Having come of age before the outcome of the war was known, he brought with him the values of the 1930s, an era when the aristocracy of the common man was being established as a replacement for the failings of the upper classes in decades previous.... There was a sense that it was possible to make great improvements if only the right action were taken.

I think that George took this to heart. He had his era's sense of faith in humanity's ability to improve itself. This, coupled with a modest background, a mother who provided an example of ebullience in the face of hard work, and an elder brother whom he admired enough to try to do as well as he did, provided George with a sense of mission about his work which he carried with him his entire life.

I'm not sure that George saw himself as a man with a mission. It was simply his way of responding to life. One of his favorite exclamations was simply "Ain't life grand?"

In a formal sense, George now lives on through his name in ways he would like: his name on the building at Cal where he gave so many of his popular freshman chemistry lectures; the scholarships and awards named for him at schools and colleges and, of course, the ACS chemical education award. But his spirit lives on in a much broader aspect of time and space. His research was concerned with the interaction of molecules, and I like to think of his spiritual molecules dispersing and recombining as ubiquitously and eternally as the water cycle, as his enthusiasm, diligence, optimism, and compassion affect those who knew him, and those that they will know. As Jan pointed out:

> Each of us has a great deal of power to effect change. Like a pebble dropped into a pond—the ripples spread out until the entire surface of the water is affected. My father somehow understood this. He touched things with love, with energy, with creativity, and with joy.

Acknowledgment The author gratefully acknowledges the help and inspiration of all George's students, colleagues, and friends who have contributed, directly or indirectly, to this article. She welcomes comments and stories about George for her personal enjoyment and the Pimentel archives.

George C. Pimentel—Professional Activities and Awards: Highlights

Prepared by Prof. Kenneth S. Pitzer, with assistance from Prof. C. Bradley Moore, Department of Chemistry, University of California at Berkeley, California, 94720

Dr. Pimentel's research was in the fields of infrared spectroscopy, chemical lasers, molecular structure, free radicals, and hydrogen bonding. His interests centered on the application of spectroscopic methods to the study of unusual chemical bonding. A major contribution was the development and exploitation of the matrix isolation method for the spectroscopic detection of highly unstable molecules. This involves stabilization of such molecules in a matrix of frozen inert gas, such as argon, at very low temperature to permit leisurely spectroscopic study. Application of this matrix isolation method led to the discovery of many unusual and highly reactive molecules that could not otherwise have been detected.

During studies of photochemical reactions, Dr. Pimentel and his students discovered the first chemically pumped laser. Flash photolysis methods on the microsecond time scale permitted the measurement, through laser emissions, of nascent population inversions produced in the normal course of a chemical reaction. A variety of chemically pumped vibrational and rotational lasers have been discovered in his laboratory, providing valuable state-to- state kinetic information.

Dr. Pimentel's pioneering development of rapid-scan techniques for infrared spectroscopy extended to the gas phase these spectroscopic studies of normally transient species. This work led to the design of a unique infrared spectrometer for the 1969 Mariner interplanetary spacecraft to determine the composition of the atmosphere of Mars.

In addition to over 200 scientific papers, Pimentel was the author or coauthor of 11 books. Three concerned his areas of research—the others, education at various levels. The book arising from the ChemStudy project for high schools, Chemistry—An Experimental Science, sold more than a million copies and was translated into 13 languages.

Dr. Pimentel received many awards. Those emphasizing education included the American Chemical Society Award in Chemical Education, now called the George C. Pimentel Award. Emphasizing research were the Wolf Prize in Chemistry that he received in 1983, the National Medal of Science in 1985, and the Robert A. Welch Award in 1986. The highest honor of the American Chemical Society is the Priestley Medal, which Pimentel received in 1989. Other honors included election to the National Academy of Sciences in 1966 and to the American Philosophical Society in 1985 and several honorary degrees.

In addition to his long service as Professor of Chemistry at the University of California, Berkeley, Pimentel was Deputy Director of the National Science Foundation in 1977–1980 and President of the American Chemical Society in 1986.

Reading Between the Lines[a]

Scientific papers are not always as objective as their authors suppose

Hugh Aldersey-Williams[b]

"C60: Buckminsterfullerene" [1] is an unusual paper. For a start, its title tells you what it is about only in the most oblique terms. As this announcement of the discovery of this 60-atom, spherically symmetric, chemically bonded molecular form of carbon proceeds, the authors engage in further games that not only reveal the facts of the discovery, but also convey a sense of its importance, and of the thrill they feel as they break their news.

Because of this, I thought it profitable to attempt what literary theorists call a deconstruction of the paper (published in *Nature* in November 1985 and written by Professor Harold Kroto of the University of Sussex and Professors Richard Smalley and Robert Curl of Rice University in Houston and their then research students Jim Heath and Sean O'Brien) in my recent popular book on the discovery of buckminsterfullerene [2]. This exercise revealed not exactly a hidden agenda, but it did show something of what was on the scientists' minds, their surprise and excitement at the discovery, their playful good humor, and their confidence in its significance.

I included this section in part because I wished to expose my readers to a "real scientific paper" and to demonstrate that such communications are not as impenetrable as is often thought. But I also wished to show that a "literary" technique could be applied to a "nonliterary" work. There is much current interest among literary theorists in applying their methods to scientific writing [3]. But, almost universally, these scholars have chosen to examine historic works of science where, thanks to the patina of age, it is easy to believe there is also literary intent.

Professor Gillian Beer of Cambridge University, for example, has made a detailed study of the writings of Charles Darwin [4], "In its imaginative consequences for science, literature, society and feeling, *The Origin of Species* is one of the most extraordinary examples of a work which included more than the maker of it at the time knew…," she writes in introduction. This over-inclusion of information is the temptation for such a project. But the extraordinariness of the chosen work is its weakness. *The Origin* is a book-length work, now more than a century old, written to be read not only by scientists but by educated people of all kinds.

For all these reasons, it is possible to see this work as literary as much as scientific in style. Beer discovers, for example, that Darwin "rearranges the elements of creation myths"—the Garden of Eden becomes the ocean, the tree of life becomes the evolutionary tree. At a time before scientific language had grown apart from everyday language, it was easy for such images and metaphors to pass from everyday parlance to scientific usage and back again, thus enhancing people's comprehension of the science and adding layers of complex meaning to the accounts of Darwin's simple observations. "His language is expressive rather than rigorous," Beer writes. But how exclusively rigorous can scientific language ever be? Expression is always there; language expresses thought. Scientific language is unavoidably full of literary devices, notwithstanding that it is sometimes the scientists who deny this.

Consider one of the most famous scientific communications of this century, the announcement in 1953 by James Watson and Francis Crick in *Nature* of the famous double helix [5]. This paper has an unprepossessing title: "Molecular Structure of Nucleic Acids." But from there on, modesty is cast aside. While many scientific papers uphold the ideal of supposed objectivity through anonymity, this one is written in the first person. Its first sentence could hardly he bolder: "We wish to suggest a structure for the salt of deoxyribose nucleic acid (D.N.A.)." This is the action of throwing down a gauntlet. They follow their opening declaration with a breathtaking understatement: "This structure has novel features which are of considerable biological interest."

The paper proceeds for the most part in the same tone, with many sentences beginning "We…." The great merit of the double-helix structure—literally its *raison d'être!*—was that it inherently solved the puzzle of self-replication. Watson and Crick knew this all too well and could not help but tease readers of *Nature* with the concluding litotes: "It has not escaped our notice that the specific pairing we have postulated immediately suggests a possible copying mechanism for the genetic material."

---

[a]*Chemical Intelligencer* 1996(4), 37–41.

[b]33 Hugo Road, London N19 5EU, UK

B. Hargittai and I. Hargittai (eds.), *Culture of Chemistry: The Best Articles on the Human Side of 20th-Century Chemistry from the Archives of the Chemical Intelligencer*, DOI 10.1007/978-1-4899-7565-2_41, © Springer Science+Business Media New York 2015

In order to further demonstrate the applicability of literary analysis to contemporary works of science, let us deconstruct the companion paper to Kroto et al.'s 1985 announcement, Wolfgang Krätschmer, Lowell Lamb, Kostantinos Fostiropoulos, and Donald Huffman's 1990 paper, once again in *Nature,* describing how C_{60} was obtained in quantity for the first time [6]. This paper has certain conscious (and some perhaps unconscious) parallels with the earlier paper as well as other elements of subtext. It is a long paper, so I have selected only parts of the paper for analysis.

In its formal composition, the title has echoes of "C_{60}: Buckminsterfullerene," although, by now, none of its mystery:

Solid C_{60} : A New Form of Carbon

The first sentence of the abstract contains a number of claims to novelty.

> A new form of pure, solid carbon has been synthesized consisting of a somewhat disordered hexagonal close packing of soccer-ball-shaped C_{60} molecules.

The bald statement comes with the words *new form.* The word *solid* gives us the first new information (the room-temperature physical state of buckminsterfullerene had been unknown and presumed perhaps to be liquid). The other chosen adjective, *pure,* is gilding the lily with tautology: a new form of any element must, by definition, be pure.

The choice of *synthesized* places a further tax on our credulity. Krätschmer et al. presumably wish to distance their facile and copious technique for making C_{60} from the hopelessly impractical method of Smalley's cluster beam apparatus. This is not a synthesis in terms that an organic chemist would recognize. But it is perhaps forgivable hubris from physicists who have made an astonishing chemical discovery. The remainder of the sentence is uncontroversial description; note the choice of the adjective *soccer-ball-shaped* in place of anything more "scientific," again an echo of the Kroto paper, in which a soccer ball was used as a photographic illustration of the proposed C_{60} structure.

The abstract continues:

> Infrared spectra and X-ray diffraction studies of the molecular packing confirm that the molecules have the anticipated "fullerene" structure. Mass spectroscopy shows that the C_{70} molecule is present at levels of a few per cent. The solid-state and molecular properties of C_{60} and its possible role in interstellar space can now be studied in detail.

In the last sentence, the *now* is significant, with its overtone of "at last." It has been a trying five-year wait to get to this stage, and it is Krätschmer and Huffman's breakthrough, as much as Kroto and Smalley's, that will permit the resolution of these larger questions.

The paper itself begins as follows:

> Following the observation that even-numbered clusters of carbon atoms in the range $C_{50}-C_{100}$ are present in carbon vapour,[1] conditions were found[2-4] for which the C_{60} molecule could be made dominant in the large-mass fraction of vapourized graphite.

The authors are scrupulous in referring to the sequence of events surrounding the discovery of C_{60}, conscious no doubt of the controversy that had erupted since 1985. Reference 1 is to work by researchers at Exxon who in 1984 had recorded mass spectra showing 60 carbon atoms as the dominant peak but had not singled that fact out for further examination. Reference 2 is to Kroto et al.'s *Nature* paper in 1985, and the third and fourth references are to further work in Smalley's laboratory establishing the stability of fullerene ions.

After sentences describing the torrent of calculation and speculation based upon the proposed truncated icosahedron structure unleashed by the 1985 discovery, Krätschmer and Huffman bring their readers back to reality and the merit of their discovery, which is to make possible measurement in place of mere theorizing:

> Until now, it has not been possible to produce sufficient quantities of the new material to permit measurement of the physical properties, to test the theoretical calculations, or to evaluate possible applications.

It is comparatively rare in scientific papers to advertise the novelty they contain so blatantly. But here, *Until now* makes it very clear what is new.

The tone of the paper up to this point has been impersonal. Those responsible for work described are not identified except by citation. In the subsequent paragraph, the tone changes abruptly to the first person:

> Some of us have recently reported evidence[21,22] for the presence of the C_{60} molecule in soot condensed from evaporated graphite.... Here we report how to extract the carrier of the features from the soot, how to purify it, and evidence that the material obtained is in fact primarily C_{60}.

These citations are to earlier papers by Krätschmer, Fostiropoulos, and Huffman in which they had tentatively advanced the notion that certain features of spectra of laboratory-produced carbon dust might be explained by the new molecule. Without the full evidence they needed, they had confined these thoughts to obscure and slow-publishing media.

Although the authors say they will proceed in effect to give the recipe for C_{60}, they are less than detailed:

> The starting material for our process is pure graphitic carbon soot...with a few per cent by weight of C_{60} molecules, as described in refs 21, 22. It is produced by evaporating graphite electrodes in an atmosphere of ~100 torr of helium. The resulting black soot is gently scraped from the collecting surfaces …

And that's it. That is the complete extent of the recipe that Krätschmer and his colleagues give—or are willing to give—for producing a material whose manufacture had frustrated groups around the world working with the most sophisticated apparatus. The brevity of the recipe emphasizes its simplicity and, in so doing, serves to magnify the scale of the physicists' break through.

There then follows a description of the purification of the C_{60}-rich soot by means of recrystallization from various solvents, a procedure that these physicists clearly view with some distaste. With relief, they add that there is a "physical" rather than "chemical" (i.e., liquid-phase) remedy:

> An alternative concentration procedure is to heat the soot to 400 °C in a vacuum or in an inert atmosphere, thus subliming the C_{60} out of the soot...

But even this requires a preparatory washing of the initial soot in ether to remove the ubiquitous hydrocarbons...

The authors press on with empirical descriptions of the behavior of the first solid C_{60}.

> Thin films and powder samples of the new material can be handled without special precautions and seem to be stable in air for at least several weeks, although there does seem to be some deterioration with time for reasons that are as yet unclear. The material can be sublimed repeatedly without decomposition. Using the apparatus described, one person can produce of the order of 100 mg of the purified material in a day.

Chemists might have speculated about reasons (oxidation? photodissociation?) for the deterioration observed. These authors, however, dwell upon the physical stability of their new substance, saying that it *can be sublimed repeatedly without decomposition.* The most significant comment comes in the last sentence of the paragraph, *Using the apparatus described...*, which lays claim to the unique efficacy of this apparatus for the production of purified C_{60}, a presage to moves to patent the process.

The following paragraph offers further description:

> Studies by optical microscopy of the material left after evaporating the benzene show a variety of what appear to be crystals—mainly rods, platelets and star-like flakes.

This poetic final phrase hints at the hexagonal symmetry:

> ...All crystals lend to exhibit six-fold symmetry. In transmitted light they appear red to brown in colour; in reflected light the larger crystals have a metallic appearance whereas the platelets show interference colours. The platelets can be rather thin and are thus ideally suited for electron-diffraction studies in an electron microscope...

And indeed, the electron diffraction (and X-ray diffraction) patterns that accompany this statement are to be the principal new evidence for the C_{60} structure both as molecule and in bulk. The following section of the paper, however, focuses on what historically has been the only undisputed means for characterizing C_{60}: mass spectroscopy. It begins:

> The material has been analysed by mass spectrometry at several facilities.

Why *several facilities?* It does no harm to duplicate one's results of course, but this is not what is going on here. This is public (albeit coded) revenge taken by Huffman over a technician who wanted his name added to the paper simply for recording the spectrum. His request was refused, his spectrum omitted, and an earlier, poorer spectrum of Krätschmer's published instead.

The results confirm the constitution of the material as substantially C_{60}. They reveal that there is about 10% of C_{70} in a typical sample and that this proportion can be reduced by different means of taking the spectra, much in line with observations made during the laser vaporization experiments of Smalley's group. The section closes with a promise that

> Further details of the mass spectroscopy of the new material will be published elsewhere.

What further details? Where? And why not now? This statement presumably reflects Huffman's wish still to publish the spectrum recorded by his technician.

With this confirmation out of the way, the authors proceed to do what has not been done before—to characterize C_{60} for the first time as a bulk material distinct from the discrete molecules made by Kroto and Smalley:

> To determine if the C_{60} molecules form a regular lattice, we performed electron and X-ray diffraction studies on the individual crystals and on the powder....From the hexagonal array of diffraction spots..., a d spacing of 8.7Å was deduced corresponding to the (100) reciprocal lattice vector of a hexagonal lattice.

The discussion continues with the deduction of the nearest-neighbor distance of the buckyballs and a calculation of the density of the bulk material. It is clear that

> ...the C_{60} molecules seem to assemble themselves into a somewhat ordered array as if they are effectively spherical, which is entirely consistent with the hypothesis that they are shaped like soccer balls.

This formulation, *is entirety consistent with the hypothesis,* rather than simply "confirms," echoes several papers of Smalley's group which were able to add layers of circumstantial evidence that C_{60} was spheroidal without quite being able to produce unequivocal proof. By now, however, there is no reasonable doubt about the shape of the individual molecules, and Krätschmer and Huffman concern themselves with the bulk material:

> In summary, our diffraction data imply that the substance isolated is at least partially crystalline. The inferred lattice constants, when interpreted in terms of close-packed icosahedral C_{60}, yield a density consistent with the measured value. Further evidence that the molecules are indeed buckminsterfullerene and that the solid primarily consists of these molecules comes from the spectroscopic results.

The following section assembles infrared and ultraviolet spectroscopic results broadly in agreement with those previously observed and predicted by theory.

There then follows a discussion under the heading

Possible interstellar dust

The original stimulus for the work[2] that led to the hypothesis of the soccer-ball-shaped C_{60} molecule, buckminsterfullerene, was an interest in certain unexplained features in the absorption and emission spectra of interstellar matter.

By citing Kroto et al. of 1985 once again at this point, these authors indicate a shared interest and motivation.

These include an intense absorption band at 217 nm which has long been attributed to small particles of graphite[31], a group of unidentified interstellar absorption bands in the visible that have defied explanation for more than 70 years[31,32], and several strong emission bands attributed to polycyclic aromatic hydrocarbons[33,34].

Reference 31 is to Huffman's own 1977 work on dust particles. It establishes his independent credentials and shows that he has a perspective on the subject that predates Kroto and Smalley's discovery. Now comes the crunch:

Based on the visible and infrared absorption spectra..., we do not see any obvious matches with the interstellar features.

This is a disappointment. The authors do what they can to leave the door open:

The ultraviolet band at 216–219 nm has a similar peak wavelength to an interstellar feature, although the other strong bands of the spectrum have no interstellar counterparts. As the influence of C_{70} absorptions on the spectrum is not yet known, a conclusive comparison with the 217-nm interstellar band is difficult. We note that the visible-ultraviolet spectrum presented here is characteristic of a solid, rather than of free molecules. In addition, these new results do not relate directly to absorption in the free C_{60}^{+} molecular ion, which has been envisaged[19] to explain the diffuse interstellar bands. Nevertheless, these data should now provide guidance for possible infrared detection of the C_{60} molecule, if it is indeed as ubiquitous in the cosmos as some have supposed.

Reference 19 is to a conjecture of Kroto's published in 1987. The supposition of cosmic ubiquity is Kroto's, too. Kratschmer and Huffman are gentle in their disparagement. They proceed to their own, down-to-earth summary:

To our method for producing macroscopic quantities of C_{60}, we have added a method for concentrating it in pure solid form. Analyses including mass spectroscopy, infrared spectroscopy, electron diffraction and X-ray diffraction leave little doubt that we have produced a solid material that apparently has not been reported previously.

The authors take care to separate the methods of production and purification, perhaps preparing the ground for patent applications. Their scientific discovery, the isolation of a new substance, they apparently regard as secondary to these technologies. They are nevertheless keen to lay ownership to the new solid material as well:

We call the solid fullerite as a simple extension of the shortened term fullerene, which has been applied to the large cage-shaped molecules typified by buckminsterfullerene (C_{60}).

The name *fullerite* is well chosen. It is easy to say. It sounds natural. Indeed, it sounds like a mineral. Others doubted that the solid bulk material warranted its own name. Unlike "fullerene," it has not caught on.

The various physical and chemical properties of C_{60} can now be measured and speculations concerning its potential uses can be tested.

In contrast to Kroto and Smalley five years earlier, Krätschmer and Huffman forbear to offer their own speculations. By now, there is no shortage of these.

References

1. Kroto, H. W.; Heath. J. R.: O'Brien. S. C.; Curl, R. F.; Smalley, R. E. *Nature* **1985**, *318*, 162.
2. Aldersey-Williams, H. *The Most Beautiful Molecule: An Adventure in Chemistry*; Aurum: London, 1994; *The Most Beautiful Molecule: The Discovery of the Buckyball*; Wiley: New York, 1995.
3. See, for example, Locke. D. *Science as Writing*; Yale University Press: New Haven, Connecticut, 1992; Gross, A. *The Rhetoric of Science*; Harvard University Press: Cambridge. Massachusetts, 1990.
4. Beer, G. *Darwin's Plots: Evolutionary Narrative in Darwin, George Eliot and Nineteenth Century Fiction*; Ark: London, 1983.
5. Watson, J. D.; Crick. F. H. C. *Nature* **1953**, *171*, 737.
6. Krätschmer, W.; Lamb, L. D.; Fostiropoulos, K.; Huffman, D.R. *Nature* **1990**, 347, 354.

Captain Nemo's Battery: Chemistry and the Science Fiction of Jules Verne[a]

William B. Jensen[b]

The Father of "Scientifiction"

My high school English teacher always insisted that the first prerequisite of a good essay was a catchy title, and I flatter myself that my choice for this literary excursion is not half bad. Thus, it is with a great deal of reluctance that I must also immediately confess that it is misleading—misleading because of the existence of two very common and widespread myths about Jules Verne (Fig. 1).

The first and most fundamental of these is the myth that Verne wrote science fiction—indeed that he not only wrote it but actually invented the genre. This misconception appears to be due to none other than Hugo Gernsback, who in April of 1926 began publication of *Amazing Stories,* America's first science fiction pulp magazine. In his introductory editorial, Gernsback explained exactly what he meant by the kind of literature that he called "scientifiction"—a rather unmelodious term that has happily disappeared from the English lexicon [1]:

> By scientifiction I mean the Jules Verne, H.G. Wells and Edgar Allan Poe type of story—a charming romance intermingled with scientific fact and prophetic vision.

Later, Gernsback would single Verne out from this trio as the "patron saint" of the genre, and the masthead of the magazine would carry a drawing of Verne's tomb at Amiens as a symbol of his everlasting "immortality."

However, as Arthur Evans has shown in his book *Jules Verne Rediscovered,* Verne never wrote science fiction—or at least not science fiction as the term is now understood [2]. There are no alien monsters, no mysterious superforces, no time travel, no magic materials, and no heroines in skimpy futuristic attire in his novels. Rather, his works are a part of a tradition of French didactic writing known as the so-called "scientific novel" and were intended as a way of painlessly popularizing science for the lay public. They use an adventure story, combined with novel but not improbable applications of existing technology, as a framework into which are inserted sizable digressions on the facts of zoology, botany, geography, astronomy, physics, and occasionally even chemistry. In the course of his life, Verne would write over 60 of these novels.

The second common myth is that Verne wrote primarily for children and young adults. This is due to the fact that most English translations of his works have been butchered, with many of the didactic digressions on science—their very raison d'être—having been either deleted or shortened to the point of becoming incomprehensible [3], It was only in the 1970s that Walter James Miller began publishing restored and annotated editions of some of Verne's classics, and it was the reading of Miller's restored edition of *Twenty Thousand Leagues under the Sea* [4] that first awoke my interest in Verne's use of chemistry.

Twenty Thousand Leagues Under the Sea

First published in 1870, the novel opens with reports of a strange sea monster that has been terrorizing shipping in both the Atlantic and Pacific oceans. The famous French scientist Professor Aronnax (Fig. 2), who has been visiting the United States accompanied by his trust servant, Conseil, agrees to join a U.S. expedition to hunt down the monster. As a result of the expedition's first encounter with the creature, Aronnax, Conseil, and a Canadian harpooner named Ned Land are thrown overboard and become the uninvited guests of Captain Nemo aboard his submarine, the *Nautilus,* which is, of course, the source of the reports of the sea monster.

The first digression on chemistry comes when Aronnax awakens after his first night as a prisoner on the *Nautilus* [4]:

> I breathed with difficulty. The heavy air seemed to oppress my lungs. Although the cell was large, we had evidently consumed a great part of the oxygen it contained. Indeed, each man consumes, in one hour, the oxygen contained in more than 176 pints of air and this air, charged with a nearly equal quantity of carbonic acid [carbon dioxide], becomes unbreathable. It became necessary to renew the atmosphere of our prison and no doubt of the whole submarine boat. That gave rise to a question in my mind. How would the commander of this floating dwelling proceed? Would he obtain air by chemical means, in getting by heat the oxygen contained in chlorate of potas [potassium chlorate], and in absorbing carbonic acid by caustic potash [potassium hydroxide]?

[a]*Chemical Intelligencer* 1997(2), 23–32.

[b]Department of Chemistry, University of Cincinnati, Cincinnati, OH 45221, USA

B. Hargittai and I. Hargittai (eds.), *Culture of Chemistry: The Best Articles on the Human Side of 20th-Century Chemistry from the Archives of the Chemical Intelligencer,* DOI 10.1007/978-1-4899-7565-2_42, © Springer Science+Business Media New York 2015

Fig. 1 Jules Verne (1828–1905) at age 76. (All illustrations are from the Oesper Collection in the History of Chemistry).

In other words, Aronnax is proposing the use of the following standard reactions as a means of maintaining the air quality aboard the *Nautilus* [5]:

$$2KClO_3(s) + heat \rightarrow 2KCl(s) + 3O_2(g) \qquad (1)$$

$$2KOH(aq) + CO_2(g) \rightarrow K_2CO_3(aq) + H_2O(l) \qquad (2)$$

In the end, however, he decides that this chemical scheme is impractical and that surfacing every 24 hours, like a whale, to replenish the air supply would be best, which is in fact exactly what Nemo does.

The most interesting digression on chemistry, however, occurs when Nemo shows Aronnax the engine room of the *Nautilus* and they discuss how the submarine is powered (Fig. 3). This is also the part of the novel that has been most flagrantly misrepresented, because in Walt Disney's 1954 movie adaptation—starring James Mason as Nemo, Paul Lukas as Aronnax, Kirk Douglas as Ned Land, and Peter Lorre grotesquely miscast as Conseil—it is implied that the *Nautilus* is powered by atomic energy and that Verne foresaw

Fig. 2 Professor Aronnax (modeled by the artist Riou on the appearance of Jules Verne as a young man).

Fig. 3 Tire engine room of the Nautilus.

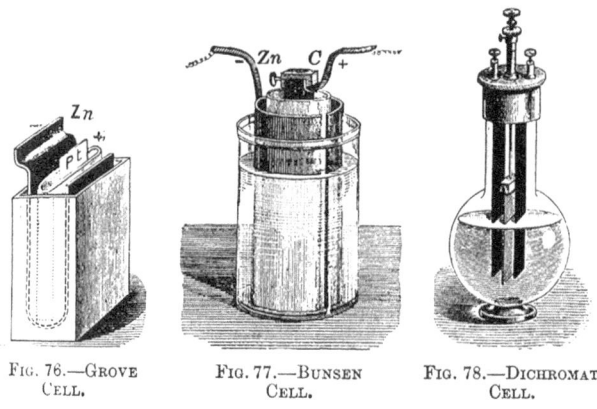

FIG. 76.—GROVE CELL.   FIG. 77.—BUNSEN CELL.   FIG. 78.—DICHROMATE CELL.

Fig. 4 From left to right: The Grove cell, the Bunsen cell, and the dichromate cell.

the nuclear age [6]. However, Nemo is very explicit about the power source of his submarine [4]:

> There is a powerful agent, obedient, rapid, easy, which conforms to every use, and reigns supreme on board my vessel. Everything is done by means of it. It lights it, warms it, and is the soul of my mechanical apparatus. This agent is electricity.

We need to remember that in 1870 electricity was the power source of the future just as atomic energy was in 1954. And what is the source of Nemo's electricity? The answer is none other than chemical storage batteries.

At the time Verne was writing, there were three important types of chemical storage battery (Fig. 4): the Grove cell, invented by the British chemist William Grove in 1839; the Bunsen cell, invented by the German chemist Robert Bunsen in 1841; and the dichromate or bichromate cell, apparently proposed by several different scientists in the period 1841–42, including Bunsen, the German physicist Johann C. Poggendorff, and the Englishman Robert Warington [7, 8]. The Grove and Bunsen cells were both based on the same chemical reactions, namely, the oxidation of zinc at the anode and the reduction of nitric acid at the cathode—the colorless nitric oxide quickly reverting to reddish-brown nitrogen dioxide on contact with air [9]:

$$2[NO_3^-(aq)+4H^+(aq)+3e^- \rightarrow$$
$$NO(g)+2H_2O(l)]] \quad +0.96\,V \quad (3)$$
$$3[Zn(s) \rightarrow Zn^{2+}(aq)+2e^-] \quad +0.76\,V \quad (4)$$
$$\overline{2NO_3^-(aq)+8H^+(aq)+3Zn(s) \rightarrow}$$
$$3Zn^{2+}(aq)+2NO(g)+4H_2O(l) \quad +1.72\,V \quad (5)$$

The sole difference was that Bunsen had replaced the expensive platinum cathode of Grove's original cell with an inexpensive one made of porous coke.

Grove or Bunsen cells would have been impractical on a submarine because of the necessity of venting the NO_2 fumes, so the best choice would have been the dichromate

cell, which substituted the reduction of the dichromate anion for the reduction of nitric acid at the cathode and in the process also increased the overall emf of the cell [10]:

$$Cr_2O_7^{2-}(aq)+14H^+(aq)+6e^- \rightarrow$$
$$2Cr^{3+}(aq)+7H_2O(l) \quad +1.33\,V \quad (6)$$
$$3[Zn(s) \rightarrow Zn^{2+}(aq)+2e^-] \quad +0.76\,V \quad (7)$$
$$\overline{Cr_2O_7^{2-}(aq)+14H^+(aq)+3Zn(s) \rightarrow}$$
$$3Zn^{2+}(aq)+2Cr^{3+}(aq)+7H_2O(l) \quad +2.09\,V \quad (8)$$

Nemo, however, has a fetish about obtaining all of his material needs from the ocean and, when Aronnax asks him where he gets the zinc for his batteries, he replies that he doesn't use zinc, but rather sodium metal extracted from seawater [4]:

> So it is this sodium that I extract from sea water, and of which I compose my ingredients…Mixed with mercury, sodium forms an amalgam which can take the place of zinc in Bunsen batteries. The mercury is never consumed, only the sodium is used up, and that is supplied from sea water. Moreover, sodium batteries are the most powerful, since their motive force is twice that of zinc batteries.

Though Verne does not cite quantitative emf values, it is interesting to note that use of modern data shows that Verne's estimate of the "idealized" relative strength of Captain Nemo's sodium cell, versus that of the conventional dichromate cell, is accurate (i.e., 4.04 V versus 2.09 V) provided one uses the reduction potential for sodium metal [11]:

$$Cr_2O_7^{2-}(aq)+14H^+(aq)+6e^- \rightarrow$$
$$2Cr^{3+}(aq)+7H_2O(l) \quad +1.33\,V \quad (9)$$
$$6[Na(Hg) \rightarrow Na^+(aq)+e^-] \quad +2.71\,V \quad (10)$$
$$\overline{Cr_2O_7^{2-}(aq)+14H^+(aq)+6Na(Hg) \rightarrow}$$
$$6Na^+(aq)+2Cr^{3+}(aq)+7H_2O(l) \quad +4.04\,V \quad (11)$$

This strongly suggests that Nemo's cell was based on an actual experimental account published in the scientific literature of the period. Though I have not been able to trace the original reference, the most likely candidates for all of Verne's information on electrochemistry, as we will see in greater detail later, are the writings of the French electrochemist Antoine-César Becquerel. In passing, it is also of interest to note that Antoine-César was the grandfather of Antoine-Henri Becquerel, best known for his discovery of radioactivity in 1896 [12].

Aronnax then raises the question of how Nemo extracts his sodium [4]:

> I can see how sodium serves your needs. And there is plenty of it in sea water. But you have to manufacture it, to extract it. How? You could use your batteries to extract it, but it seems to me you would need more sodium for such equipment than it would be extracting. I mean, would you not consume more than you produce?

In short, Aronnax is suggesting that Nemo use his batteries to electrolyze seawater:

$$\text{electricity} + 2\text{NaCl}(aq) \rightarrow 2\text{Na}(\text{Hg}) + \text{Cl}_2(g) \qquad (12)$$

though he immediately realizes that such a process would violate the conservation of energy. Nemo replies:

> No, I do not use batteries, at least not for the extraction process. I use heat generated by coal.

This answer is ambiguous but probably refers to the production of sodium via the carbon reduction of sodium carbonate, which was the standard method of manufacture in the 1870s [13]:

$$\text{Na}_2\text{CO}_3(s) + 2\text{C}(s) \rightarrow 2\text{Na}(g) + 3\text{CO}(g) \qquad (13)$$

In keeping with his theme of "all from the sea," Nemo implies that he mines his coal at the bottom of the ocean. However, the need to convert NaCl into Na_2CO_3, which requires use of either the Leblanc or the Solvay process, as well as the necessity of manufacturing the sulfuric acid and potassium dichromate required for the cathode reaction, strongly suggest that Nemo must have a land base somewhere to carry on these processes, and, as we will see later, this is indeed the case.

Note that not only is Nemo's claim of "all from the sea" chemically weak in this case, it is also geologically weak, as the coal that he mines on the bottom of the ocean is certainly not a product of the ocean itself but the result of the submergence of conventional land-based coal deposits formed from the decomposition of prehistoric land-based plant life. Indeed, in a later chapter entitled "The Submarine Coal Mines," Verne as much as admits that this is the case, though Nemo's submerged coal deposits are rather improbably located in the crater of an extinct volcano which is connected to the ocean via an underwater system of caves. Prior to large-scale industrialization and mining of coal in the eighteenth century, chunks of coal were often found along ocean beaches, where they were collected by women and children. Though this material was actually broken off from submerged shore-line coal outcroppings and washed ashore by wave action, it appeared to the common man to be a product of the ocean and was consequently known as "sea coal." This incorrect association was still prevalent among the uneducated classes in the nineteenth century and is exploited by Nemo in the course of his discussion with Aronnax.

One final point of interest. When Nemo takes Aronnax for a walk on the ocean floor in one of his special diving suits, Aronnax asks Nemo what he uses to light his way in the blackness of the ocean abyss (Fig. 5). Nemo replies that he uses one of his special sodium batteries and a "Ruhmkorff apparatus" (i.e., an induction coil) connected to a special lantern [4]:

> In this lantern is a spiral glass which contains a small quantity of carbonic gas [carbon dioxide]. When the apparatus is at work this gas becomes luminous, giving out a white and continuous light.

Fig. 5 Nemo's electric carbon dioxide lanterns at work.

What Nemo is describing is, of course, a Geissler tube—a sort of crude precursor of the fluorescent light (Fig. 6)—and this same contrivance is used to light the interior of the *Nautilus*. H. W. Meyer, in his book *A History of Electricity and Magnetism*, describes a similar device [14]:

> About the year 1895, D. McFarlan Moore of the United States began experimenting with long glass tubes filled with carbon dioxide gas, which gave off a good quality white light when a current of electricity was sent through them at relatively high voltage. Beginning about the year 1904, many installations of such tube lighting were made, especially in stores.

So it would appear that Verne was prophetic about new applications of existing technology after all!

The Mysterious Island

This brings us to the sequel to *Twenty Thousand Leagues under the Sea*, the three-part novel *The Mysterious Island*, which was published in 1874, four years after *Twenty Thousand Leagues* [15]. Set during the American Civil War, the story involves a group of Union prisoners held in Richmond, Virginia, who escape the city in March of 1865 in a Confederate observation balloon in the midst of

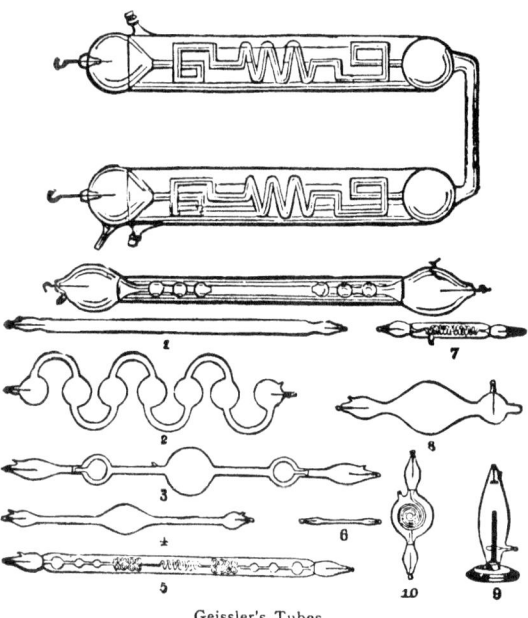

Geissler's Tubes.

Fig. 6 Typical nineteenth-century Geissler tubes.

Fig. 7 *Cyrus Harding, the engineer-hero of* The Mysterious Island.

a violent storm. The storm blows them west across the United States and out into the Pacific Ocean, where they crash on an uncharted island. Events eventually reveal that this island is one of Captain Nemo's land bases, hinted at in *Twenty Thousand Leagues*, but it is the first two parts of the novel that are of most interest to us.

Unlike the castaways in Johann Wyss's famous novel *The Swiss Family Robinson,* who have access to the cargo of their wrecked ship and are amply supplied with tools, provisions, guns, and domestic animals, the castaways in Verne's novel have only the clothes on their backs, the knowledge in their heads, and a single match—the wrecked balloon having been blown back out to sea [16]. What follows might be appropriately called "The Chemical Swiss Family Robinson." It is a paean to the now defunct advertising phrase "better things for better living through chemistry" and a celebration of the engineer as hero. The engineer in question is one Cyrus Harding (Fig. 7), and the worship of his fellow castaways is apparent from the beginning of the novel [15]:

> The engineer was to them a microcosm, a compound of every science, a possessor of all human knowledge. It was better to be with Cyrus Harding on a desert island than without him in the midst of the most flourishing town in the United States. With him they could want nothing; with him they would never despair.

The island itself, which the castaways name "Lincoln Island" in a display of patriotism, is of volcanic origin and is particularly rich in minerals. What follows is a partial chronology of the rise of "chemical man" on Lincoln Island, and

it goes without saying that in each instance Verne inserts a short digression painlessly describing for the reader the chemistry involved.

Within eight days of their arrival (i.e., by the 31st of March), Cyrus Harding has discovered pyrites, clay, limestone, and coal deposits on the island. These materials are quickly put to use. Between the 2nd and 13th of April, the castaways manufacture bricks from fired clay and make mortar from stone and lime, the latter being produced by thermally decomposing limestone and slaking the resulting quicklime with water [13, 15]:

$$CaCO_3(s) + CaO(s) \rightarrow CO_2(g) \tag{14}$$

$$CaO(s) + H_2O(l) \rightarrow Ca(OH)_2(s) \tag{15}$$

The bricks and mortar are then used to construct a pottery kiln (Fig. 8), which the castaways use to fire crude pots, dishes, etc.

By the 17th of April, Harding has added niter (KNO_3) and iron ore to his mineralogical discoveries and begins the construction of a large bellows using sealskin and clay pipe

Fig. 8 The castaways make pottery.

Fig. 9 The castaways make iron.

manufactured in the pottery kiln. Between the 21st of April and the 5th of May, the resulting forced-air furnace is used to produce iron and steel (Fig. 9):

$$Fe_3O_4(s) + 2C(s) \rightarrow 3Fe(s) + 2CO_2(g) \qquad (16)$$

which the castaways use to manufacture crude saws, hammers, nails, axes, hatchets, chisels, spades, and pickaxes.

Between the 7th and 18th of May, Harding extracts green vitriol $\left(FeSO_4 \cdot 7H_2O\right)$ and alum from schistose pyrites and soda (Na_2CO_3) from the ashes of marine plants. He then uses the soda to produce soap and glycerin by saponifying the fat of a dugong that has been mysteriously killed after attacking the castaway's pet dog, Top:

$$Na_2CO_3(s) + H_2O(l) \rightarrow$$
$$2Na^+(aq) + OH^-(aq) + HCO_3^-(aq) \qquad (17)$$

$$C_3H_5(OOCR)_3(s) + 3OH^-(aq) \rightarrow$$
$$C_3H_5(OH)_3(l) + 3RCOO^-(aq) \qquad (18)$$

$$RCOO^-(aq) + Na^+(aq) \rightarrow Na(OOCR)(s) \qquad (19)$$

Given that Harding has already manufactured slaked lime (Eqs. 14 and 15), it is surprising that he doesn't use it to convert his soda into caustic soda (NaOH):

$$Ca(OH)_2(aq) + Na_2CO_3(aq) \rightarrow$$
$$2NaOH(aq) + CaCO_3(s) \qquad (20)$$

since this would make a much more effective saponifying agent:

$$C_3H_5(OOCR)_3(s) + 3NaOH(aq) \rightarrow$$
$$C_3H_5(OH)_3(l) + 3Na(OOCR)(s) \quad (21)$$

On the 20th of May, Harding, using chemical apparatus made in the pottery kiln, manufactures sulfuric acid via the destructive distillation of the green vitriol:

$$2FeSO_4 \cdot 7H_2O(s) \rightarrow$$
$$H_2SO_4(l) + Fe_2O_3(s) + 13H_2O(l) + SO_2(g) \quad (22)$$

He then uses this, along with the niter discovered earlier, to produce concentrated "azotic acid" (nitric acid):

$$H_2SO_4(l) + 2KNO_3(s) \rightarrow$$
$$K_2SO_4(s) + 2HNO_3(l) \qquad (23)$$

Fig. 10 Cyrus Harding makes nitroglycerin. The original caption reads "It's nitroglycerin!".

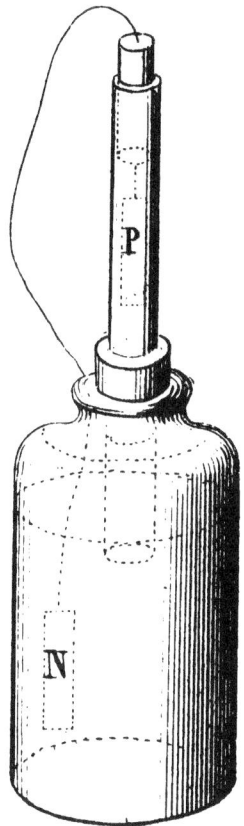

Fig. 11 Becquerel's original acid-alkali cell.

and this, in turn, is used, in combination with the sulfuric acid and glycerin, to make nitroglycerin (Fig. 10), which is subsequently used for various large-scale engineering projects on the island:

$$C_3H_5(OH)_3(l) + 3HNO_3(l) \rightarrow$$
$$C_3H_5(NO_3)_3(l) + 3H_2O(l) \quad (24)$$

On the 5th of June, Harding manufactures candles from seal fat, lime, and sulfuric acid and, finally, to round out their first year on the island, he extracts sugar from a local variety of the maple tree on the 25th of August.

In early January of their second year on the island, Harding uses his supply of sulfuric and nitric acids, in combination with native plant cellulose, to manufacture pyroxylin or guncotton. On the 28th of March, he makes glass using sand, chalk produced from limestone, and soda extracted from seaweed, and in January of their third, and last, year on the island, he decides to build an electric telegraph in order to facilitate communication between the various outposts that the castaways have established. This brings Verne back to the subject of electricity and chemical batteries. In this case, his choice is an unusual acid/alkaline battery invented in 1820 by the French physicist and electrochemist Antoine-César Becquerel, whom we met earlier in connection with Captain Nemo's sodium cell. Verne describes Becquerel's cell in great detail and in terms which strongly suggest that he has read Cyrus Harding, after mature consideration, decided to manufacture a very simple battery...in which zinc only is employed [obtained from the lining of a sea chest in which Captain Nemo has anonymously left supplies for the castaways]. The other substances, azotic [nitric] acid and potash [potassium carbonate], were all at his disposal.

The way in which the battery was composed was as follows, and the results were to be attained by the reaction of acid and potash on each other. A number of glass bottles were made and filled with azotic acid. The engineer corked them by means of a stopper through which passed a glass tube, bored at its lower extremity, and intended to be plunged into the acid, by means of a clay stopper secured by a rag. Into this tube, through its upper extremity, he poured a solution of potash, previously obtained by burning and reducing to ashes various plants, and in this way the acid and potash could act on each other through the clay [see Figs. 11 and 12].

Cyrus Harding then took two slips of zinc, one of which was plunged into the azotic acid, the other into the solution of potash. A current was immediately produced, which was transmitted from the slip of zinc in the bottle to that in the

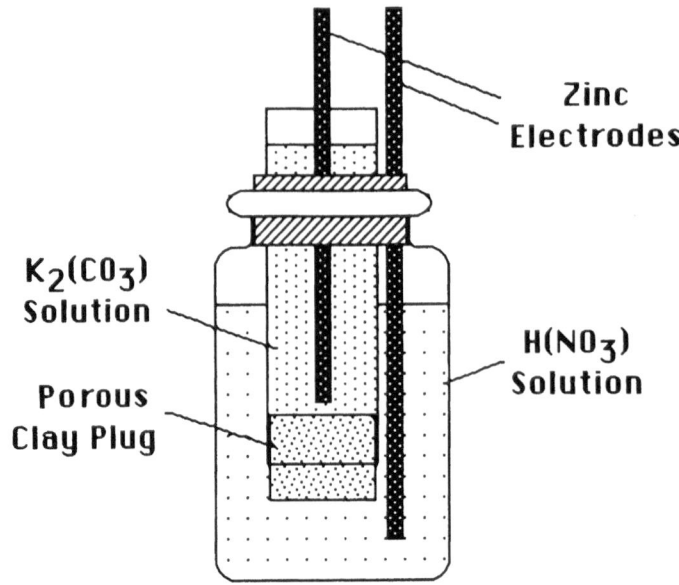

Fig. 12 *A reconstruction of the acid-alkali cell described by Cyrus Harding in* The Mysterious Island.

tube, and the two slips having been connected by a metallic wire, the slip in the tube became the positive pole and that in the bottle the negative pole of the apparatus. Each bottle, therefore, produced as many currents as, united, would be sufficient to produce all the phenomena of the electric telegraph.

As in the case of Verne's paraphrase, Becquerel's own account of his cell tells us little about its chemistry other than the fact that dioxygen gas is generated at the anode, probably via the reaction (18):

$$4CO_3^{2-}(aq) + 2H_2O(l) \rightarrow$$
$$4HCO_3^-(aq) + O_2(g) + 4e^- \quad -0.62 \text{ V} \quad (25)$$

while Benjamin, who refers to it as the "Becquerel Oxygenated Gas Cell" in his 1893 treatise on the voltaic cell, claims that the cathode reaction corresponds to the reduction of the concentrated nitric acid to ammonium nitrate [19]:

$$10H^+(aq) + NO_3^-(aq) + 8e^- \rightarrow$$
$$NH_4^+(aq) + 3H_2O(l) \quad +0.88 \text{ V} \quad (26)$$

As can be seen, these half-reactions give us a thermodynamically favorable net emf of +0.26 V for the cell at unit activities. I must confess, however, to having certain reservations about representing the cathode reaction in terms of Eq. 26, since Latimer reports that the reduction of nitric acid to nitric oxide, as observed in the case of the Grove cell, is slightly more favorable [20]:

$$4H^+(aq) + NO_3^-(aq) + 3e^- \rightarrow$$
$$NO(g) + 2H_2O(l) \quad +0.96 \text{ V} \quad (27)$$

This would give us a favorable net emf value of around +0.34 V at unit activity. In his original account, Becquerel used platinum, rather than zinc, for his electrodes, and a quick replication of the cell in my laboratory, using a saturated potassium carbonate solution and 16 M nitric acid, gave a potential of around +0.87 V, provided that one used either platinum or nichrome wire electrodes, though I could observe no gas evolution at either electrode. This is not bad agreement, given the enormous deviations from unit activities. Unfortunately, Verne's substitution of zinc in place of platinum for his electrode material appears to be the source of a serious defect in his scheme, since I found that all attempts to use zinc for the electrode in the nitric acid halfcell led to its rapid destruction, regardless of how dilute the acid [21].

Though Verne does not explicitly spell out his reasons for choosing this rather unusual cell, it appears to be related to the fact that the castaways have access to a continuous supply of only one metal—iron. As a consequence, they are unable to construct batteries based on the chemical difference between two metal electrodes, since that would lead to the net consumption of their strictly limited supply of zinc from the lining of the sea chest. Verne emphasizes this circumstance when he discusses their substitution of iron for lead in making shot for their guns and iron for copper when making the wires for their telegraph. If this is, in fact, the true reason for Verne's choice of a cell having two identical metal electrodes, then it is elegant testimony to the care with which he planned the scientific details of his novels, even though, as already indicated, he negated this advantage via his ill-advised substitution of zinc in place of platinum.

Based on his comments in both *Twenty Thousand Leagues under the Sea* and *The Mysterious Island,* there is little doubt that electricity was Verne's favorite choice as the power source of the future. Nevertheless, he was not unaware of society's ultimate dependence on fossil fuels, and at one point in *The Mysterious Island* he has the castaways discuss the possibility of a future energy crisis. A castaway by the name of Gideon Spilett begins this discussion by asking Harding how he is able to square the consequences of such a crisis with his habitually optimistic view of mankind's technological future [15]:

> But now, my dear Cyrus, all this industrial and commercial movement to which you predict a continual advance, does it not run the danger of being sooner or later completely stopped...by the want of coal, which may justly be called the most precious of minerals.

Harding agrees but is not upset, as he foresees a future in which coal will be replaced by an alternative fuel [15]:

> Water...but water decomposed into its primitive elements...and decomposed, doubtless, by electricity which will then become a powerful and manageable force, for all great discoveries, by some inexplicable law, appear to agree and become complete at the same time. Yes, my friends, I believe that water will one day be employed as fuel, that hydrogen and oxygen which constitute it, used singly or together, will furnish an inexhaustible source of heat and light, of an intensity of which coal is not capable. Some day the coal-rooms of steamers and the tenders of locomotives will, instead of coal, be stored with these two condensed gases, which will burn in the furnaces with enormous caloric power. There is, therefore, nothing to fear. As long as the earth is inhabited it will supply the wants of its inhabitants, and there will be no want of either light or heat as long as the productions of the vegetable, mineral, or animal kingdoms do not fail us. I believe that when the deposits of coal are exhausted we shall heat and warm ourselves with water. Water will be the coal of the future.

In this selection Verne exploits the known difference in the heats of combustion per unit mass of dihydrogen gas versus carbon:

$$H_2(g) + \frac{1}{2} O_2(g) \rightarrow H_2O$$
$$\Delta H = -142.93 \text{ kJ} / \text{gH}_2 \quad (28)$$

$$C(s) + O_2(g) \rightarrow CO_2(g)$$
$$\Delta H = -32.79 \text{ kJ} / \text{gC} \quad (29)$$

but unhappily fails to tell us how we are going to generate the electricity necessary to electrolyze all of this water in the first place.

Several of Verne's other novels also contain digressions on chemistry [22]. Thus, in his famous account of space travel, *From the Earth to the Moon* (1865), Verne discusses the manufacture of aluminum using the Deville process, the manufacture of guncotton, and various chemical schemes for generating dioxygen gas and for absorbing carbon dioxide aboard his proposed spacecraft. In his equally famous novel

Journey to the Center of the Earth (1864), he discusses various chemical theories of volcanism and makes use of the same carbon dioxide Geissler lamps later used by Captain Nemo in *Twenty Thousand Leagues.* Likewise, Captain Nemo's sodium cell makes a second appearance in the 1886 novel *The Clipper of the Clouds*—this time as the power source for a lighter-than-air craft called the "Albatross," which is commanded by a Nemo-like clone by the name of Robur. In the short story *Dr. Ox's Experiment* (1874), Verne once more returns to the subject of various alternative methods of generating dioxygen gas and the effects of increased dioxygen concentrations on the physiological and psychological behavior of living organisms, while in the novel *The Southern Star Mystery* or *The Star of the South* (1884), he deals with the synthesis of artificial diamonds. But none of these digressions come close to rivaling the chemical versatility of the castaways in *The Mysterious Island.*

In February of 1873, while still in the process of planning the details of *The Mysterious Island,* Verne wrote a letter to his friend and publisher Pierre-Jules Hetzel, in which he referred to his new project as a "roman chimique"—a chemical romance—and confessed that he had been spending his time doing background research "among Professors of Chemistry and in chemical plants" [23]. Surely, it is time that chemists and teachers of chemistry return the compliment and spend some time with Verne enjoying what must surely be the only known example of that most elite of literary genres—the *roman chimique.*

Notes and References

1. Quoted in Gunn, J. *Alternate Worlds: The Illustrated History of Science Fiction;* A&W Visual Library: Englewood Cliffs, NJ, 1975; p 120. See also Moskowitz, S. *Explorers of the Infinite: Shapers of Science Fiction;* Hyperion Press: Westport, CT, 1974; Chapter 19.
2. Evans, A. B. *Jules Verne Rediscovered: Didacticism and the Scientific Novel;* Greenwood Press: Westport. CT, 1988. Though intended for adults and far better researched, Verne's novels actually have more in common with such juvenilia as the *Tom Swift series* than with modern science fiction.
3. A good example of a butchered version of *Twenty Thousand Leagues* can be found in *Jules Verne, Classic Science Fiction: Three Complete Novels;* Russell, A. K., Ed.; Castle Books. Secaucus, NJ, 1981. Though care was taken in this collection and in its earlier companion volume (*The Best of Jules Verne: Three Complete Novels;* Russell, A. K., Ed.; Castle Books: Secaucus, NJ, 1978) to reproduce all of the nineteenth-century French illustrations to the novels, most of which had been deleted from the original English editions, similar care was not taken with the text. As a consequence, the version of *Twenty Thousand Leagues* which it reprints is missing virtually all of the technical passages found in the version given in Ref. 4 and which I quote in this article.
4. Miller, W. J. *The Annotated Jules Verne: Twenty Thousand Leagues under the Sea*; Crowell: New York, 1976; pp 54–55, 75–77, 100.
5. Verne refers to these reactions again in a later chapter entitled "Want of Air" in which the *Nautilus* becomes trapped under the ice

and the crew is caught in a frantic struggle to free her before they die of suffocation.

6. In addition to the 1954 Disney film, there are two earlier silent-film versions of *Twenty Thousand Leagues under the Sea,* one made by Universal Studios in 1916 and the other by the French director George Méliès in 1907.

7. For historical background on these batteries, see (a) Dunsch, L. *Geschichte der Elektrochemie;* Deutscher Verlag für Grundstoffindustrie: Leipzig, 1985; pp 56–57; (b) Ritter von Urbanitzky, A. *Electricity in the Service of Man: A Popular and Practical Treatise on the Applications of Electricity in Modern Life;* Cassell: London, 1886; pp 106–115; (c) Lowry, T. *Inorganic Chemistry;* Macmillan: London, 1931; pp 208–209; (d) Benjamin, P. *The Voltaic Cell: Its Construction and Capacity;* Wiley: New York, 1893; and (e) Slock, J. T. "Bunsen's Batteries and the Electric Arc", *J. Chem. Educ.* **1995,** *72,* 99–102.

8. There seems to be considerable confusion in the literature as to who actually invented the dichromate cell. Kevin Desmond, in his reference work *A Timetable of Inventions and Discoveries* (Evans: New York, 1986), claims that it was invented by Heinrich Ruhmkorff in 1855; Dunsch (Ref. 7a) attributes it to Robert Bunsen in 1842; Benjamin (Ref. 7d) claims that it was developed by Johann C. Poggendorff the same year; and Stock (Ref. 7e) claims that it was mentioned in passing by Bunsen in 1841 and discussed in detail by Robert Warington in 1842. Of these four authors, only Stock provides original literature citations to support his claims.

9. The emf values quoted correspond to unit activities. In practice, the nitric acid in the cells was much more concentrated, and Benjamin (Ref. 7d) reports actual operating values of 1.60–1.90 V for the Grove cell and 1.93–1.96 V for the Bunsen cell, whereas lowry (Ref. 7c) reports values of 1.80–1.96 V for the Grove cell.

10. The emf values quoted correspond to unit activities. In practice, the solutions were much more concentrated. Benjamin (Ref. 7d) reports actual operating value of 1.92–2.2 V for the dichromate cell, whereas Lowry (Ref. 7c) reports a value of 2.0 V.

11. The "idealized" emf values quoted are based on unit activities. The actual voltage of Nemo's sodium cell would depend on the activity of the sodium in the amalgam. The value of 1.957 V for the sodium amalgam half-cell reported by Dietrick et al. (Dietrick, H.; Yeager, E.; Hovorka, F. *The Electrochemical Properties of Dilute Sodium Amalgams*; U.S. Office of the Naval Research, Technical Report 3; Western Reserve University: Cleveland OH, 1953) would give an overall value of 3.29 V for the cell, which is only about one and half times that of the dichromate cell. This suggests that Nemo was using very concentrated amalgams in his cells.

12. The only nineteenth-century estimates reported by Benjamin (Ref. 7d) for the emf values of sodium amalgams are those originally reported by Antoine-César Becquerel and his son, Alexandre-Edmond Becquerel, in their 1855 volume *Traité expérimental de l'électricité et du magnétisme*. These range from 2.303 to 2.334 V, which again gives an overall emf value that is only about one and half times that of the standard dichromate cell.

13. A nineteenth-century account of the chemistry underlying all of the manufacturing processes described by Verne in both *Twenty Thousand Leagues under the Sea* and *The Mysterious Island* can be found in the *Encyclopaedia of Chemistry, Theoretical, Practical and Analytical as Applied to the Arts and Manufacturers* (2 volumes; Lippincott: Philadelphia, 1879), under the entries for sodium, pottery, cement, iron, steel, soap, alum, sulfuric acid, nitroglycerin, candles, sugar, guncotton, and glass.

14. Meyer, H. W. *A History of Electricity and Magnetism*; MIT Press: Cambridge, MA, 1971; pp 174–175.

15. Verne, J. *The Mysterious Island*; A. L. Burt : New York, no date; pp 57, 90, 94–95, 96, 107–113, 124–126, 126–127, 145–146, 162–163, 225–226, 234–235, 250–252, 311–312. This is one of many inexpensive rip-off editions of Verne. Other English translations of this novel use the name Cyrus Smith, rather than Cyrus Harding, for the engineer-hero. Unfortunately I have been unable to examine a French edition and so cannot tell which rendition is the correct one. The *Mysterious Island* has been filmed three times: once by MGM in 1929, and twice by Columbia Pictures, in 1951 and 1961, respectively. Captain Nemo was played by Lionel Barrymore in the 1929 production and by Herbert Lom in the 1961 production.

16. Later in the novel, the resources of the castaways are further supplemented by accidental finds of supplies that have apparently washed ashore from wrecked ships but which are, in fact, provided by Captain Nemo, who has been secretly observing their progress.

17. There are some obvious confusions in this quote, which again reflect the low quality of most English translations of Verne's novels emphasized by Miller in Ref. 4.

18. Becquerel, A. C. *Traité de physique considérée dans ses rapports avec la chimie et les sciences naturelles*; Didot: Paris, 1844; Vol. 2, pp 300–301.

19. Benjamin (Reference 7d), p 267.

20. Latimer, W. *Oxidation Potentials*, 2nd ed.; Prentice-Hall: Englewood Cliffs, NJ, 1952; p 93. For consistency, this reference has been used to calculate all other thermodynamic values cited in the article.

21. Verne may have been misled by Becquerel's remark in Ref. 18 that the emf of his cell could be increased by substituting zinc for platinum at the anode. He may not have realized that a similar substitution would not work for the cathode, nor that the increase in the emf is due, in the case of the anode substitution, to oxidation of the zinc.

22. I have discussed several of these in greater detail in the essays "Sir Humphry Davy and the Hollow Earth: The Geochemistry of *Journey to the Center of the Earth*" and "Tom Swift Among the Diamond Makers: Synthetic Diamonds in Fact and Fiction" to be published in future issues of *The Chemical Intelligencer*.

23. Quoted in Martin, C. N. *La vie et l'oeuvre de Jules Verne*; Michel de l'Ormeraie: Paris, 1978; p 200.

William B. Jensen is Oesper Professor of the History of Chemistry and Chemical Education at the University of Cincinnati. Founder and former editor of The Bulletin for the History of Chemistry, he also serves as curator of both The Oesper Collection of Books and Prints in the History of Chemistry and the departmental apparatus museum. In addition to his activities in chemical education arid history, his research interests include the theory and application of the Lewis acid-base concepts, the development of empirical structure-reactivity sorting maps, chemical periodicity, inorganic crystal chemistry, and qualitative bonding theory.

Across Two Oceans onto the Reef—The Genesis of Marine Natural Products[a]

A Personal Account

Paul J. Scheuer[b]

The Editor's invitation to write an article for *The Chemical Intelligencer* "about your career and your pioneering work in marine natural products" was flattering and challenging; it proved to be rewarding. For the first time in my life, I have taken the time to examine the unlikely and meandering route that led me from a middle-class upbringing in southern Germany to an exciting academic adventure in the Hawaiian Islands.

The journey which you will share had a slow and uncertain beginning, taking as it did nine years from high school graduation to a B.S. degree in chemistry and another seven years to a Ph.D. But an assistant professorship at the University of Hawai'i revealed a new world with unexpected challenges and rewards. Previously unexplored rain forests and coral reef organisms offered unlimited opportunities for natural products research. For a person who had never studied botany or zoology and whose natural products research in graduate school began with a pure crystalline solid, it was a true beginning. It was a beginning in another respect as well. The Chemistry Department at the University of Hawai'i was not equipped for research, physically or intellectually. Rarely has any beginning assistant professor been able to experience the excitement and frustrations of starting with a tabula rasa. But the end result, as you will see, has been gratifying.

Education and *Wanderjahre*

Primary public education in the Weimar Republic (1919–1933) embraced a democratic element that had been absent in pre-World War I Germany. All six-year-olds were enrolled in identical elementary schools (Volkschule) for four years, at the end of which a mandatory qualifying examination determined whether a student was eligible for a nine-year high school or a vocationally oriented middle school. Another decision point for high school students came after six years, when only those who aimed at a university education would continue for the last three years, which culminated in another comprehensive examination. Those who had left high school after six years would begin careers in industry or commerce. Presumably, that mid-point career decision was heavily influenced by the student's parents.

When I graduated from the Heilbronn Realgymnasium in the spring of 1934, the Weimar Republic had died. The national elections of 1933 made the National Socialists (Nazis) the majority party and Adolf Hitler the Chancellor, who rapidly became the dictator. Although I graduated first in a tiny class of 11 students, the racial laws of the Third Reich had eliminated my option to attend a university.

If university attendance had been allowed, I have no idea what I would have studied. It decidedly would not have been chemistry. Although I had had two years of high school chemistry, my classmates and I were more fascinated by the antics of the instructor than by the subject that he taught (badly). There were few role models in a family of merchants. An uncle who was an attorney would occasionally invite me to attend civil trials, which I found utterly boring. In a way then, it was fortunate that I gained time before deciding on a career. The subsequent events, which eventually culminated in a professorship in the field of natural products, were serendipitous and remarkably devoid of deliberate planning.

During my teens I belonged to a youth group which combined elements of scouting with discussions of current events. One of the group leaders, a Ph.D. in economics, was working in his family's leather tannery. He asked me whether I would like to be a tannery apprentice, an offer which I accepted. Following an old German tradition of apprenticeship, I started my career, not with sweeping the floor of the workshop, but with unloading raw cowhides onto trucks from freight cars parked on a factory siding. Since the factory specialized in making leather for shoe soles, the hides were big and heavy. It was the hardest physical labor I had ever done!

[a]*Chemical Intelligencer* 1997(2), 46–57.

[b]Department of Chemistry, University of Hawai'i at Manoa, Honolulu, HI 96822, USA (deceased)

Over the course of 20 months, I gradually worked my way through the entire operation from raw hides to finished leather, but I was left with only a vague idea of the actual process that turns a cowhide into leather. My mentor believed that I should also learn how fine leather is made. He suggested a second apprenticeship in a tannery in Hungary, which was owned by a family friend. In December 1955, I took the train to the provincial town of Pécs in southwest Hungary, leaving Germany, as it turned out, for good, except for a brief visit in 1937 for my mother's funeral and a longer visit, courtesy of the U.S. Army, from May 1945 to June 1946.

I stayed in Pécs for a year before moving on to another tannery in Simontornya, a very small Hungarian town. It was there that I became interested in chemistry. The technical director of the tannery, a Ph.D. chemist, decided that I should learn some basic theory that would teach me something about the tanning process, historically one of the earliest examples (and a highly successful one) of biotechnology. Almost daily, after work—what else was there to do?—he tutored me in what might be called simplified classical biochemistry. The chemistry of the tanning process—whether accomplished by chromium salts or by the complex phenols of tree bark—remained a mystery, but I became fascinated with chemistry as an intellectual challenge. It was then and there that I decided to become a chemist.

Short stints in tanneries in Slovenske Konjice and finally Downton, Wiltshire, tiny towns in Yugoslavia (now Slovenia) and England, respectively, lasted through the fall of 1938. During the spring and summer of 1938, war in Europe became increasingly probable, and my thoughts of leaving the old country became more frequent. Ever since my mother's oldest brother, who had emigrated to the United States in the 1890s to avoid military service and had lived in California, had visited Germany when I was a teenager, I had been fascinated by his stories of vast distances, tall buildings, and luxurious trains. Uncle Josef Neu had died by then, but a cousin of my mother's, a New York attorney, was willing to sponsor me for an immigration visa to the United States. After Hitler and Chamberlain had made their agreement that would guarantee peace, war appeared more imminent than ever. And so, in September of 1938, I went to the U.S. Consulate in London, obtained a visa, and booked passage on the first available ship. The *Queen Mary* brought me into New York harbor in mid-October. Cousin Ben Herzberg met me at the Hudson River pier and suggested that I call a client of his, a leather wholesaler in lower Manhattan near City Hall.

A job sorting and packaging calf and sheep leather materialized. By January 1939, I was a foreman in a small tannery in Ayer, Massachusetts, which the owner of the New York leather company had recently purchased. Once I was settled, I looked for an opportunity to continue my study of chemistry. Northeastern University in Boston, a small engineering-oriented college, had the most extensive night school program. By the fall of 1939, I was enrolled in freshman chemistry. Two evenings a week, I drove my newly acquired two-seater '32 Ford to Boston, a 65-mile round trip. With the invasion of Poland by German troops in September of 1939, World War II had begun.

During the course of the year, it became clear to me that commuting to night school was perhaps not the best way to obtain a university education or to achieve a better understanding of the processes that change a cowhide into a piece of leather. (I was still naive enough to believe in that goal.) I decided to quit my job, move to Boston, go to school full time, and try to make a living from part-time jobs, nights and weekends. In the fall of 1940, I became a day student in the College of Liberal Arts at Northeastern University, majoring in chemistry. Savings from my job in Ayer paid the first year's tuition. I earned my living expenses by taking part-time jobs that ranged from pin boy in a bowling alley (no automatic pin setters in 1940!) to taxi driver. After six years as a migrant tanner, I was eager to get on with my education.

In Northeastern's cooperative education plan, a full-time freshman year was followed by 4 years of alternate study and work periods. I had been able to get by on part-time jobs and did not relish the thought of a five-year degree. I decided on a program of full-time study. Because the curricula were planned for coop students, schedules were at times rocky, but it worked. By the time I took organic chemistry, I had completed all of my math, physics, and analytical chemistry; it was immediately obvious to me that this was the chemistry I would want to study. Even by 1941 standards, the course was old-fashioned, but carbon was the most interesting element in the periodic table.

In December of 1941, the United States entered the war and everybody's life changed. As a German national, I was an enemy alien and had a few travel restrictions. Like everybody else, I registered for the draft, but I was rejected when I was first called.

In April 1943, I graduated from Northeastern with a B.S. in Chemistry. I no longer recall when the idea of an advanced degree occurred to me. It may well have been during the summer of 1942, when I served as the "maid" for the chairman of the chemistry department in return for room and board. I remember well that I intended to apply to MIT and Harvard, a natural choice for a Bostonian. The application forms arrived, and MIT was quickly eliminated when I saw the detailed (and rather irrelevant) questions that had to be answered: full names and titles of all textbooks for all chemistry courses, pages covered in each text, etc. I applied and was accepted at Harvard. I even received a tuition scholarship.

During the summer of 1943, I drove a Checker Cab, but strict gas rationing, which gave rise to some small-time racketeering at the pump, took the fun out of navigating the streets of Boston, which I had used to enjoy.

Graduate school started for me in the fall of 1943. The Harvard chemistry faculty was small, as was the number of graduate students. G.P. Baxter, the Chairman, was everybody's informal and willing adviser. Of the three organic faculty members, Louis Fieser, Paul Bartlett, and Bob Woodward, only the latter two represented viable choices as thesis advisers because Fieser was rarely in residence as he was heavily engaged in war-related research. I chose Woodward (R.B.).

R.B. had been elevated from Junior Fellow to Instructor when Patrick Linstead returned to his native England for war-related activities. Several of R.B.'s graduate students, Jack Chanley, Dick Eastman, and Bob Loftfield among them, were Linstead holdovers. Some, for example, Harry Wasserman, had started with R.B. but were in military service. Of original R.B. students, I recall only Betsy Clarke. Her husband John (grandson of Max Planck) was a student at MIT. Group meetings were held in a small conference room in the basement of Converse; they were informal and lacked the sharply competitive atmosphere of the postwar years.

My research during the fall semester of 1943 consisted of many unsuccessful attempts to add ketene to α-vinylpyridine, probably in connection with the synthesis of quinine. In the spring of 1944, the Department needed an instructor for a short introductory chemistry course to be taught to about 30 GIs. Professor Baxter handed me a text, a class list, and a course schedule. There I was in front of a class in an ancient classroom in Sever Hall. I thought that I had prepared a good first lecture. I managed to say it all in 10 minutes. Class dismissed.

I met some of the same GIs again later in the spring at Fort Devens, Massachusetts. I had been drafted—by then even "enemy aliens" were eligible—and while waiting for orders to report for basic training at Camp Sibert, Alabama, I was on KP. I was scrubbing garbage cans in back of the mess hall when a group of my former students walked by. It made them feel good to see their former Harvard instructor on KP while they were goldbricking.

Although I was assigned to the Chemical Warfare Service, the two years and four months I spent in the U.S. Army were yet another hiatus in my chemistry education. My most spectacular chemical experience was the firing of WP (white phosphorus) shells from 4.2″ chemical mortars in the Alabama hills. A two-month course in quantitative analysis at Edgewood Arsenal, Maryland, rounded out the chemistry. While at Edgewood I became a U.S. citizen in the U.S. District Court in Baltimore. By January of 1945, I had left the Chemical Warfare Service and was assigned to Camp Ritchie, Maryland, for training in military intelligence. After two months at Ritchie and six weeks at the University of Pennsylvania to learn basic German ("wo ist der Bahnhof?"), I was ready to join the war effort. In early May 1945, a few days before VE Day, we assembled in Fort Totten, New York, were flown to Paris with fuel stops in Bermuda and Gander, Newfoundland, were issued our jeeps, and traveled to Third Army Headquarters in Bavaria. There our team of 12 CIC agents received orders—to establish a post in Eslarn, a village a few miles from the German-Czech border. Aside from temporary duty at the Nuremberg war crimes trials, my 14 months as a Special Agent were uneventful. Return to the United States aboard the S.S. *Aiken Victory* was distinctly unglamorous—12 days in a hold with six tiers of bunks during very rough weather. I received my discharge at Fort Devens, Massachusetts, having risen to the rank of Staff Sergeant. After a stopgap job on the assembly line in a plastics extrusion plant, I was ready to resume my education in September 1946, now to be financed under the G.I. Bill.

Postwar Harvard was a far cry from its skeletal existence during the war years. As was true all over the country, returned veterans together with the normal enrollment swelled the student ranks. The new organic instructors included Gilbert Stork and Morris Kupchan. I moved into a dorm, Perkins Hall, where my several roommates included Aksel Bothner-by. Though this was my first dorm experience, at the age of 31, what with a fireplace in a large living room and maid service, it bore no resemblance to Camp Sibert, Alabama, where two "space heaters" served the whole company.

Discovering Natural Products

The Woodward research group had become large. It occupied many widely dispersed labs in Converse and Mallinckrodt. Despite the group meetings, which now were held in the evening and rarely ended before midnight, it was difficult to know everybody. Dick Eastman was the only holdover from the war years. He had graduated but stayed on as postdoc to manage the group while R.B. drove his old jalopy to Reno to be divorced from his first wife. Structure and synthesis of strychnine and steroid synthesis were the dominant projects. Somehow, my unrewarding ketene experiments of 1943 had been forgotten. Instead, R.B. handed me a 500-gram bottle of crystalline strychnine (**1**) and asked me to follow a published procedure to prepare *neo*-strychnine (**2**) [1]. It was to be an intermediate that might resolve the principal difference between the Woodward (**1**) and Robinson (**3**) expressions, a six-membered versus a five-membered ring VI. In *neo*-strychnine the double bond of ring VII had isomerized to ring VI, thereby setting the stage for cleavage and reclosure of ring VI, which would provide direct evidence for the size of that ring. In time, the *neo*-strychnine route was abandoned. Warren Brehm's research [2] had provided conclusive evidence for the Woodward structure (**1**). Synthesis plans were in full swing, and useful relay compounds were in demand. So it became my task to establish an unassailable structure for the hitherto elusive oxostrychnine [3, 4]. I accomplished the project, but R.B. maintained only a marginal interest. When I needed advice, I consulted Gilbert Stork, as did most Woodward students who were not working on the hot project of the hour.

1

2

3

The course entitled "The Chemistry of Natural Products" was R.B.'s teaching assignment, also inherited from Patrick Linstead—complete with lecture notes. (I can still hear RB.'s chuckle at the mention of it.) According to the university catalog, it met Tuesday, Thursday, and, at the pleasure of the instructor, on Saturday at 11. In his first lecture, R.B. made it clear that it was rarely the pleasure of the instructor to lecture on Saturday. In fact, there never were any Saturday lectures. But every single lecture was an intellectual summit and an aesthetic pleasure. Woodward had, of course, no notes. Colored chalk was the single visual aid. At the end of the hour, the single large blackboard would be covered with beautifully drawn structural formulas, each in a place that would relate it perfectly to its antecedents, contemporaries, and descendants. No eraser was ever used. Each lecture strengthened my desire to become involved in the study of natural products.

At the impressive and exciting commencement exercises in Harvard Yard, I (with hundreds of others) was welcomed into the society of scholars; this event, however, could not obscure the fact that I did not have a job in early June of 1950. I had interviewed at an industrial laboratory in upstate New York, an unexciting prospect at best. Bernd Witkop was moving to the National Institutes of Health (NIH) and had suggested that I might want to join him there. That was a viable option when Gilbert Stork, on the morning of June 24, called the lab to inform us of a visit by a chemistry department chairwoman from the University of Hawai'i, who was on a faculty recruiting trip. The interview with Leonora Bilger resulted on July 13 in an offer of an assistant professorship at an annual salary of $4,140 plus a monthly territorial "bonus" of $48. My ignorance of Hawai'i and of its university was profound. I had not even seen a catalog. Consulting fellow graduate students only revealed that my lack of knowledge was not unique. Only one of my contemporaries had visited Honolulu during his World War II naval service. But for me, somehow, the absence of facts only heightened the sense of adventure. Not to mention my resolve in the 1943 winter, when I dragged chemical mortars through muddy Alabama hills, never to be cold again once the war was over. I was ready to accept the offer and my fiancée, Alice Dash, who had been in my organic lab section during her senior year at Radcliffe, was willing to share this somewhat nebulous future. Alice, though born and raised on the eastern seaboard, knew more about Hawai'i than anyone else around, but even she knew nothing about the university.

Destination: Honolulu

On September 5, Alice and I were married in the Harvard Chapel, and the ceremony was followed by a reception in the Mallinckrodt conference room. On the following day we flew to San Francisco. On September 8 we boarded the S.S. *Lurline* for the trip to Hawai'i. At sunrise on September 13, the *Lurline* slowed near Makapu'u Lighthouse to take on the harbor pilot and scores of women laden with fragrant plumeria (*Plumeria acuminata*) leis. We slowly steamed toward a green mountainous island in a deep blue sea. It was an unforgettable beginning of a new life in a part of the world that I had never even read about.

We were met at the pier by two members of the Chemistry Department, who drove us to Faculty Housing on the university campus. A studio apartment, one of four in a converted army barracks, had been reserved for us. It rented for $35/month plus $6 for utilities. It was furnished with a refrigerator, a stove, two army cots, a table, a chest of drawers, and four chairs. The University farm, cows, chickens, and pigs were our nearest neighbors. If we wanted adventure, we surely were having it!

In addition to Mrs. Bilger and her husband, there were two veteran faculty members. One of these was on leave of absence. The remaining six faculty members were a sophomore (who would leave at the end of his second year), a recently retired visiting professor, three new Ph.D.'s, and a part-time instructor, who doubled as the Department's sole Ph.D. candidate. This situation apparently was not unique. Faculty turnover was rapid, with Mrs. B. in firm control.

One of the attractions of the University of Hawai'i (UH) position was a new chemistry building, which was still under construction. Before it was ready to be occupied, my quarters were in—guess what?—a reconfigured army barracks. This precluded my starting my own research, but I rapidly had two M.S. students to supervise. Shortly before the start of the semester, Mrs. B. informed me that I was to supervise a second-year M.S. student. My joy over this unexpected good fortune was dampened when I was informed that all M.S. candidates must complete a research project and receive their degrees after two years' residency.

I knew that I wanted to do research in natural products, but I had not formulated a single specific project. And, clearly, time was of the essence. My first two M.S. students, Seiji Sakata ('51) and Kiichi Ohinata ('52), did their research on projects that had occurred to me in the course of my strychnine studies. This gave me much needed breathing space to learn about teaching, about UH, and about Hawai'i, its people, its culture, and its natural resources. I rapidly became acquainted with scientists in the plant, marine, and agricultural sciences. I soon discovered that Hawai'i would be an ideal place for natural products research. In fact, the two major agricultural industries, sugar and pineapple, maintained excellent institutes, where applied as well as some fundamental research was conducted.

Because of the isolation of the Hawaiian islands, 3,800 km from the nearest substantial landmass, Hawai'i has a rich endemic flora which no chemist had ever studied. Early on, I audited Harold St. John's (Harvard'14, Ph.D.'17) course in Hawaiian botany. It included weekly field trips, which taught me some Polynesian botany and opened my eyes to the incredible natural beauty and biological diversity of the island of O'ahu.

Based on archaeological findings, the earliest Polynesian settlers arrived in Hawai'i from the Marquesas Islands in their voyaging canoes around 650 A.D. Several hundred years later, a second group reached Hawai'i, probably from Tahiti. Captain James Cook is credited with the first Western contact in 1778, which is well documented [5]. Although no evidence of any earlier European visitors to Hawai'i has ever been discovered, a 1579 Dutch map shows a group of islands, "Los Volcanoes," at the correct latitude but 1600 km east of the correct location. Spaniards en route from Mexico to the Philippines may well have known of the Hawaiian islands in the sixteenth century but apparently made no contact. The error in longitude is not surprising, since only after ships' chronometers were invented in the second half of the eighteenth century was it possible for sailors to log their east-west progress accurately.

Terrestrial Natural Products

When the Polynesian settlers arrived in Hawai'i, they brought with them 24 food and fiber plants [6] as well as pigs, dogs, and fowl. The only endemic mammals in Hawai'i were a monk seal and two species of bat. In addition to essential food staples, among them taro, breadfruit, sweet potato, coconut, banana, and sugarcane, there was 'awa (or *kava*, *Piper methysticum*), which furnishes the ceremonial beverage of that name. *P. methysticum* is pan-Pacific and has been used by native peoples from Indonesia to Hawai'i. It aroused the interest of the early explorers and its chemical literature dates back to the nineteenth century. Early reports of its intoxicating and paralytic action proved to be unfounded but, like many firmly held misconceptions, died but slowly [7]. Perhaps these accounts reflect sensations experienced by Europeans unaccustomed to squatting or sitting on the ground for long periods of time during an 'awa ceremony. Rudolf Hänsel of the Freie Universität, Berlin, was the first to demonstrate the soporific and nonaddictive properties of a major 'awa constituent, dihydromethysticin (**4**) [8]. My own interest in 'awa chemistry [9] brought Professor Hänsel to my laboratory in 1961, where he had an opportunity to collect 'awa on the island of O'ahu and to establish a connection with 'awa farmers on the island of Hawai'i. Over the past

Die typische altpolynesische Kava-Zubereitung geschah folgendermaßen:

Der Wurzelstock von PIPER METHYSTICUM wurde von den Eingeborenen gesäubert und in kleine Stücke zerteilt. Diese wurden von jungen Leuten mit guten Zähnen zerkaut und in einer Kava-Schale gesammelt. Dann wurde Wasser zu der Masse gegeben und kräftig verrührt. Mit einem Hibiskusbast-Bündel wurden alle Wurzelpartikelchen entfernt. Damit war der Kava-Trank fertig zur Austeilung. Heute wird die Kava-Wurzel nicht mehr gekaut, sondern mit einem Stößel zerrieben. Das haben Missionare, die diese Zubereitungsmethode als unhygienisch empfanden, veranlaßt.

Fig. 1 From a brochure on 'awa' by the German pharmaceutical company Krewel Meuselbach GmbH.

35 years, medicinal use of 'awa in Germany as a nonaddictive sedative and soporific, resulting from Rudi Hänsel's research, has become well established (Fig. 1).

Learning of the Ocean

An even more dramatic and new experience than Polynesian rain forests was the ocean—warm, blue, and rich in animals and plants of which I was totally ignorant. As I became acquainted with faculty members in marine biology and learned of their research interests, I soon discovered a vast and virtually untapped resource for natural products research.

Sea urchins, colored red to purple, many covered with calcareous spines, are conspicuous sessile invertebrates on Hawai'i's rocky shores. I first saw members of this exclusively marine phylum of animals at Hanauma Bay, a submerged crater on O'ahu's southeast shore and one of the island's spectacular scenic attractions. Discussions with a faculty member in the Zoology Department suggested that little research had been done on the pigments of sea urchin shells and spines. A look into the chemical literature readily confirmed this. European and Japanese workers had studied sea urchin pigments, but only a single structure, that of echinochrome A (**5**), had been confirmed by synthesis [10]. A Friedel-Crafts reaction in an AlCl$_3$-NaCl melt at 180 °C produced the desired compound in 1.5–2 % yield. In an age when elemental analysis, melting points, and UV–visible spectra were the only accessible

physical parameters, confirmation by synthesis was vital. I was ready to explore the void in our chemical knowledge of sea urchin pigments.

So it came about that a graduate student, Piobert Amai, my wife, and I, armed with screwdrivers, descended on Hanauma Bay to collect *Echinometra oblonga* [11]. (The Bay has been a marine preserve since 1967 and a City and County Park since 1978.) The research did not result in an unambiguous structure proof. It served to highlight the inadequacy of the separation method that the workers in the field practiced. The accepted technique was chromatography on a column of calcium carbonate. Colored bands did indeed develop, but—not surprisingly—nothing would move. After all, when the pigments are part of the live animal, they are tightly adsorbed on calcium carbonate of the shells and spines. A way had to be found to separate these closely related compounds, derivatives of naphthazarin or juglone differing only in minor structural features. A systematic and comprehensive investigation by Clifford Chang [12] provided the answer: chromatography on severely deactivated (washed with 0.5 N HCl, followed by air drying) silica gel separated the individual pigments beautifully and finally opened up research on echinoderm pigments. Hand in hand with the discovery of how to separate effectively these acidic naphthaquinone derivatives was the acquisition of a Vartan A-60 NMR spectrometer. At last, there was a new and vital

Fig. 2 *The author and R.E. Moore looking at the Palythoa toxica tide pool near Hana, Maui, in May, 1963.*

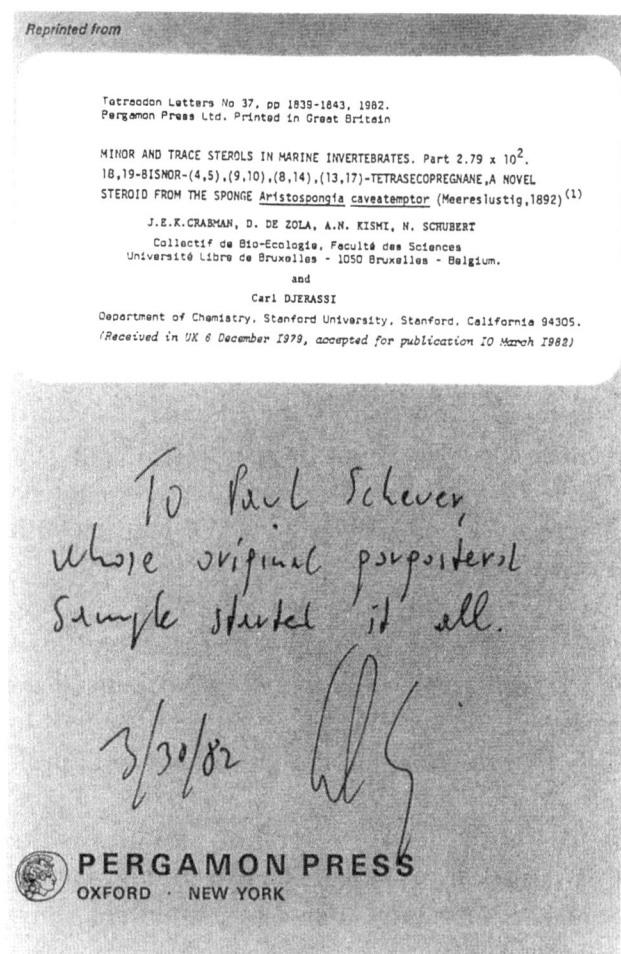

Fig. 3 *A spoof* Tetrahedron Letters *reprint produced in 1982 to celebrate Carl Djerassi's 1000th publication.*

tool, which finally made possible a definitive structure of spinochrome M (**6**) [13].

A decisive influence on the direction of my future research came from my friendship with the late A.H. (Hank) Banner, a marine biology professor. While stationed in the Pacific during World War II with the U.S. Air Force, Banner became aware of ciguatera fish poisoning. In 1957 he assembled a multidisciplinary research team to study ciguatera and invited me to participate. The major features of this involvement have been reported, as have some of the inadvertent by-products of ciguatera research [14], Palytoxin was perhaps the most significant and spectacular ciguatera offspring [15] (Fig. 2).

Chemotaxonomic relationships have long been a valuable diagnostic feature of natural products research of terrestrial flowering plants. Our comparative study of echinoderm pigments [16] led us to the work of Bergmann, who since the early 1930s had used sterols as chemotaxonomic markers in his study of marine invertebrates [17]. It was indeed a challenge if one considers the state of the art in chromatography and spectroscopy. In 1943 he isolated from a Caribbean gor-

gonian, *Plexaura flexuosa,* an unusually high-melting (180–182 °C) sterol, with likely composition $C_{30}H_{50-52}O$, which he called gorgosterol [18]. Twenty-five years later, Bergmann's student Ciereszko encountered gorgosterol in a number of coelenterates and secured the formula of $C_{30}H_{50}O$ by mass spectrometry [19]. My student Kishan Gupta isolated gorgosterol from a *Palythoa* sp., where it was present in a mixture of five sterols [20]. It had a long retention time on GLC, where it could be separated as its volatile trimethylsilyl ether. An authentic sample from Ciereszko confirmed identity. Our Varian A-60 NMR instrument was unable to probe its structural details. We turned to Carl Djerassi and his 100-MHz NMR instrument. He was able to solve the structure of gorgosterol (**7**) [21] and—as a veteran sterol chemist—became hooked on the study, particularly the biosynthesis, of marine sterols. When Djerassi's co-workers helped him celebrate his publication millennium in 1982, they created a spoof communication disguised as a *Tetrahedron Letter* (Fig. 3).

I received a copy with the dedication "To Paul Scheuer, whose original gorgosterol sample started it all."

Chemical Marine Ecology

My interest in ecology was piqued by my acquaintance with Bob Johannes, then a graduate student in the Zoology Department. He had added a nudibranch, *Phyllidia varicosa,* to his aquarium and discovered that all of his fish and shrimp soon died [22]. He traced the cause to an unpleasant-smelling skin secretion of the nudibranch, which he discovered by stroking the animal. He showed by dialysis and heating to 100 °C that the secretion was a small, heat-stable molecule. Although we spread the word—including in the local newsletter of the malacologists—that we would like to study the *P. varicosa* skin secretion, progress was slow. Once we received a call from Maui reporting a sighting of a school (?) of the mollusks in Ma'alaea Bay. Two co-workers promptly flew to Maui and returned with bags of Maui's famous potato chips, but without mollusks. At last, in the summer of 1973, Jay Burreson spotted the animal with its

characteristic yellow and blue coloration at Pupukea Bay—on O'ahu's north shore. He rapidly verified Johannes's observations. More significantly, he discovered that within a few days in captivity the skin secretion of the mollusk would dwindle to zero. An inquiry to a noted malacologist brought no inkling of the nudibranch's diet. It would be another year, another dive season at Pupukea, which is inaccessible during the winter, when Burreson was in luck: he observed the animal feeding on an off-white sponge subsequently identified as *Ciocalypta* sp. (Fig. 4). The sponge indeed was the source of the *P. varicosa* secretion. Moreover, the chemistry was also rewarding: 9-isocyanopupukeanane (**8**) was a sesquiterpene with a new skeleton and a rare isocyano function [23]. Isocyano natural products, most of them terpenoids of marine origin, have blossomed during the past 20 years into a fruitful area of research. The biological activities of isocyano compounds include antifouling and antimalarial activities [24]. The trivial name pupukeanane for this new sesquiterpene skeleton continues my practice, begun in 1961, of paying tribute to the rich flora and fauna of Hawai'i by coining names that use the sonorous Hawaiian language.

Fig. 4 *Nudibranch* Phyllidia varicosa *feeding on a sponge,* Ciocalypta *sp.*

7

8

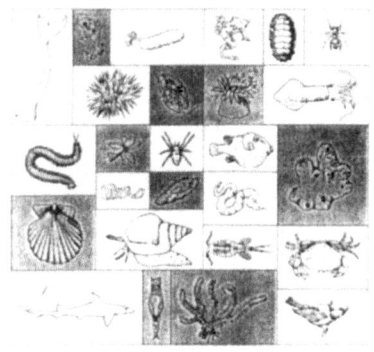

Chemistry of Marine Natural Products

PAUL J. SCHEUER

ACADEMIC PRESS

Fig. 5 *Dustcover of the author's 1973 book.*

During the 1960s my research emphasis gradually gravitated from terrestrial to marine natural products. Although I published terrestrially oriented papers through the mid-1970s, the idea of writing a monograph on marine natural products began to germinate in the late 1960s. *Chemistry of Marine Natural Products* was published in 1973 (Fig. 5) and was the first book on the subject in any language [25].

Pupukea Bay, with its underwater caves and lava tubes, has remained an attractive dive spot (Fig. 6). I appreciated it only after 1981, when, on the persuasive urging of my graduate student Gary Schulte, I took scuba lessons and became a certified diver. Gary, while an undergraduate researcher in Bill Fenical's group at the University of California at San Diego, had become interested in mollusks. On one of our night excursions to Pupukea in the late 1970s, he collected some mollusks in the tidepools. They seemed to appear with the incoming tide to feed and mate. They were identified as *Philinopsis speciosa,* members of the gastropod order Cephalaspidea. According to conventional wisdom of the time, the anticipated metabolites were of polypropionate biogenesis. This prediction was borne out: we isolated two unexceptional polypropionate derivatives [26] and an alkyl-pyridine [27] reminiscent of an alarm pheromone constituent of a closely related mollusk, *Navanax inermis* [28]. More significant than these results proved to be an observation by Steve Coval that the polar extract of *P. speciosa* appeared to contain a peptide, isolation of this peptide and determination of its structure and stereochemistry proved to be an enduring experience. In the light of the long time lapse between our

Fig. 6 *Pupukea Bay, O'ahu.*

first collections and a definitive peptide structure, we named the peptide kulolide (**9**) from the Hawaiian word *kulo* = taking a long time.

Two factors contributed prominently to the longevity of the kulolide project: instrumentation and ecology. Our 300-MHz NMR instrument in the early 1980s proved to be inadequate to resolve the complicated spectra of a $C_{43}H_{63}N_5O_9$ (794 Da) molecule, which existed in two conformations in all solvents over a wide temperature range. Acquisition of a 500-MHz instrument in 1989 resolved this difficulty. Attempts to determine the absolute stereochemistry of the two component acids, 3-phenyllactic and 2,2-dimethyl-3-hydroxy-7-octynoic acid, were frustrated by our dwindling supply of kulolide and by our failure to collect more animals.

Knowledge about the organismic biology and ecology of many marine invertebrates is rather limited. Alison Kay's authoritative treatise [29] lists no specific food source for *P. speciosa;* the family Philinacea to which it belongs are carnivores, feeding on foraminiferans, worms, and mollusks. A potential lead appeared in the work by Cimino and co-workers, who had linked the metabolites of the Mediterranean *Philinopsis depicta* (syn. *Aglaja depicta*) to its prey, the cephalaspidean mollusk *Bulla striata* [30]. According to Kay, the Hawaiian Bullidae are subject to dramatic fluctuations in population, ranging from virtually zero to a thousand shells at a single location [29]. During many years of collecting at Pupukea, we in fact never saw more than a few *Bulla* specimens on any one occasion.

9

10 R = Br

11 R = H

As luck would have it, ecology came to our rescue, and chemistry, in turn, solved the ecological puzzle. Both 1994 and 1995 seasons brought large populations of *P. speciosa* to Pupukea, which allowed us to re-isolate kulolide (9) and

resolve the remaining stereochemical lacunae, many years after our first collection [31]. The bountiful harvest of *P. speciosa,* followed by the tedious re-isolation of kulolide (9), provided a clue to the food source of the animal. During chromatography of the crude extract, it became apparent that kulolide was the major, but by no means the only, peptide in *P. speciosa.* In addition to several kulolide-related cyclic depsipeptides, there was a linear tetrapeptide, which we called pupukeamide [32]. It proved to be the key that would answer the question, "What do *P. speciosa* eat?" The structure of pupukeamide was instantly reminiscent of the majusculamides, which Moore and co-workers had isolated from the marine blue-green alga *Lyngbya majuscula* [33]. But since *P. speciosa* is a carnivore, what is the missing link?

As luck would have it, the answer was found in my own research records. *L. majuscula* had been implicated as the source of a contact dermatitis, "swimmers' itch" [34, 35]. We had briefly investigated it as a possible origin of ciguatera fish poisoning, with negative results [14]. Watson's research into the toxins of sea hares (Notaspidean mollusks) was another potshot and unproductive approach to trace the ciguatera toxin [36]. While the original purpose failed, Watson's "ether-soluble toxin" seemed a worthwhile research objective. In time, we succeeded in determining the structure of aplysiatoxin (10) and debromoaplysiatoxin (11) from *Stylocheilus longicaudus,* a small but relatively abundant sea hare [37] (Fig. 7). It was the last major piece of research in my laboratory carried out without the benefit of HPLC or high-field NMR instrumentation on 12 g of crude toxin. Several years later, Moore and co-workers isolated debromoaplysiatoxin (11) from *Lyngbya majuscula,* the same blue-green alga that had yielded the majusculainides. Now it was crystal clear: *Stylocheilus,* who feed on *Lyngbya,* are in turn eaten by *Philinopsis.* In fact, we had noted *Stylocheilus* in the Pupukea area but disregarded the animals, which had been studied long ago. But now we have a new puzzle: Why are there no aplysiatoxin-like compounds in *Philinopsis?* Like old soldiers, good research projects never die!

Drug Discovery

Dramatic biological activity and structures of natural products have formed an interwoven fabric since the science began, although emphasis has fluctuated. After a few decades in which the study of natural products took a back seat to laboratory synthesis and theory, natural products once again are looked upon as important drug leads. Not only is this an obvious liaison, but in an age when most funding agencies look for societal benefits even while advancing basic knowledge, it has become a *sine qua non.* New anticancer agents have been the targets of choice ever since the establishment of the National Cancer Institute as a visible symbol of Richard Nixon's "War on Cancer." One of our candidates currently in preclinical trials against lung and colon cancers is

Fig. 7 *Sea hare* Stylocheilus longicaudus.

Fig. 8 *Sacoglossan mollusk* Elysia rufescens.

kahalalide F (**12**) [38]. It was an unexpected isolate from a herbivorous sacoglossan mollusk, *Elysia rufescens* (Fig. 8), as the chemical and biological literature had led us to anticipate terpenoid constituents. The animal feeds on a green alga, *Bryopsis* sp., from which we were able to isolate kahalalide F as well as an acyclic analogue as the major constituent.

12

Another significant product from my laboratory, while not a potential drug, has become a commercial chemical for use in molecular biology. Okadaic acid (**13**), first isolated from a sponge, *Halichondria okadai* [39], selectively inhib-

its phosphatases 1 and 2A. It is widely used for the study of cellular mechanisms.

What of the Future?

What with genetic engineering, combinatorial drug design, and detailed mapping of receptor sites, are we not witnessing the last hurrah of natural products research? I think not. There will never be a substitute for discovering new molecular architecture and for learning in greater detail the molecular composition of living organisms. As we all have experienced in our lifetime, technological advances ensure that we continue to learn more from ever smaller samples. What about marine research? Is it not true that with increasing frequency we isolate known compounds? Indeed, this is the case, but two developments promise continued success. In parallel with the history of terrestrial natural products research, when microorganisms succeeded flowering plants as prime targets, marine bacteria and fungi are increasingly being scrutinized for new natural products. And their numbers are vast. A second exciting development is the use of rebreathing technology. This will allow ocean exploration at greater depths and for longer periods of time. Although mini-submersibles are wonderful inventions—and I cherish the memories of my first sight of bioluminescent gorgonian corals or suddenly viewing the wreck of a World War II dive-bomber (ours) off Makapuu—they are not well suited for collecting, and they are expensive to operate. Rebreathing apparatus will allow a pair of human, rather than mechanical, hands to collect, say, bryozoans or encrusting sponges, which are beyond the capacity of scuba. Rebreathers will also provide access to a depth zone which up to now was out of reach. And finally, as we begin to study and understand the biosynthesis and genetics of marine organisms, exciting new natural products will be produced in quantity by fermentation technology.

13

Scheuer family, May 1995. Front row (l to r), Tim Carlson, Debbie, Paul, Alice, Jonathan; back row, David, Elizabeth Carlson.

Acknowledgment I have indeed been fortunate to have traveled from the tanning vats in a German leather factory to the coral reefs of the "loveliest fleet of islands that lies anchored in any ocean" [40]. This journey has been made possible and rewarding by the constant support of my wife Alice and our children Elizabeth, Deborah, David, and Jonathan. Without the talent and devotion of my co-workers, there would be little to write about. Among the funding agencies that have supported my work, I must mention Research Corporation, which gave me a start in 1951 with a $2,800 grant. Continued funding over the years has come from many sources, but most steadfastly from the National Science Foundation, the Sea Grant College Program, and PharmaMar, S.A.

References

1. Kotake, M.; Yokoyama, M. *Sci. Pap. Inst. Phys. Chem. Res. (Tokyo)* **1937**, *31*, 321.
2. Woodward, R. B.; Brehm, W. J. *J. Am. Chem. Soc.* **1947**, *70*, 2107.
3. Scheuer, P. J. Ph.D. Dissertation. Harvard University, 1950.
4. Scheuer, P. J. *J. Am. Chem. Soc.* **1960**, *82*, 193.
5. Kippis, A. *A Narrative of the Voyages round the World Performed by Captain James Cook*; Leavitt and Allen: New York, 1855; Vol. II, p 111.
6. Hillebrand, W. *Flora of the Hawaiian Islands;* Williams and Norgate: London, 1888; p. XIII ff.
7. *The Merck Index*, 12th Ed., p 1047 (Merck & Co.: Whitehouse Station, NJ, 1996) still refers to 'awa as an intoxicating beverage.
8. (a) Hansel, R.; Beiersdorf, H. U. *Naturwissenschaften* **1958**, *45*, 573; (b) Hänsel, R. Z. *Phytother.* **1996**, *17*, 180.
9. Scheuer. P. J.; Horigan, T. J. *Nature* **1959**, *184*, 979.
10. Wallenfels, K.; Gauhe, A. *Chem. Ber.* **1943**, *76*, 325.
11. Amai, R. L. S. M.S. Thesis, University of Hawai'i, 1956.
12. Chang, C. W. J. Ph.D. Dissertation, University of Hawai'i, 1964.
13. Chang, C. W. J.; Moore, R. E.; Scheuer, P. J. *J. Am. Chem. Soc.* **1964**, *86*, 2959.
14. Scheuer, P. J. *Tetrahedron* **1994**, *50*, 1.
15. Moore, R. E.; Scheuer, P. J. *Science* **1971**, *172*, 495.
16. Singh, H.; Moore, R. E.; Scheuer, P. J. *Experientia* **1967**, *23*, 624.
17. Bergmann, W.; Domsky, I.I. *Ann. N.Y. Acad. Sci.* **1960**, *90*, 906.
18. Bergmann, W.; McLean, M. L.; Lester, D. *J. Org. Chem.* **1943**, *8*, 271.
19. Ciereszko, L. S.; Johnson. M. A.; Schmidt, R. W.; Koons, C. B. *Comp. Biochem. Physiol.* **1968**, *24*, 899.
20. Gupta, K. C. Ph.D. Dissertation, University of Hawai'i, 1967.
21. Hale, R. L.; Leclercq, J.; Tursch, B.; Djerassi. C.; Gross, R. A., Jr.; Weinheimer, A. J.; Gupta, K.; Scheuer, P. J. *J. Am. Chem. Soc.* **1970**, *92*, 2179.
22. Johannes R. E. *Veliger* **1963**, *5*, 104.
23. Burreson, B. J.; Scheuer P. J.; Finer, J.; Clardy, J. *J. Am. Chem. Soc.* **1975**, *97*, 4763.
24. Chang, C. W. J. "Naturally Occurring Isocyano/Isothiocyanato and Related Compounds', *Prog. Chem. Nat. Prod.*, in press.
25. Scheuer, P. J. *Chemistry of Marine Natural Products;* Academic Press: New York, 1973.
26. Coval, S. J.; Schulte, G. R.; Matsumoto, G. K.; Roll, D. M.; Scheuer, P. J. *Tetrahedron Lett* **1985**, *26*, 5359.
27. Coval, S. J.; Scheuer, P. J. *J. Org. Chem.* **1985**, *50*, 3025.
28. Sleeper, H. L.; Fenical, W. *J. Am. Chem. Soc.* **1977**, *99*, 2367.
29. Kay, E. A. *Hawaiian Marine Shells,* Bishop Muscum Spccial Publication 64(4); Bishop Museum Press: Honolulu, HI, 1979; pp 430–431.
30. Cimino, G.; Sodano, G.; Spinella, A. *J. Org. Chem.* **1987**, *52*, 5326.
31. Reese, M. T.; Gulavita, N. K.; Nakao, Y.; Hamann, M. T.; Yoshida, W. Y.; Coval, S. J.; Scheuer, P. J. *J. Am. Chem. Soc.* in press.
32. Nakao, Y.; Yoshida, W. Y.; Scheuer, P. J. *Tetrahedron Lett.*, in press.
33. Marner, F. J.; Moore, R. E.; Hirotsu, K.; Clardy, J, *J. Org. Chem.* **1977**, *42*, 2815.
34. Banner, A. H. *Hawaii Med. J.* **1959**, *19*, 35.
35. Moikeha, S. N.; Chu, G. W.; Berger, L. R. *J. Phycoil.* **1971**, 7.
36. Watson, M. *Toxicon* **1973**, *11*, 259.
37. Kalo, Y.; Scheuer, P. J. *J. Am. Chem. Soc.* **1974**, *96*, 2245.
38. Hamann, M. T.; Scheuer, P. J. *J. Am. Chem. Soc.* **1993**, *115*, 5825.
39. Tachibana, K.; Scheuer, P. J.; Tsukitani, Y.; Kikuchi, D.; Van Engen, D.; Clardy, J.; Gopichand, Y.; Schmitz, F. J. *J. Am. Chem. Soc.* **1981**, *103*, 2469.
40. Clemens, S. L. (Mark Twain). On his visit to Hawai'i in 1866 as a reporter for the Sacramento *Weekly Union.*

Paul J. Scheuer was born in Heilbronn, Germany, in 1915. He emigrated to the United States in 1938 and graduated from Northeastern University, Boston, with a B.S. in Chemistry in 1943. His graduate studies at Harvard, interrupted by service in the U.S. Army, led to a Ph.D. in organic chemistry in 1950, when he was appointed Assistant Professor at the University of Hawai'i. Since 1985 he has been Professor Emeritus. He continues to direct a research group investigating natural products—originally of terrestrial plants, but now exclusively of marine organisms. He has directed the research of 33 M.S. and 25 Ph.D. students, over 100 postdoctoral associates, and numerous undergraduates. More than 260 publications, including 1 authored and 12 edited books, have resulted from this work. In 1992 his former students initiated a Paul J. Scheuer Award in Marine Natural Products, of which he was the first recipient. In 1994 he received the Ernest Guenther Award of the American Chemical Society and the Research Achievement Award for the American Society of Pharmacognosy. Northeastern University recognized him with a Distinguished Alumni Award in 1984. He was cited in 1972 for Excellence in Research by the Regents of the University of Hawai'i.

The "Wall of Fame" in the Chemistry Department at the Technion, Haifa[a]

After years in a temporary wooden building in the center of Haifa, the chemistry department at the Technion moved in 1964 to new larger quarters at Technion City, the Mount Carmel campus. The new location consisted of three inter-connected structures: the research wing (Canada Building), an administration and student laboratory wing (Compton Building), and a classroom wing (Bernstein Building). Structural settling, a common occurrence with new build-ings, resulted in a large, ugly crack over the blackboards across the front wall of the main lecture theater in the Bernstein Building. Consultation with the man responsible for the aesthetics of interior decoration brought forth a sug-gestion that the offending area be covered with a large, read-able periodic table. This did not satisfy David Ginsburg [1], the founding father of modern chemistry at the Technion and the then department chairman; it was too commonplace. It led, however, to the idea of a human periodic table bearing the names of the leaders of chemistry through the ages.

The area was divided into separate interfitting name plates made of Formica on which the appropriate names would be inscribed and which would then be arranged in alphabetical order with sufficient blank spaces to allow for future addi-tions. The size of each plate was 19 cm × 39.5 cm so that each name would be clearly seen (if the signature was decipher-able) even in the last row of the room. Committees were set up in the various areas of chemistry to propose names for inclusion on the wall. The names of earlier chemists were presented in printed letters [2], and those of living chemists were represented by enlarged reproductions of their signa-tures. The letter below was sent to about 100 people asking each to send us a signature.

Letter from David Ginsburg
May 1964

It may interest you to learn that in the main lecture hall of our new Chemistry Building we are planning to cover the wall behind the lecturer with facsimile signatures of great chemists who have made important contributions to various fields of chemistry. I am sending out this letter to about one hundred chemists the world over asking for a facsimile signature. We then intend to add about 2–3 additional names annually to this chemical hall of fame.

We should be honored if you would be kind enough to send us a facsimile of your signature, written on ordinary paper with a thick pen and about twice the size of your usual signature. Our students will then be in the fortunate position of learning, also by this means, the names of the leaders of chemistry in the world today.

Reminders were hardly necessary, and we even had some good wishes (two, in fact) expressed for the future of the department occupying this building. Figure 1 shows a gen-eral view of the Wall as it appears at the present time (note the small periodic table on the left side of the room). In addi-tion, a small reproduction appears on the back wall of the lecture hall with the names (all in print) and locations and, in the case of Nobel prizewinners, the year in which the Nobel Prize was awarded. This small version bears the inscription (in Hebrew and English) from Daniel 1:4 "Cunning in knowledge and knowing science." Close-up views of the signatures of the four Zurich diners mentioned below appear in Fig. 2.

In the years immediately following installation of the Wall, the annual fall announcement of the Nobel Prize in chemistry was the occasion for some tension in Haifa. Would the name of the current year's awardee(s) already be on the Wall or had we goofed? With one or two embarrassing exceptions, it turned out for a good many years that the name was already on the Wall. Now, more than 50 years later, the practice is to periodically update the Wall with the names of the latest Nobel laureates. Discretion overpowered valor from the beginning, and the rule is that the only Israeli chem-ists to appear on the wall will be Nobel laureates. There is still room on the wall, and eventually it could be expanded to the side walls of the lecture hall if future generations so desire.

In time, the Wall acquired the name "The Wall of Fame," and chemists whose names appeared on it were sometimes irreverently referred to as "wallflowers." We have had much fun from the Wall. It is the first stop on any tour of the chemistry department, particularly with a visitor whose name appears there. It is obviously educational for students and faculty. It is rewarding when lecturing to students to be able to add,

[a] *Chemical Intelligencer* 1997(3), 44–49.

[b] Department of Chemistry, Technion—Israel Institute of Technology, Haifa, Israel (deceased)

B. Hargittai and I. Hargittai (eds.), *Culture of Chemistry: The Best Articles on the Human Side of 20th-Century Chemistry from the Archives of the Chemical Intelligencer*, DOI 10.1007/978-1-4899-7565-2_44, © Springer Science+Business Media New York 2015

229

Fig. 1 View from the rear of the main lecture hall at the Technion Chemistry Department.

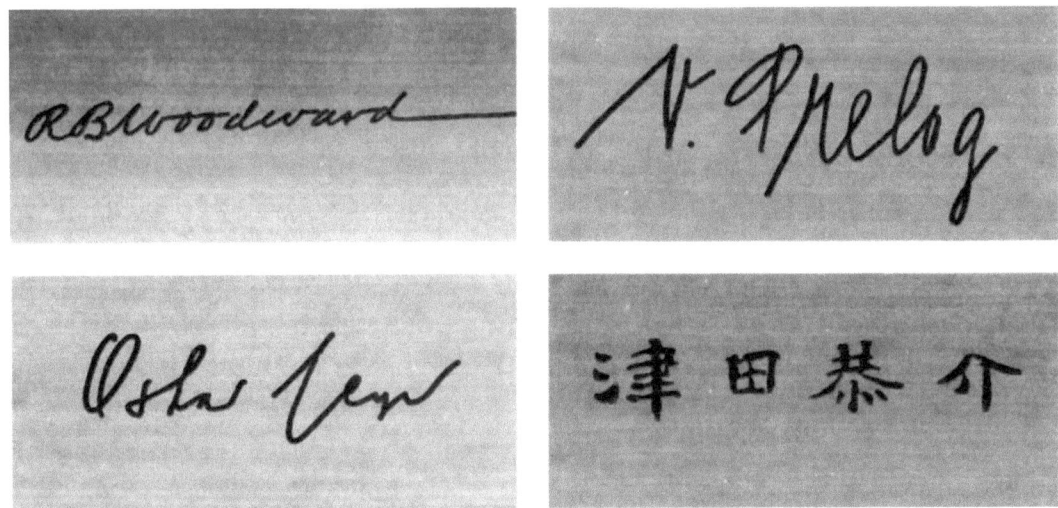

Fig. 2 Signatures of the four Zurich diners on the Wall. Across: R.B. Woodward, V. Prelog, Oskar Jeger, K. Tsuda.

"there is his name on the wall behind me." Our students have had an April 1st custom of adding names of unlikely candidates; fortunately, this has always been done with erasable writing materials. We keep the projection screen up except during lectures so that the entire wall is visible; only the sides appear when the screen is down. Prior to a lecture (and sometimes during), it is instructive to while away some time with thoughts on the names appearing on the wall.

The first name is that of Anatole Abragam, a physicist who was included because of his pioneering work in nuclear magnetic resonance. His reply to the letter above follows [3]:

May 1964
Dear Professor Ginsburg,
Your very kind proposal to have my name on a wall as a great chemist demonstrated that either you do not know me or that I do not know what a chemist is. My ignorance of chemistry makes me a laughing stock among my young collaborators and if the gravity of your functions did not put you above suspicion, I would have suspected that someone was pulling my leg.
If you define chemistry as the study of properties of matter in bulk I might perhaps qualify as a chemist, but is not that a little farfetched. However, if in the light of what I have said you still wish to keep me on your wall, here is my signature more than life-size.

A. Abragam

One year after the Wall was established, four "wallflowers" dined together in Zurich. The following correspondence is the result of that meeting.

Zürich,
June 4, 1965
Dear Professor Ginsburg:
It happened on Thursday, June 3rd in Zürich (Switzerland) that four chemists, Robert B. Woodward (Harvard University), Kyosuke Tsuda (Tokyo University), Vlado Prelog and Oskar Jeger (ETH, Swiss Federal Institute of Technology) were dining together and dividing the world into chemical spheres of influence. In the course of an animated conversation the subject of which institution, Harvard or ETH, had more name plaques on the wall of honor in the big lecturing theater in the chemistry department of your respected institution came up. This vigorous discussion evoked the betting spirit of Vlado Prelog and Robert B. Woodward which resulted in the wager of £1 of the former against 10 SFr. of the latter. The action was duly witnessed by the senior chemist of Tokyo and the undersigned is holding the bet in escrow until the problem is resolved by the honorable head of the chemistry department of Technion, Haifa. Respectfully submitted by your faithful friend, the bet secretary.

Oskar Jeger

Professor Ginsburg (Fig. 3) went to the main lecture hall and counted up the ETH and Harvard names and on June 9th replied to Professor Jeger that there were 6 (six) from the ETH and 11 (eleven) from Harvard. This brought forth the following pained response from Jeger to Woodward.

June 21, 1965
Der Chronist dieses (wissenschaftlichen) Kampfes hat schwere Bedenken, ob der richterliche Entscheid von Carmel unparteiisch war. Es ist ihm unverstaendlich wie die Harvard Individualisten auf die enorm hohe Zahl von 11 Ehrungen an der

Fig. 3 David Ginsburg (1920–1988). chairman of the Chemistry Department of the Technion until 1968.

Wand des Ruhmes kommen konnten. Geschah dies in fairem Kampf oder sind unerlaubte Lobby-Methoden angewandt worden? Kann der Richter von Carmel richtig bis 11 zaehlen? Hat er vielleicht auch die Namen der Fieser'schen Katzen mitgezaehlt? (A rough English translation: "It is incomprehensible that the Harvard chemists could reach the enormous number of 11 names on the wall. Was the arithmetician of the Carmel impartial? Can he even count all the way to eleven? Perhaps he was including Fieser's cats [4].")

Woodward had the last word:

Cambridge, Mass.
June 26, 1965
Dear Oskar:
Naturally I took much pleasure in receiving the official surrender of the Zurich forces engaged in our recent joust—a pleasure in no way diminished through the outcome having lacked any element of surprise, it having been at all moments a foregone conclusion. May I also take this opportunity to let you all know how much I have admired the courage, if not the wisdom, of your forces in being willing to sally forth so boldly against such formidable opponents.
The concession was only somewhat marred by a touch of Nixonian spirit. The veiled references to possible unfairness and to lobbying, and in particular the aspersions cast on the flawless arithmetician of Haifa do their authors little honor, and I am sure will be forgotten when the heat of the battle cools, since the dominant note of the document is clearly the more creditable one of a real desire for self-improvement. Indeed to me the most satisfying aspect of my participation in the contest is my realization that I have been able to do something which has helped imbue the Zürich forces with a spirit which may well one day win them a place in the first rank.

List of names on the Wall of Fame. Vertical rows represent horizontal rows on the Wall

A. Abragam	M. BERTHELOT	E. BUCHNER	M. CURIE
R. Adams	C.L. BERTHOLLET	R.W. BUNSEN	T. CURTIUS
S. Altman	J.J. BERZELIUS	A. BUTLEROV	J. DALTON
C.B. Anfinsen	J. Bigeleisen	J.I.G. Cadogan	R. Daudel
D. Arigoni	J.M. Bijvoet	M. Calvin	H. DAVY
ARISTOTELES	A.J. Birch	S. CANNIZZARO	P.J.W. Debye
O.I. Arnon	N.J. BJERRUM	W.H. CAROTHERS	J. Deisenhofer
S. ARRHENIUS	J. BLACK	H. CAVENDISH	DEMOCRITUS
F.W. ASTON	F. Bloch	T. Cech	B.V. Deryagin
K. VON AUWERS	K. Bloch	E.B. Chain	P. Deslongchamps
A. AVOGADRO	W. Bodenstein	J.A.C. CHARLES	O. DIELS
R. BACON	H. BOERHAAVE	M.E. CHEVREUL	C. Djerassi
L.H. BAEKELAND	N. BOHR	L. CLAISEN	W. von E. Doering
A. VON BAEYER	L. BOLTZMANN	S.G. Cohen	E.A. Doisy
J.C. Bailar, Jr.	K. BOSCH	A.C. Cope	P. Doty
P.D. Bartlett	R. BOYLE	E.J. Corey	A. Dreiding
D.H.R. Barton	S.F. Boys	C.F. Cori	P.L. DULONG
A.R. Battersby	J. Brachet	J.W. Cornforth	J.B.A. DUMAS
R.P. Bell	W.L. Bragg	C.D. Coryell	J.D. Dunitz
P. Berg	A.E. Braunstein	F.A. Cotton	G.M. Edelman
F. BERGIUS	J. BREDT	C.A. Coulson	P. EHRLICH
M. BERGMANN	P.W. BRIDGMAN	D.J. Cram	
T. BERGMAN	J.N. BRØNSTED	B.L. Crawford, Jr.	
J.D. Bernal	H.C. Brown	F.H.C. Crick	
M. Eigen	L.J. GAY-LUSSAC	R.D. Haworth	C.K. Ingold
E.L. Eliel	C. GERHARDT	W.N. HAWORTH	V.N. IPATIEFF
P.J. Elving	F.F. Giaque	M. Heidelberger	F. Jacob
H.J. Emeleus	J.W. GIBBS	E. Heilbronner	O. Jeger
W.A. Engelhardt	J.R. GLAUBER	I.M. HEILBRON	F. JOLIOT-CURIE
H. Erdtman	V.M. GOLDSCHMIDT	D. Herschbach	I. JOLIOT-CURIE
E. ERLENMEYER	M. GOMBERG	G. Herzberg	E.R.H. Jones
R. Ernst	C. GRAEBE	G.H. HESS	J. LENNARD-JONES
A. Eschenmoser	T. GRAHAM	G. Hevesy	W.S. Johnson
H. Eyring	D.E. Green	J. Heyrovsky	N.O. Kaplan
M. FARADAY	V. GRIGNARD	C.N. Hinshelwood	J. Karle
F. Feigl	E.A. Guggenheim	J.O. Hirschfelder	P. Karrer
E. FISCHER	C.M. GULDBERG	J.W. HITTORF	A. KEKULE
E.O. Fischer	I.C. Gunsalus	D. Crowfoot Hodgkin	E.C. Kendall
H. FISCHER	H.G. Gutowsky	J.H.VANT HOFF	J.C. Kendrew
L.F. Fieser	P. HABER	R. Hofmann	J.A.A. Ketelaar
P. Flory	G. Hagg	A.W. VON HOFMANN	M.S. KHARASCH
G. Fodor	O. Hahn	R.W. Holley	H.G. Khorana
J. FRANCK	L.P. Hammett	R. HOOKE	F.S. KIPPING
E. FRANKLAND	G.S. Hammond	B. Horecker	J.G. KIRKWOOD
K. Freudenberg	A.R. HANTZSCH	R. Huber	G.B. Kistiakowsky
H. FREUNDLICH	A. HARDEN	E. Hückel	J. KJELDAHL
K. Fukui	O. Hassel	C.S. HUDSON	A. Klug
R.M. Fuoss	H. Hauptman	R. Huisgen	E KNOEVENAGEL
			F. KOHLRAUSCH
			I.M. Kolthoff

List of names on the Wall of Fame. Vertical rows represent horizontal rows on the Wall (continued from previous page)

A. Kornberg	W.N. Lipscomb. Jr.	B. Merrifield	M. Nirenberg
H.A. Krebs	M. LOMONOSOV	K.H. MEYER	R.G.W. Norrish
A. Kuppermann	F.A. Long	L. MEYER	J.H. Northrop
A. LADENBURG	H.C. Longuet-Higgins	V. MEYER	R.S. Nyholm
1. LANGMUIR	K. Lonsdale	O. MEYERHOF	S. Ochoa
A. LAPWORTH	P.D. Lowdin	A.I. Meyers	G. Olah
A. LAURENT	S.E. Luria	A. MICHAEL	L. Onsager
A.L. LAVOISIER	A. Lwoff	L. MICHAELIS	A.I. Oparin
J.A. LEBEL	F. Lynen	H. Michel	W. OSTWALD
H. LECHATELIER	PJ. MACQUER	P. Mitchell	J.Th.G. Overbeek
J. Lederberg	M. Magal	E.A. MITSCHERLICH	PARACELSUS
E. Lederer	ALBERTUS MAGNUS	W. MOFFITT	R.G. Parr
Y.T. Lee	R. Marcus	J. Monod	L. PASTEUR
R.J.W. LeFevre	H.F. Mark	S. Moore	L. Pauling
J.M. Lehn	V.V. MARKOVNIKOV	H. MOISSAN	C. PEDERSEN
L. Leloir	A.J.P. Martin	R.S. Mulliken	W.H. PERKIN
N.J. Leonard	C.S. Marvel	K. Mullis	M.F. Perutz
P.A.T. LEVENE	J.C. MAXWELL	G. Natta	A.T. PETIT
G.N. LEWIS	J.E. Mayer	W. NERNST	K.S. Pitzer
W.F. Libby	M. Goeppert Mayer	A.N. Nesmeyanov	M.PLANCK
J. VON LIEBIG	E.M. McMillan	C. Neuberg	J. Polanyi
J.W. Linnett	H. Meerwein	A. Neuberger	J.A. Pople
R.P. Linstead	L. Melander	H. Neurath	G. Porter
F. Lipmann	D. MENDELEEV	JA. NIEUWLAND	R.R. Porter

F. PREGL	F. Sanger	T. Svedberg	O. WARBURG
V. Prelog	A.M. SAYTZEV	R.L.M. Synge	J.D.Watson
J. PRIESTLEY	C.W. SCHEELE	A. Szent-Gyorgyi	A. WERNER
1. Prigogine	H. Schmid	H. Tamiya	F.H. Westheimer
J.L. PROUST	R. SCHOENHEIMER	H. Taube	F.C. WHITMORE
G. Quinkert	E. SCHRODINGER	H.S. Taylor	H. WIELAND
E. Racker	G.M. Schwab	H. Theorell	G. Wilke
W. RAMSAY	G. Schwarzenbach	L. THIELE	G. Wilkinson
F.M. RAOULT	G.T. Seaborg	A. TISELIUS	R. WILLSTÄTTER
R.A. Raphael	N.N. Semenov	Lord Todd	E.B. Wilson, Jr.
T. Reichstein	D. Seebach	L.A. TSCHUGAEV	A. WINDAUS
O.A. Reutov	J.C. Sheehan	K. Tsuda	S. Winstein
F.O. Rice	M.M. Shemyakin	H.C. Urey	J. WISLICENUS
T.W. RICHARDS	ABU ALI IBN SINA	V. du Vigneaud	G. Wittig
J. RICHTER	P.S. Skell	A.L. Virtanen	F. WOHLER
E. Rideal	J.C. Slater	E. Vogel	R.B. Woodward
D. Rittenberg	E.F. SMITH	P. WAAGE	A. WURTZ
J.D. Roberts	M. Smith	J.D. VAN DER WAALS	L. Yaffe
J.M. Robertson	C.P. Smyth	G. Wald	L. Zechmeister
R. Robinson	F. SODDY	P. VON WALDEN	K. Ziegler
E. RUTHERFORD	F. Sorm	O. WALLACH	R. ZSIGMONDY
L. Ruzicka	G.E. STAHL		
P. SABATIER	S. Spiegelman		
L. Salem	W.M. Stanley		
	W.H. Stein		
	G. Stork		
	J.B. SUMNER		

I have graciously acted upon the generous suggestion of the vanquished Colonel Vlado, and have transmitted the spoils of war to Haifa, Of course, lest any element of taint (Mod. Am., payola) be attached to the funds. I have had to effect the transferal under rigorously prescribed conditions [5].

Consequently, I have instructed the authorities at Haifa to use the monies for the foundation of a fund for the special polishing, on a regular schedule, of the ETH names inscribed on the wall of fame—or for any other charitable purpose they may deem justified. Since I am sure you will agree that there could be no more certain and effective way of adding luster to your names, I sincerely hope they will choose the former alternative.

May I conclude with my most generous good wishes to all of you for ultimate success in your striving for a place in the sun.

Yours,

Bob

Woodward's exhortations to the ETH have had some effect over the years. In 1996 there were 14 names from Harvard and 10 from the ETH compared with 11 to 6 in 1964. We are still polishing the appropriate names.

A list of the names as they appear on the Wall are included. Those in lower case on the list appear as enlarged facsimiles of signatures of the individual; the names in capital letters are those of earlier chemists. Like most human efforts, the selection is undoubtedly imperfect. Time has dimmed the luster of some of the names as fashions change. Other chemists whose names would be included today were only small children at the time the Wall was established and now must achieve the Nobel Prize in order to join.

References and Notes

1. For an obituary, see: Rubin, M. B. *Tetrahedron* **1989,** *45,* iii.
2. Professor Ginsburg decided that he did not wish to spend the effort in collecting more ancient signatures, which presumably could be traced, and Roman capitals were used for these.
3. Opposite p 158 in *Reflections of a Physicist* by Anatole Abragam; Clarendon Press: Oxford, 1986; translation from the French by Ray Freeman.
4. Cf. any of the numerous books by Louis and/or Mary Fieser.
5. The cash which arrived was the yen equivalent of £1 and SFr. 10! Evidently, Professor Tsuda used the opportunity to change some of his home currency.

146 Semesters of Chemistry Studies[a]

In Memory of Vladimir Prelog (1906–1998)

Kurt Mislow[b]

With the passing of Vlado Prelog—the name by which his friends knew him and under which he published the first of his many scientific triumphs, the synthesis of adamantane [1]—we mourn the loss of one of the giants of twentieth-century organic chemistry. The founder of modern stereochemistry, it was he who initiated and intellectually invigorated the current renaissance in this field, a feat duly recognized by the awarding of the 1975 Nobel Prize in chemistry to Prelog for his research on the stereochemistry of organic molecules and reactions. As William Klyne remarked on this occasion [2]: "van 't Hoff is considered as the founder of 'Chemistry in space'; Prelog has been the High Priest of this cult for the past 20 years."

Vladimir Prelog's ex libris by Hans Erni.

Kurt Mislow was the first recipient of the Vladimir Prelog Medal and the first Prelog Lecturer at the ETH Zürich in 1986, honoring Vladimir Prelog's 80th birthday. On that occasion, Professor Duilio Arigoni of the ETH characterized the scientific interaction of Prelog and Mislow in the following way: "Their scientific interaction developed over the years into a kind of father-son relationship with all the advantages and occasionally some of the disadvantages of such a situation. They have fought together many battles, mostly on the same side of the barricade" [*Chimia* **1986**, *40*, 394–595]. The Editor is grateful to Professor Mislow for having accepted the invitation to write this account.

[a]*Chemical Intelligencer* 1998(3), 51–54.

[b]Department of Chemistry, Princeton University, Princeton, NJ 08544, USA

Throughout his long and illustrious career, Prelog maintained a consuming interest in the chemistry of natural products, an area to which he made major contributions, including the elucidation of the structures of nonactin, boromycin, ferrioxamines, and rifamycins. The spatial disposition of atoms is an integral part of the structural information content, and considerations of stereochemistry therefore became ineluctable. Thus, as early as 1944, we find papers by Prelog entitled "Über die Konfiguration der asymmetrischen Kohlenstoffatome 3, 4 und 8 der China-Alkaloide" [3] and "Über die Konfiguration von (−)-3-Methyl-4-äthyl-hexan" [4]. The same year also saw the report on the resolution of Tröger's base by chromatography on a column of lactose hydrate [5], a stereochemical landmark that has achieved textbook status [6].

Just a few years later, in 1947, Prelog's acyloin synthesis of medium-sized rings [7] opened the way to a comprehensive study of many-membered ring systems. In the first Centenary Lecture of the Chemical Society [8], which he presented in 1949, Prelog employed, for the first time, the methodology of conformational analysis to rationalize the physical and chemical properties of medium-membered ring compounds. This groundbreaking work on conformationally dependent steric effects, which Prelog elaborated in subsequent reviews [9–11], was completed independently of and prior to Derek Barton's epochal publication on the conformational analysis of fused cyclohexane systems [12]. During that time, Prelog also succeeded in formulating an empirical rule governing the relationship between the configurations of chiral alcohols and the sign of rotation of atrolactic acid obtained by saponification of the product from the reaction of phenylglyoxylate esters of such alcohols with Grignard reagents [13]. The underlying idea behind this rule, now known as Prelog's rule, was that the relative steric bulk of the groups attached to the carbinol carbon was responsible for the predominant direction of attack by the Grignard reagent on the carbonyl group. It is commonly assumed nowadays that the course of asymmetric syntheses is more often than not controlled by differential nonbonded interactions. Prelog's pioneering concept therefore represented a breakthrough of major proportions.

B. Hargittai and I. Hargittai (eds.), *Culture of Chemistry: The Best Articles on the Human Side of 20th-Century Chemistry from the Archives of the Chemical Intelligencer*, DOI 10.1007/978-1-4899-7565-2_45, © Springer Science+Business Media New York 2015

By the end of the 1950s, stereochemistry had clearly become the dominant theme in Prelog's work. In the field of natural products, the structure of the macrotetrolide antibiotic nonactin—the first natural product with S_4 symmetry— was being worked out by Prelog and his colleagues [14] in what has since been characterized as "a classic application of stereochemical logic and chemical analysis" [15]. Work was continuing apace on conformational analysis and on asymmetric syntheses, including studies of stereoselectivity in microbial and enzymatic reactions. At the same time, the problem of stereochemical notation had begun to capture Prelog's interest. In 1953, together with Barton, Odd Hassel, and Kenneth Pitzer, he proposed the axial/equatorial terminology to describe the spatial arrangement of cyclohexane bonds [16]. In 1956, together with Robert Cahn and Christopher Ingold, he proposed the *R/S* terminology, now familiar to everyone as the CIP system, to specify the configuration of stereoisomers [17]. And in 1960, together with Klyne, he proposed the syn/anti clinal/periplanar terminology to describe steric relations across single bonds [18]. The last two papers were destined to become "Citation Classics"—in 1982, the paper with Cahn and Ingold had been cited in over 660 publications since 1961 [19], and in 1984, the paper with Klyne had been cited in over 675 publications since 1960 [20].

Prelog's work with Cahn and Ingold was to have a profound effect on the direction of his research in later years. Before returning to this subject, however, I need to digress in order to indulge in some personal reminiscences. It was my good fortune that during those heady days I found myself spending a year as guest of the Organischchemisches Laboratorium at the ETH Zürich. It was a memorable experience; 40 years have passed, yet I still remember this *annus mirabilis* as though it were yesterday. The air at that time was electric with the excitement generated by a constant stream of galvanizing discoveries, by the race to beat the scientific competition, and by the hubbub of incessant scientific argument and counterargument. This wonderfully stimulating atmosphere was due entirely to Prelog's nurturing influence and charisma. As genial *primus inter pares*, he presided over a stable of brilliant young Dozenten, Duilio Arigoni, Jack Dunitz, and Albert Eschenmoser, three upward mobile stars who were soon to achieve world-class renown. Conviviality in this hothouse of ideas was assured by the affection, admiration, and respect that the denizens of the Laboratorium felt for each other and was promoted by Prelog's kindness, warmth, and self-deprecating modesty and, not least, by his pointed sense of humor and by his genius as a raconteur. As a witness to these doings and a more than willing participant, I happily immersed myself in this ferment of intellectual activity.

Two joint research projects were completed during this time: the absolute configuration of a chiral binaphthyl was determined by use of the asymmetric atrolactic acid synthesis [21], and a kinetic isotope effect was demonstrated in the solvolysis of alkyl sulfonates deuterated in the α-position [22]. Instructive as these researches had been, their significance to me paled in comparison with the stimulation afforded by personal discussions with the master. With fatherly benevolence, Prelog had taken me under his wing and had made clear to me, in innumerable lectures invariably enlivened with jokes and apposite anecdotes, that uppermost in his mind was the need to impose a logical structure on stereochemistry so as to bring reason and order to a field dominated by rank empiricism, pragmatism, and intuitive thinking. The time was right, in short, to provide stereochemistry with a conceptually sound foundation.

This was to become Prelog's chief endeavor from the early 1960s on. Its roots lay in the work on the CIP system. Up until about 1950, configurational notation had been in disarray. Thus, in a letter to *Nature* in 1951 [23], C. Buchanan had noted that

> standard textbooks…[are] about equally divided in naming dextrorotatory tartaric acid D(+) or L(−). This confusion arises because the relationship of the tartaric acids to the reference compound D(+)-glyceraldehyde has been established in two different ways from which opposite conclusions have been drawn…Any attempt to classify all optically active compounds into D- and L-series is obviously absurd…Although the relative configurations of many optically active compounds can be established, it is impossible to allocate the compounds to D- and L-series without ambiguity.

In a note added in proof, however, he remarked: "Cahn and Ingold [24] have proposed a new convention, for the specification of configuration, which is free from the ambiguities of the current system." What had happened was that Cahn, as editor of the Chemical Society, had become involved in the problem of configurational notation and had consulted Ingold, who was chairman of the Publications Committee. Their collaboration gave birth to the sequence rule and the "steering wheel" convention that lie at the heart of the CIP system. In 1954 Prelog met with the duo at a conference in Manchester and, after some discussion, was invited to join them in writing an article on the subject. The ensuing paper [17], which significantly modified and extended the Cahn-Ingold paper, received general acceptance because, according to Prelog, "it satisfied the need for an unambiguous, general system for specification of the innumerable stereoisomers of an organic compound of known constitution" [19].

Ten more years were to pass before a fully elaborated version of the CIP system [25] was to be presented to the world. But this paper, "which it is said Ingold wrote, Prelog criticised and Cahn mediated" [26], and whose writing entailed what Prelog in retrospect delicately characterized as "many fraternal arguments" [27], left important questions unanswered. With regard to nomenclature, these were ultimately addressed in a proposed revision of the CIP system [28]. Of equal if not greater significance, however, was the fact that the CIP paper provided no hint as to how to predict novel types of stereoisomers yet to be discovered. In that sense the work was incomplete, for a system designed to specify the configuration of stereoisomers cannot be general and complete unless all possible types of stereoisomers are known—a daunting task at best. In brief, what was clearly needed was a systematic classification of stereoisomers, that is, a way of cataloging the "innumerable stereoisomers of an organic compound of known constitution." It was a challenge that Prelog tackled with enthusiasm and with the uncompromising intellectual rigor that was the hallmark of all of his work.

Prelog's work on novel types of stereoisomers—cyclostereoisomers [29], vesperenes [30], and molecules possessing "elements of pseudoasymmetry" [31]—underscored the perceived need for such a catalog. Building on the work of van 't Hoff, Prelog, by constructing models of stereoisomers from various combinations of simplexes, developed what he called the geometric foundation of stereoisomerism and, at other times, chemical topology. He eloquently propagated his vision in plenary lectures [32–35], culminating in his Nobel Lecture [36]. It was the well-deserved pinnacle of his extraordinary career.

On the occasion of Prelog's 80th birthday, Dunitz [37] wrote:

> Prelog gave us the following advice from his store of folk wisdom: "If you want to be happy for an hour, buy a bottle of wine; if you want to be happy for a week, slaughter a pig; if you want to be happy for a year, get married; if you want to be happy for your life, enjoy your work." Prelog is a man who radiates happiness. It seems that he has the good fortune to have followed his own advice.

Indeed, Prelog included the motto *Studium chymiae nec nisi cum morte finitur*, which he found in an old book on alchemy [38], as a subtitle of his autobiography, "My 132 Semesters of Chemistry Studies" [39], a sure sign that the study of chemistry was at the core of his happiness.

To the end of his days, Prelog remained true to his first love. "I have always been interested in natural compounds chemistry [and it] will always remain the most fascinating area of chemistry," he confided in an interview on March 17, 1995 [40]. There had been those who had held a different opinion. Saul Winstein, with whom Prelog spent several days in the fall of 1951, bluntly asked: "Why do you waste your time working on natural products?" [41]. While Prelog professed to be "deeply shocked" by Winstein's question, it failed to deter him from his pioneering pursuits. And so, in the

years that followed, the tree of stereochemistry planted in the soil of natural products chemistry blossomed, flowered, and grew mighty under the care of the talented gardener whose memory I have tried to honor in this piece.

References

1. Prelog, V.; Seiwerth. R. *Ber. Deutsch. Chem. Ges.* **1941**, *74*, 1644, 1769.
2. Klyne. W. *Nature* **1975**, *258*, 96.
3. Prelog, V.; Zalán, E. *Helv. Chim. Acta* **1944**, *27*, 535.
4. Prelog, V.; Zalán, E. *Helv. Chim. Acta* **1944**, *27*, 545.
5. Prelog, V.; Wieland, P. *Helv. Chim. Acta* **1944**, *27*, 1127.
6. For example, see: Quinkert, G.; Egert, E.; Griesinger, C. *Aspects of Organic Chemistry: Structure*; VCH Publishers: New York, 1996; p 58.
7. Prelog, V.; Frenkiel, L.; Kobelt, M.; Barman. P. *Helv. Chim. Acta* **1947**, *30*, 1741.
8. Prelog, V. *J. Chem. Soc.* **1950**, 420.
9. Prelog, V. "Bedeutung der vielgliedrigen Ringverbindungen für die theoretische organische Chemie." In *Perspectives in Organic Chemistry;* Todd, A., Ed.; Interscience; New York, 1956; p. 96.
10. Prelog, V. *Butt. Soc. Chim. Fr.* **1960**, 1433.
11. Prelog, V. *Pure Appl. Chem.* **1963**, *6*, 545.
12. Barton, D. H. R. *Experientia* **1950**, *6*, 316.
13. Prelog, V. *Helv. Chim. Acta* **1953**, *36*, 308. See also: Prelog, V. *Bull. Soc. Chim. Fr.* **1956**, 987.
14. (a)Dominguez, J.; Dunitz, J. D.; Gerlach, H.; Prelog, V. *Helv. chim. Acta* **1962**, *45,* 129; (b) Gerlach, H.; Prelog, V. *Liebigs Ann. Chem.* **1963**, *669*,121.
15. Bartlett, P. A.; Meadows, J. D.; Ottow, E. *J. Am. Chem. Soc.* **1984**, *106*, 5304.
16. Barton, D.H.R.; Hassel, O.; Pitzer, K. S.; Prelog, V. *Nature* **1953**, *172*, 1096.
17. Cahn, R. S.; Ingold, C. K.; Prelog, V. *Experientia* **1956**, *12*, 81.
18. Klyne, W.; Prelog, V. *Experientia* **1960**, *16,* 512.
19. *Current Contents*(PC&ES), December 13, **1982**, p 18.
20. *Current Contents*(PC&ES), July 23, **1984**, p 12.
21. Mislow, K.; Prelog, V.; Scherrer, H. *Helv. Chim. Acta* **1958**, *41*, 1410.
22. Mislow, K.; Borcic, S.; Prelog, V. *Helv. Chim. Acta* **1957**, *40*, 2477.
23. Buchanan, C. *Nature* **1951**, *167*, 689.
24. Cahn, R. S.; Ingold, C. K. *J. Chem. Soc.* **1951**, 612.
25. Cahn, R. S.; Ingold, C. K.; Prelog, V. *Angew. Chem. Int. Ed. Engl.* **1966**, *5*, 385.
26. Robinson, F.A. *Chem. Brit.* **1982**, 359.
27. Prelog, V. "From Configurational Notation of Stereoisomers to the Conceptual Basis of Stereochemistry." In *Van't Hoff-Le Bel Centennial*; Ramsay, O. B., Ed.; American Chemical Society Symposium Series No. 12; American Chemical Society: Washington, DC, **1975**; pp 179–188.
28. Prelog, V.; Helmchen, G. *Angew. Chem.Int. Ed. Engl.* **1982**, *21,* 567.
29. (a)Prelog, V.; Gerlach, H. *Helv. Chim. Acta* **1964**, *47*, 2288; (b) Gerlach, H.; Owtschinnikow, J. A.; Prelog, V. *Helv. Chim. Acta* **1964**, *47*, 2294.
30. Hass, G.; Prelog, V. *Helv. Chim. Acta* **1969**, *52*, 1202.
31. (a) Prelog, V.; Helmchen, G. *Helv. Chim. Acta* **1972**, *55*, 2581; (b) Helmchen, G.; Prelog, V. *Helv. Chim. Acta* **1972**, *55*, 2612.
32. Prelog, V. "Das asymmetrische Atom, Chiralitat und Pseudoasymmetrie," Lecture held at the 3rd Van't Hoff Commemoration of the Division of Sciences of the Royal Neatherlands Academy of Sciences and Letters, January 27, 1968.

See *Koninkl. Nederl. Akad. Wetensch.*—Amsterdam Proceedings, Series B, **1968**, *71*, 108.

33. Prelog, V. "Problems in Chemical Topology," Robert Robinson Lecture, Dublin, April 3, 1968. See *Chem. Brit.* **1968**, *4,* 382.

34. Prelog, V. "Problems in Chemical Topology", 21st National Organic Chemistry Symposium of the American Chemical Society, Salt Lake City, Utah, June 15-19, 1969. See Abstracts , p 72.

35. Prelog, V. "Die geometrischen Grundlagen der Stereoisomerie," Hauptversammlung der Gesellschaft deutscher Chemiker, Hamburg, September 1969.

36. Prelog, V. "Chirality in Chemistry" Nobel Lecture, Stockholm, Sweden, December 12, 1975. See *Science* **1976**, *193*, 17.

37. Dunitz, J. *Chem. Brit.* **1986**, *22*, 606.

38. Prelog, V. *Naturwiss. Rundsch.* **1985**, *38*, 259.

39. Prelog, V. "My 132 Semesters of Chemistry Studies: Studium chymiae nec nisi cum morte finitur." In *Profiles, Pathways, and Dreams*; Seeman, J.I., Ed.; American Chemical Society: Washington, DC, **1991**.

40. *The Chemical Intelligencer* **1996**, *2*(2), 16.

41. Prelog, V. *Croat.Chem. Acta* **1985**, *58*, 349.

A Ceramist's View of Chemistry in Art[a]

Jane W. Larson[b]

The vision of "nature as energy" was formulated by poets and artists in Victorian England about the time managers were learning to harness nature for the industrial revolution [1]. It set the stage for Van Gogh's writhing iris and Cézanne's feverish red grounds—a heightened attention by oil painters to environmental energy. Thus, I have found it only a step farther, using the ceramic medium's powdered rocks and mineral melts, to manipulate inorganic chemistry and fire to that same end—depicting nature's energy. And in so doing, I have opened a new window, I believe, on art between Gaia—this great biosphere on which we live [2]—and life itself.

For some 30 years, I have explored the ceramic medium's high-temperature clays and earthy glaze melts, mixing formulas that are responsive to oxygen manipulation as well as heat, hoping to end up with a color palette from which I can convincingly evoke the natural world. I press plant materials and other imagery into clay canvases which I liken to fossil shales. Besides the many browns and blues from durable iron and cobalt melts that withstand any kiln fire, I struggle with anciently famous reduction glazes—the oxblood reds and peach bloom pinks originally developed for me by chemistry professor Clarence Larson. The reds and pinks are elusive copper carbonate mixes that evaporate easily and dramatically change color, basically from green to red, with an infinite variety of shades and tones in between. We have yet to tackle purple glazes but have developed varieties of white, gray, and black as well as orange and a rare yellow. These all have to be adjusted, of course, so as to mature at the same temperature in the kiln, and all have to be oxygen-reduction friendly—no mean feat, I've found. Glazes have subtle changes in color depending on their chemistry: does the feldspar emphasize sodium or potassium compounds? How finely pulverized is the silica? Is talc going to be compatible, and if so, how much? Kiln atmospheres add subtle characteristics to the work, depending upon moisture in the air, drafts, smoke, and the length of time it takes the temperature to rise. All of these variables, when studied long enough, however, can eventually produce an entire rainbow of color so natural, so full of the energy of nature's own colors, that it fits pre-cisely—eerily, even—into images of nature pressed into the clay canvas.

Embedding imagery in imitation of the embedding of fossils in shale, it seems to me, is essential to this work. When pressing imagery into the soft clay canvas, tiny shoulders of clay rise up around the image to protect it and allow an eventual buildup of melt to coat tiny cavities that could not otherwise hold color. Glaze melts then become a substitute for what might have been an organism's debris, left in the fossil cavity millions of years ago. And how precise is an impression in clay, down to that of the little mayfly, complete with every thread-fine leg intact! When chemistry provides the inorganic melt for embedded images, and a high kiln fire the final paroxysm, the combined techniques provide pieces of art that I have come to call bedding planes.

Glaze color in the ceramic medium comes from small amounts of metallic oxides cradled in a mix of silica, the glass former, clay for adhesion to the clay body, and a flux to bring the melt to manageable levels. But, given slight deviations, the glaze color variations are a matter of fortune favoring the prepared mind, for the artist is finally only midwife to the fire. Iron oxide as colorant, for example, provides many kinds of tan and brown, or it can provide a strong black or so-called iron red. However, if oxygen is removed from the kiln atmosphere at the right moment (by closing off air intake and damper), the iron melt will turn toward green or gray green, still perhaps with brown mixed in depending on how "variable" is the reduction in oxygen.

It is clear that building and using a palette from inorganic chemistry and the fire is not like wielding brushes and fielding ready-made colors. But the rewards are great. The brown and green iron-colored melts are of the precise earthy shades one finds in nature. They are the colors of the locale, i.e., woods and trees, fields and landscapes, to which we all, until urban landscapes began to overwhelm us, have been instinctively tuned. Copper red melts, too, have amazing affinities for expressing not only autumn colors and the pinks of a cold spring, but human emotions, the fury of strong light, the energy in a visualized landscape, you name it. When the shades and tones captured are matched to a visual scene that can absorb them, the results can be startling.

[a] *Chemical Intelligencer* 1998(3), 48–50.

[b] 6514 Bradley Blvd, Bethesda, MD 20817, USA (deceased)

B. Hargittai and I. Hargittai (eds.), *Culture of Chemistry: The Best Articles on the Human Side of 20th-Century Chemistry from the Archives of the Chemical Intelligencer,* DOI 10.1007/978-1-4899-7565-2_46, © Springer Science+Business Media New York 2015

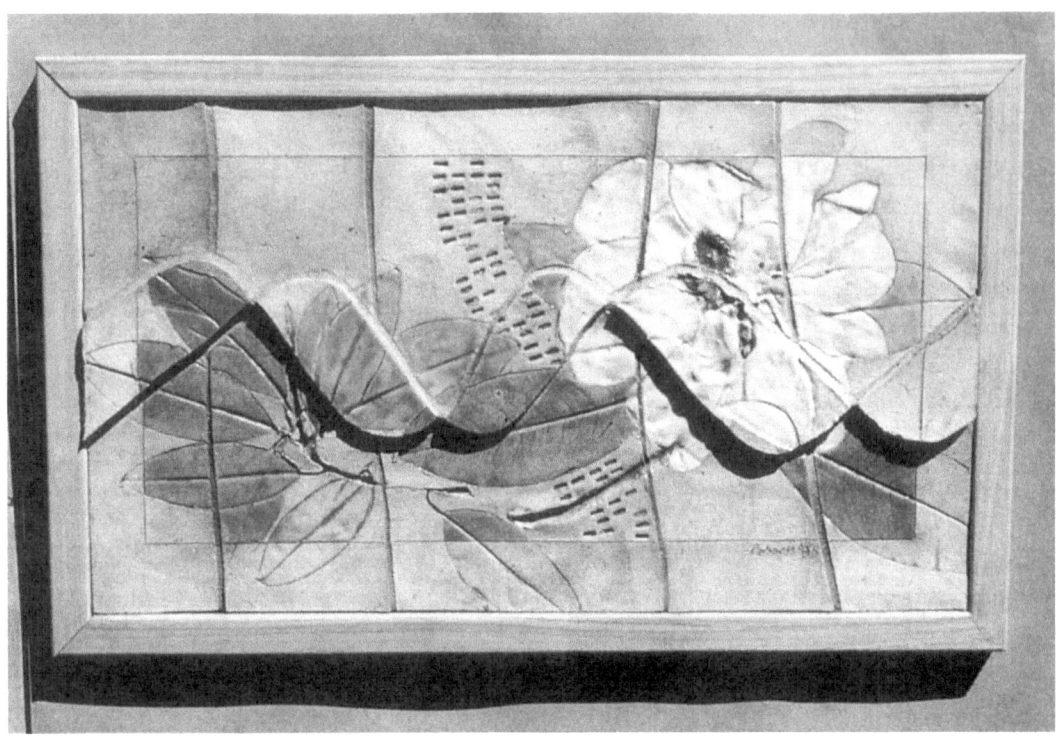

Fig. 1 *Ancient Magnolia and DNA*, 1995 (13.5″ height × 24.5″ width).

Recently, I have been reading scientific studies that seem to suggest that minerals from the inorganic world such as I use are found in organic fossil settings, too. This is an area I would like to see studied more assiduously, for I believe the likes of soft-bodied fossil finds 20 to 30 million years old incorporate ingredients that I perhaps am approaching in the studio with variable reduction melts [3].

Since my intent with imagery over time became strongly subject to the kiln fire's final achievement, interpretations have had to come after the firing, not before. One can see this exhibited in *'Ancient Magnolia and DNA'* (Fig. 1), where a red blush in the growing tip of a green branch points up the current pull of life, even while the clay is signaling the availability of ancient fossil magnolia DNA, with genes to be read. I did not plan that blush. I did not even think of emphasizing that the magnolia still grows strongly today and doesn't necessarily want to be known for ancient access to its genes! In the baseball plaque *'Where's the Ball?'* (Fig. 2), an upstart female outfielder was to carry the eye as main attraction, but remained pale. Instead the batter, with a firmly heightened oxblood profile, is obviously "up." So much for my ideas. I always anticipate something beyond my own design, now, and am ready to second-guess in all kinds of ways.

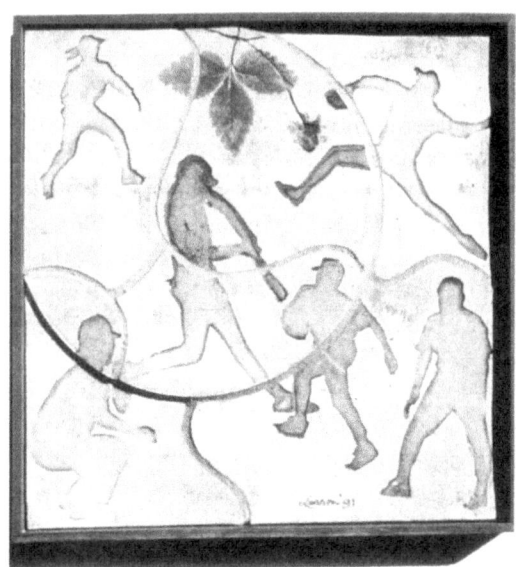

Fig. 2 *Where's the Ball?*, 1992 (12″ × 13″).

The University of Maryland Chemistry Department mural *'Ten Molecules That Shaped the World'* (Fig. 3) is art of that sort. The tulip poplar tree, a member of the magnolia

Fig. 3 *Ten Molecules That Shaped the World*, 1997 (4′×8′ long). The ten molecules are chlorophyll, heme, urea, adenosine triphosphate (ATP), DNA, glucose, cellulose, penicillin G, progesterone, and cisplatin. In addition, there is the suggestion of water and a "mystery molecule".

family, was featured because it bequeathed to us 20 million-year-old leaf fossils with colors intact, found in an Idaho dig [4]. How much more dramatic are its branches than I had the wit to plan! Sheep in this mural were to be "mutton-fat" white, a pertinent ancient glaze perfected by my husband, but the fire instead decreed that live sheep are not white but—as one finally sees them. It seems to me a new kind of aesthetic experience, a pictorial art in a rock canvas that uses inorganic earth to more deeply extract organic life.

Very fine clay is now being implicated, by some scientists, in the origin of life [5]. It is suspected of being a matrix or template for the first replicating molecules. People in South America sometimes eat clay, in faith that it will heal and strengthen them. I like the theory of Gaia (developed by a NASA scientist) for it suggests that our blue ball—Earth as it looks from space—is in some ways the largest living organism. It suits my experience. I find the theory compatible with a chemistry of inorganic earth materials which, when subjected to high heat and oxygen manipulation, develop along lines similar to those followed by organic materials. In bedding clay and subjected to the high temperatures of the kiln fire, inorganic earth melts can spill forth into a durable art form illustrating the thrust of life.

References

1. Bronowski, J. *The Ascent of Man*; Boston: Little, Brown, 1973; p 282.
2. Lovelock, J. *The Ages of Gaia: A Biography of Our Living Earth*; W. W. Norton, New York 1988.
3. Niklas, K. J.; Giannasi, D. E. "The Green Fossil," *Garden Magazine*, November–December, 1977. p 7.
4. Gould, S. J. "Magnolias from Moscow," *Natural History Magazine*, September 1992, p 10.
5. Cairns-Smith, A. G. *Seven Clues to the Origin of Life*; Cambridge University Press, Cambridge, U.K. 1985; Chapter 11.

The photo shows Jane Larson and her husband, Professor Clarence Larson, in December 1997, in front of the mural at the Department of Chemistry, University of Maryland, College Park.

Jane W. Larson studied at Swarthmore College and at the University of Rochester and earned a B.A. with honors and a Phi Beta Kappa key in 1943. She studied sculpture with Anna Mahler at the University of Southern California in 1957–58 and earned an M.F.A. in ceramics from Antioch College in 1982. She has traveled extensively to Europe, Egypt, Japan, Indonesia, and Mexico as well as in the United States. She worked from 1943 to 1957 as a science reporter, technical editor, and technical librarian, worked on her designs in a Los Angeles pottery factory in 1951–52, and taught ceramics at the American University between 1985 and 1988 and in 1994. Between 1975 and 1996 she worked as cameraperson for a project directed by her chemistry professor husband, Clarence Larson, which involved the videotaping of interviews with some 50 pioneers of science and technology. Her public commissions include murals at the Residence Inn, Bethesda City Center, Maryland, at the Federal City Shelter and the AAAS Headquarters in Washington, D.C., at the Beckman Center, Irvine, California, at the Community Art Center in Oak Ridge, Tennessee, and. the latest, at the Chemistry Department of the University of Maryland, College Park, Maryland. Her works are exhibited in several permanent collections.

Chaim Weizmann[a]

Hugh Aldersey-Williams[b]

Statesmen called him the greatest Jew since Moses. When the State of Israel formally came into existence 50 years ago, Chaim Weizmann became its first president. He got there through his science.

At the age of 30, Weizmann left Switzerland to come to Manchester. It was a place, he felt, where a Jew might work on merit. It was also the center of chemical industry in Britain. "Here he became the founder of what we now call biotechnology, which was to enable him to become the founder of the State of Israel," according to the British Nobel laureate, Max Perutz, in his essays on science and scientists [1].

The story of Chaim Weizmann and his dual achievement shows how intertwined the pursuit of science can become with other interests. He would later write in his autobiography, *Trial and Error* [2],

"The tug-of-war between my scientific inclinations and my absorption in the Zionist movement has lasted throughout my life. There has never been a time when I could feel justified in withdrawing, except temporarily—and even then in a sort of strategic retreat only—from the Jewish political field. Always it

seemed that there was a crisis, and always my conscience forbade me to devote more than a part of my time—usually the smaller—to my personal ambitions. The story of my life will show how, in the end, my scientific labours and my Zionist interests ultimately coalesced, and became supplementary aspects of a single purpose."

Weizmann was born in Russia, in the village of Motol near Pinsk (today it is in Belarus), in 1874 and received his scientific education in Germany and Switzerland. At first, his interest in science dominated; distracted by his studies in Fribourg, he missed the long-awaited First Zionist Congress convened in 1897 not far away in Basel.

Chemistry and Commerce

In Berlin, he worked in the laboratory of Karl Liebermann at Charlottenburg Technical College, one of the pioneers of synthetic dyestuffs, one of the great achievements of nineteenth century chemistry and the bedrock of the German chemical industry. The German programme of research was set up in competition with William Perkin in Manchester who in 1857 had made mauve, one of the first synthetic dyes. The experience was formative for Weizmann in establishing his belief that science could be done for profit and for the national good.

Soon he was receiving a regular income in the form of royalties on the patent for a novel dyestuff synthesis. By the time he obtained a position as Privatdozent (an unpaid position that relies on fees from teaching) at the University of Geneva in 1900, he could afford to lecture without pay. (Ironically, the company to which he sold his patent was IG Farbenindustrie, the German chemicals combine that later became infamous for making the gas used in the Nazi death camps and was broken up after the Second World War into the pharmaceutical giants Bayer, BASF and Hoechst. Weizmann wrote in *Trial and Error*, published in 1949, that his dealings with the company caused him no concern: "Hardly anyone thought of it then as the focus of German military might and of German dreams of world conquest. But it gives me a queer feeling to remember that I too, like many another innocent foreign chemist, contributed my little to the power of that sinister instrument of German ambition" [2].)

Weizmann stamp commemorating the Balfour Declaration. (All photos in this article are courtesy of the The Weizmann Institute of Science)

[a]*Chemical Intelligencer* 1999(1), 33–37.

[b]33 Hugo Road, London N19 5EU, UK

B. Hargittai and I. Hargittai (eds.), *Culture of Chemistry: The Best Articles on the Human Side of 20th-Century Chemistry from the Archives of the Chemical Intelligencer*, DOI 10.1007/978-1-4899-7565-2_47, © Springer Science+Business Media New York 2015

Albert Einstein and Chaim Weizmann in 1921.

At the dedication of The Weizmann Institute of Science in Rehovot, Israel, November 2, 1949: Prime Minister David Ben-Gurion, Chaim Weizmann.

Despite his comparatively comfortable position, Weizmann felt the want of money to fund his Zionist activities. However, it was not long before he received an invitation from Samuel Shriro, a Lithuanian-born oil magnate working in Baku. He could speak there on the Zionist cause and, at the same time, undertake some research into ways that waste products of oil refining might be put to use. Aromatic compounds similar to those used to make synthetic dyes are abundant in oil waste, so Weizmann was well qualified for this work. Shriro introduced him as a promising chemist and provided more funds for him to travel to Zionist conferences. It was the first of many occasions when Weizmann would combine his scientific skills with the promotion of Zionism.

Weizmann came to Britain in 1904. He chose Manchester over London for its new laboratories; it was also a hub of Zionist activity. Here was a place where he could attempt to unify the Eastern and Western strands of Zionism. He was taken on by Professor William Perkin, the son of the William who had made the mauve dye. For the next several years, Weizmann achieved some success in chemistry. He began to diversify into areas of biology, working on proteins and physiological chemistry with spells at the Pasteur Institute in Paris.

One problem of the day was to create a synthetic alternative to rubber, whose price was then rising rapidly. Weizmann and a colleague at the Pasteur Institute explored the possibility of using bacteria to ferment potato starch. This process generated two important products, acetone and butyl alcohol. The latter could be reduced in stages to provide the starting materials for many chemical syntheses. The unsaturated hydrocarbon butadiene, for example, is an important precursor in the synthesis of more complex organic chain and ring (aromatic) molecules. In polymerised form butadiene and similar compounds become rubber-like. The investigation was suspended when the price of raw rubber fell back. (When the price went up again during and after the war, it was German industry that took most advantage of the synthesis Weizmann had developed.)

The Government Emissary

When the First World War broke out, the process developed by Weizmann assumed enormous importance. The solvent acetone is needed in the production of smokeless cordite and other explosives. It is conventionally made by distillation, but Weizmann's "Bacillus BY" (the B for bacterium, the Y for Weizmann, it had the scientific name *Clostridium acetobutylicum*) enabled the rapid production of larger quantities. At the Admiralty, Winston Churchill presented Weizmann with a demand for 30,000 tons of the liquid. After explaining the difficulties of scaling up from laboratory to industry, Weizmann sought advice from the people who were already performing bacterial fermentation on an industrial scale, the brewers.

Seconded to the Ministry of Munitions, and made aware by David Lloyd George of Britain's poor preparation in explosives, Weizmann embarked upon a series of travels, visiting scientific specialists in friendly countries and assessing the potential for manufacturing acetone in various parts of the British Empire. In the meantime, the long task to develop and build an acetone plant and to convert requisitioned breweries for this purpose continued. By the time that an acetone plant was feasible, all Britain's starch crops were needed for food. The scene of the acetone production moved to North America, using maize as the raw material. Weizmann had succeeded in introducing one of the first bacterial processes in industry; it was this achievement that earned his reputation as the father of biotechnology.

Dr. Weizmann with Harry Truman in Washington after Truman was elected President of the United States.

In lieu of a fee for his services, Weizmann came to an arrangement with the British government for his compensation after the war. These earnings Weizmann invested in his Manchester friends' new business, Marks and Spencer.

Weizmann himself was now in London, where the government had furnished him with a laboratory. Here he sought to develop uses for butyl alcohol, which others, including Perkin, had regarded as waste. Ultimately, its derivatives were to prove more valuable than the acetone process, especially in the aviation and automobile industries, then in their infancy.

In London and when travelling as a government emissary, Weizmann was now able to build on his contact with the statesmen and government officials whom he met. Unlike many Zionists, Weizmann thought that a statement from the British government could cement the solidarity of nations with significant Jewish populations, in particular the United States and Russia, which was then in the throes of revolution. In the event, the Balfour Declaration—agreed on by the Cabinet on the last day of October 1917—did not achieve these ends, but it was seen to legitimise the Jews' claim to a "national home" in Palestine. For Weizmann, the famously delphic Declaration could "mean as much or as little as the Jewish people made of it" [3].

Lloyd George saw Weizmann's apparent offer of scientific expertise for favors towards the Zionist cause as a straight exchange. Weizmann disputed this view, pointing out that his overtures on the matter had begun as early as 1906, before Britain's hour of need.

Seeds of the Weizmann Institute

The momentum that the Balfour Declaration gave to the Zionist cause meant that Weizmann spent little time in the laboratory during the 1920s. He envied the advancements made by former Manchester colleagues such as the physicist Ernest Rutherford and the chemist Robert Robinson (later President of the Royal Society). He could not find the time for further work on synthetic rubber or to develop his interest in the potential for making synthetic foods.

However, this period did see the beginnings of what would turn out to be Weizmann's most significant and lasting contribution to science. His dream, which he had nurtured since before the war, was to establish a Jewish university. As ever with Weizmann, science and Zionism were linked inextricably. "There were two factors which urged me on," he wrote: "first, my intrinsic relation to science, which had been part of my life since boyhood; second, my feeling that in one way or another it had something to do with the building of Palestine" [2].

Exactly what it had to do with nation-building was a subject of some debate. Some spoke of the need for general education, others specifically of science. There were arguments as to whether the intake should be Jewish or international, whether the focus should be on teaching or research, and if on research, whether that should be in the pure or applied sciences.

In 1933, when Hitler came to power in Germany, Weizmann renewed his fund-raising efforts, this time petitioning the Marks and Spencer brothers-in-law Simon Marks and Israel Sieff. The Sieffs saw this as a way to commemorate their son, Daniel, who had recently committed suicide, as well as to benefit the Jewish people and Palestine.

The same year, Weizmann was able to found his research institute adjoining an agriculture experimental station at Rehovot, 15 miles south of Tel Aviv. It was clear that staffing it would not be a problem: Among those forced to flee Germany were many top scientists, chief among them Weizmann's old friend Richard Willstätter, the organic chemist who had won the Nobel Prize for his study of chlorophyll.

Fritz Haber, the half-Jewish chemist who had been responsible for the use of poison gases in the trenches of the First World War, was also obliged to leave Germany. Calling on Weizmann in London, Haber seemed a broken man. "It must have been particularly bitter for him to realize that his baptism, and the baptism of his family, had not protected him. It was difficult for me to speak to him; I was ashamed for myself, ashamed for this cruel world, which allowed such things to happen, and ashamed for the error in which he had lived and worked throughout all his life" [2]. Weizmann tried to persuade Haber to come to Rehovot, but he died before he could make the journey. Albert Einstein refused Weizmann's entreaties. Even Willstätter didn't make it, preferring the retired life in Switzerland.

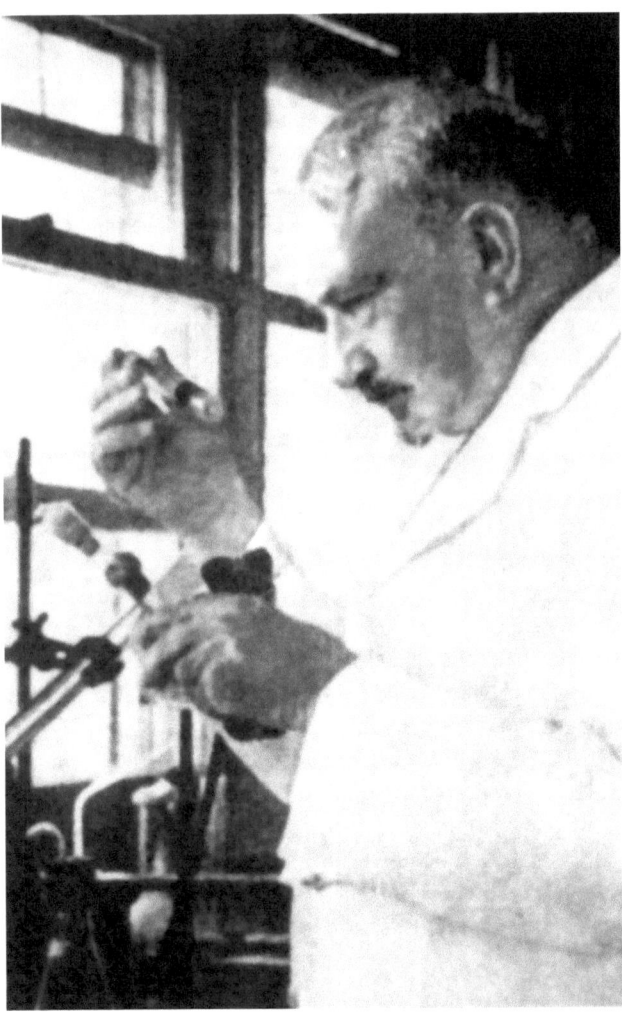

Painting by Moshe Mokady of Chaim Weizmann, first President of The Weizmann Institute of Science and of the State of Israel, in his laboratory.

Science and Zionism in the War's Dark Days

As war loomed again, Weizmann once more saw that his scientific expertise might win favours for the Zionist cause. "[H]e dreamed once again of making a Jewish contribution to the protection of England," according to one biography [3]. "In the First World War his scientific work had put him on the road to the Balfour Declaration. If the worst happened again, it could bring him to the Jewish State."

Again he met with Churchill, who was once more at the Admiralty, and again he was appointed an advisor to the government and given a laboratory. In 1940, he travelled to the United States with the task of presenting a scheme for using grain surplus for the production of butadiene for synthetic rubber. (Britain's supply of natural rubber from the Far East had been cut off.) The hidden agenda was that Weizmann's offer of scientific help would bring support for Zionism. The mission was not entirely successful. His being a scientist opened the door to the Oval Office, but unfortunately President Roosevelt saw him as a scientist and no more. The oil companies—which controlled conventional methods for the extraction of butadiene from petroleum—and American scientists met his ideas with scepticism which required all his considerable political skills to surmount.

Weizmann's method was based on a process developed at the Daniel Sieff Research Institute for the fermentation of molasses and conversion of agricultural waste into aromatic compounds. One of these was isoprene, from which synthetic rubber could be made. Other compounds related to acetone could be made into high-octane fuels. During the darkest period of the war, Weizmann spent 15 months in the US working on these projects and furthering his political ends. "I divided my time almost equally between science and Zionism" [2].

Nevertheless, the Daniel Sieff Research Institute opened its doors in the spring of 1934. Its 10 scientists concentrated on problems directly relevant to the development of the local economy, working in organic chemistry and biochemistry on matters to do with citrus, dairy, silk and tobacco production, as well as the synthesis of chemicals with potential use as pharmaceuticals. Weizmann's own interest in synthetic foods—he had devised a "blitz broth" using yeast to convert otherwise inedible vegetable matter into food during the First World War—gained new urgency in the arid surroundings. All too soon, however, this pioneering research community was to be diverted from these activities into providing medicines for allied forces in the Middle East.

Time for Reflection

As the war approached its end, Weizmann's 70th birthday present from Jewish leaders was a commitment to expand the Rehovot centre into a world-class institute of science.

Weizmann was a man of action. He pursued his career in science and statecraft, seldom stopping to consider their mutual influence. Upon the expiry of the British Mandate on 15 May 1948, his reward was to be chosen as the first president of Israel, in which capacity he served until his death on November 9th, 1952. Frustrated by the mainly ceremonial role, he at last had time to muse. In a final chapter of *Trial and Error* Weizmann concluded that his weaving together of science and politics was not purely a personal matter. The matter of natural resources force them together. "It is part of a general question of raw materials, which has been a preoccupation with me for decades, both as a scientist and a Zionist; and it had always been my view that Palestine could be made a centre of the new scientific development which would get the world past the conflict arising from the monopolistic position of oil. Not that our scientific work would be dedicated solely to that purpose; but it certainly would be one of its main enterprises" [2].

Weizmann predicted that conflict could arise in the future over access to resources other than oil, such as water or sunlight. It cannot be coincidental that some of the research at the Weizmann Institute is concerned with making the most of these resources, the one scarce, the other abundant, in Israel.

Chaim Weizmann himself showed time and again throughout his long career how science can be used for broader ends. In turn, his political ambitions to some extent determined the science he did. Without his cause to advance, he might have made quite different scientific discoveries. But without his science, he surely would not have achieved his political goal.

References

1. Perutz, M. F. *Is Science Necessary? Essays on Science and Scientists*; Barrie & Jenkins: London. 1989.
2. Weizmann, C. *Trial and Error;* Hamish Hamilton: London, 1949.
3. Litvinoff, B. *Weizmann, Last of the Patriarchs;* Hodder & Stoughton: London, 1976.

Hugh Aldersey-Williams is a freelance journalist and author. He is the design critic of the *New Statesman*, and writes on science tor the *Independent on Sunday* and other newspapers. His first popular science book. *The Most Beautiful Molecule* (Aurum Press 1994/Wiley 1995), describes the story of the discovery of buckminsterfullerene, and was a finalist for the *Los Angeles Times* Book Prizes. He is at work on two new books for Granta Publications, *Findings: Hidden Stories in First-Hand Accounts of Scientific Discovery*, which deconstructs the most significant scientific papers of the century, and a series of biographical sketches, including Weizmann, of science and nationalism.

Gibbs and Amistad[a]

Bart Kahr[b]

The current fascination with the Amistad incident—the revolt of West African captives aboard a Portuguese slaver, their subsequent incarceration, and ultimate vindication by the United States Supreme Court—provides an opportunity to learn and teach from the life and work of J. Willard Gibbs (1839–1903). The unlikely glue that binds the inventor of statistical thermodynamics to the twisted voyage of the *Amistad* is the poet Muriel Rukeyser (1913–1980), who draws these disparate themes together in the first book-length biography of Gibbs [1]. Here, we tell how an account of the Amistad incident becomes a life of Gibbs in a review of Rukeyser's 1942 biography, *Willard Gibbs*, while trying to carry forward many of the themes that she introduced in this most unusual book. In order to promote a newfound understanding of Rukeyser as scientific biographer and American historian, we offer a fresh conclusion to *Willard Gibbs* that we hope she would approve of and that restores a measure of coherence to the ambitious exercise she undertook during World War II.

It is mainly Steven Spielberg's movie *Amistad*, released after Rep. Tony Hall (D.-Ohio) introduced the controversial resolution that Congress apologize to African Americans whose ancestors "suffered as slaves under the Constitution and laws of the United States," that has brought this chapter of American history back into national consciousness. Spielberg illustrated the capture of West Africans, their transport in shackles—powerfully rendered—through the treacherous middle passage to Cuba, and sale in a market to Ruiz and Montes, who herded them aboard the *Amistad*. Cinque broke free of his chains and led the Africans in a bloody rebellion whereby they wrested control of the ship. Ruiz and Montes were commanded to sail back to Africa but deceptively steered northwest after dark. In this way, the ship reached Long Island, where the Africans were captured again, this time by an American sailor, Lieutenant Gedney, who took them to Connecticut, a slave state. Imprisoned in New Haven, the birth place of Willard Gibbs, in 1839 the year of his birth, the Africans languished. Only after several lengthy courtroom battles were the Amistad captives freed.

Their case was dramatically presented by former president John Quincy Adams (1767–1848). "When you come to the Declaration of Independence," he said, "that every man has a right to life and liberty, an inalienable right, this case is decided" [2].

Apology

The Declaration also asserts that when "it becomes necessary for one people to dissolve the …bonds which have connected them with another…they should declare the causes which impel them to the separation." This declaration—in English or any other European language—from the kidnapped Africans was impossible. It was this impasse that connected them to Gibbs through his father, Josiah Willard Gibbs Sr. (1790–1861, Fig. 1), Professor of Sacred Literature at Yale University. The elder Gibbs played a key role in the defense of the prisoners by learning bits of the West African Mendi language and locating sailors fluent in both Mendi and English; only then were the events aboard the *Amistad* told from the captives' perspective.

An apology is owed to Josiah Gibbs, who is portrayed in Spielberg's movie as a buffoon, a professor pretending to translate what he does not understand. Josiah Gibbs, like his son, was a scholar of the first rank and a most unlikely fraud, remembered "for profound scholarship, for unusual modesty, and for the conscientious and painstaking accuracy which characterized all of his published work" [3]. To know him, one should simply look at his writings "where we meet only with naked, laboriously classified, skeleton-like statements of scientific truth" [3]. Josiah Gibbs approached his role as translator for the Amistad captives in earnest. He published comparisons of several West African vocabularies [4]. His analysis of the prisoners' names proved that they must have been born in Africa and were therefore transported illegally. He rejected the Spanish names foisted upon them: "Four of these unfortunate Africans lie buried in an obscure corner of the New Haven burying ground. Their only living memorial is their name. But this is a sacred relic. It bears the impress of the nation to which they belonged, and will yet carry to their countrymen and friends the tale of their wrongs" [5]. The Africans were equally respectful. Kinna (Fig. 1) wrote to Josiah Gibbs: "dear friend I wish to

[a]*Chemical Intelligencer* 1999(2), 24–31.

[b]Department of Chemistry, University of Washington, Box 351700, Seattle, WA 98195-1700, USA

Fig. 1 Left: Traditional silhouette of Josiah Gibbs [39]. Right: Phrenological view of his friend Kinna [34].

write to you a letter because you have been so kind to me and because you love Mendi people I think of you very often…my good love to your wife and all your family I love them very much…" [1].

Science and Poetry

Of course, Rukeyser looked beyond the secondhand associations that connect Willard Gibbs with the Amistad case. She aimed to show that his work and the defense of the Amistad captives were emblems of a new nation's coming of age culturally and spiritually. With Gibbs's scientific achievements, we see the full flowering of America; with the defense of the Amistad captives, we see hope that the United States may finally fulfill the promise of the Declaration of Independence. The complementarity of science and liberty was established by Rukeyser in her discussion of John Quincy Adams as science advocate. He championed the creation of the Smithsonian Institution, inaugurated America's first astronomical observatory, and appealed for a uniform system of weights and measures. Sadly, his report to the Senate arguing for the adoption of the metric system was met with derision. Tracing the history of measurement and dealing with philosophy as much as with physics, his report challenged the scientific world "that cannot believe that a work may be sound and yet literary, artistic, and historical" [1]. The banner was picked up by young Willard Gibbs whose first scientific paper began: "A uniform system of weights and measures…is an acknowledged desideratum. I suppose that

we all hope & expect, that the world will not consent to do without so great a convenience" [6]. Of this, Rukeyser said, "He was following in a great tradition. He was going to move through it and out of it. He needed to wait and work" [1].

Rukeyser first wrote a poem about Gibbs in which she tells of a scientist largely confined to three square blocks in New Haven, drawing himself up during the Civil War, "when all of Yale is disappearing South," in preparation for future labors. She continues [7]:

> Condense, he is thinking. Concentrate, restrict. This is the state permits the whole to strand, the whole which is simpler than any of its parts. And the mortars fired, the tent-lines, lines of trains, Earthworks, breastworks of war, field hospitals…

This stanza hints at Rukeyser's larger purpose in the subsequent biography, "a footnote to the poem" [8]. She said that her role as a writer during World War II was nothing less than to fight fascism, of which she was keenly aware, having been in Spain at the start of the Civil War. Gibbs's story is "that of the pure imagination in a wartime period. War and after-war are filled with hatred, and this hatred turns against the imagination, against poetry, against structure of any kind." The act of a poet writing the biography of a mathematical physicist during wartime was for Rukeyser a challenge to so-called "reactionaries of the imagination" [1], a protest! For as long as "forms of imagination are not only separate, but exclusive" we can never have a society "in which peace is not lack of war, but a drive toward unity" [9]. If poets and scientists cannot come to an understanding, what hope is there for Nationalists and Republicans? Blacks and whites?

Rukeyser's Offense

After setting the stage with the Amistad incident, Rukeyser moves on to describe a young New Haven and the childhood of Gibbs. Not for 150 pages, until the account of Gibbs's sojourn in Europe, are we given the background that is necessary for understanding his science. Here, we meet Carnot and his engine, Count Rumford boring cannons in defense of Munich, Joule and Mayer fighting for priority over the statement of the first law of thermodynamics, and the second law of thermodynamics as devised by Kelvin, and by Clausius: "Die Entropie der Welt strebt einem Maximum zu." Maxwell was Gibbs's first champion, Ostwald his second, Boltzmann his alter ego. On his return to New Haven, Gibbs worked quietly until 1873, when he published two papers presenting more useful geometric representations of thermodynamic quantities (Fig. 2). A famous third paper established a general theory of thermodynamic equilibrium from the principle that entropy takes a maximum value at the quiescent state [3].

Admirers of Gibbs, parsimonious in his use of words adjunct to mathematical expressions, may be repelled by Rukeyser's prose that flows on and around Gibbs and his science. A reviewer in the *Journal of Chemical Education* cautioned that "readers accustomed to…complete sentences will not enjoy Miss Rukeyser's repetitive style and round-robin chapters" [10]. The economist Paul Samuelson described her writing as "excitable" and "gushing" [11]. Kertesz, author of the definitive critique of Rukeyser's work, brushed away such criticism: "A reader of *Willard Gibbs* must admit the validity of a prose that moves like poetry and has poetic coherence" [12]. But she is not a scientist. Notices from scientists were mean-spirited and prejudicial. The *Journal of the American Chemical Society* predicted that *Willard Gibbs* would fail to win approval of the scientific "fraternity" [13]. Gibbs's student Edwin Bidwell Wilson apparently said that her writing about Gibbs, given her ancestry (Jewish), "was as bad as for a Negro to be writing about a Southern gentleman" [9]. Samuelson lends some credence to this charge by questioning whether his teacher objected more to Rukeyser's gender, profession, or hometown (New York) [11]. Rukeyser was not cowed, even by Einstein's declination of a request to contribute a foreword [12]:

> There is but one way to bring a great scientist to the attention of the larger public: it is to discuss and explain, in language which will be generally understood, the problems and the solutions which have characterized his life-work. This can only be done by someone who has a fundamental grasp of the material … Otherwise, the result is banal hero worship, based on emotion and not on insight. I have learned by my own experience how hateful and ridiculous it is, when a serious man, absorbed in important endeavours, is ignorantly lionized.

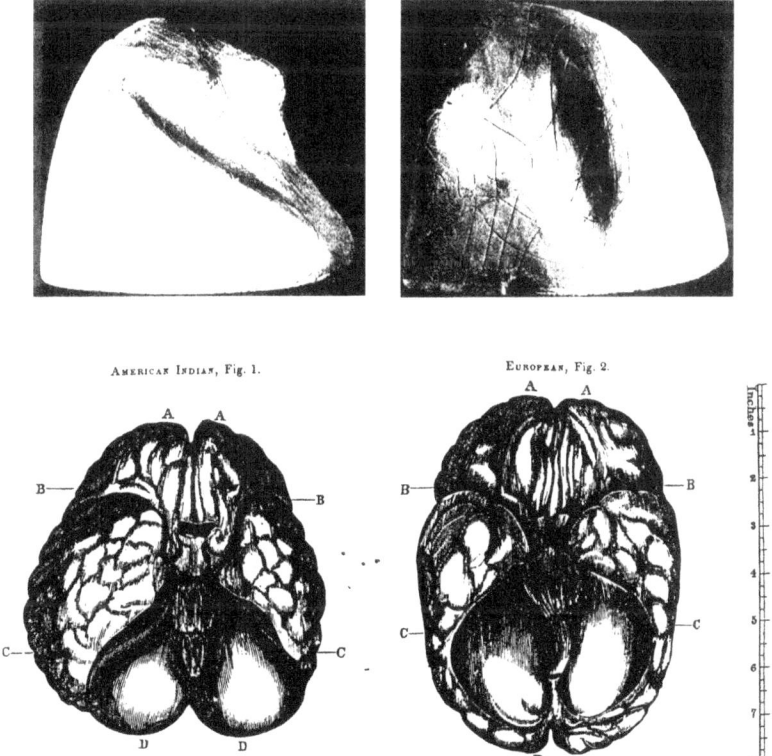

Fig. 2 The use and misuse of shape. Top: Maxwell's famous model of the Gibbs internal energy-entropy-volume surface for water [6]. Bottom: Brains from Morton and Combe's *Crania Americana* [28].

What of Rukeyser's ignorance? She has been condemned for technical inaccuracies, but the mistake that is invariably raised is her confusion of a reference to the physicist August Beer for the beverage with the same name, surely a forgivable, funny peccadillo [13, 14]. Still, we do not learn much about thermodynamics by reading *Williard Gibbs*. According to Kazin [15], "Gibbs, who declared that 'mathematics is a language,' cannot easily be touched and defined by a language that is not mathematical...He cannot be adequately restored by a prose language that beats upon him, seeking, seeking, yet can convey only our wonder, not the universality of his equations."

At the urging of Wilson and Gibbs's descendants, who were angered by Rukeyser's book, Lynde Phelps Wheeler, another former student, was persuaded to write an authorized biography. He referred to Rukeyser only once, in order to scold, and in the process damned chemists: "...it seems to me that it has been more the lack of suitably prepared mental soil than the want of perspicuity that has been involved in the difficulties which undoubtedly have been experienced by many of Gibbs' readers, particularly among the chemists" [6]. Straightforward in its approach, Wheeler's book is the place to go for an accessible chronology of Gibbs's life. However, it is to be argued whether Wheeler is more successful at explaining thermodynamics. Translating the fundamental equation into words as he does, *"Change in Energy = Temperature × change in Entropy − Pressure × change in Volume + Potential × change in Mass,"* is not the solution.

Rule of Phase Beyond Science

Any biographer of Gibbs—who wrote sparingly about his work and that of others—is confronted with a challenge. In order to fill the silences in her narrative left by Gibbs, Rukeyser made use of a loquacious contemporary, John Quincy's grandson Henry Adams (1838–1918), who wrote extensively about himself and most everyone else, even Gibbs. While a number of commentators might have served this function, she chose Adams because of the balance that he gave the narrative: "As Josiah Gibbs had given John Quincy Adams his means of communication in the Amistad case by finding an interpreter, so Adams' grandson Henry was to be the first to bring Willard Gibbs into communication with the world beyond his science" [1]. In an effort to construct a science of history, Adams identified with Gibbs and the phase rule by suggesting that human thought, considered as a substance advancing from the premedieval period, through the Renaissance, to modern times, can be likened to water passing from ice to steam. More particularly, Adams saw phases of human activity accelerating according to the law of squares, bringing "thought to the limit of its possibilities in the year 1921" [16]. This was the end of the world that

Adams foolishly assumed was cast in the second law of thermodynamics and supported by his pessimistic view of a degenerate modern society. Jordy [17] seized upon Adams's misunderstanding: "Gibbs was concerned with the coexistence of phases in equilibrium, Adams with their succession as a result of dis-equilibrium. Indeed it would be difficult to see how Adams could possibly have distorted Gibbs' Phase Rule more completely." Adams never did pretend to understand Gibbs. He lamented in *The Education* that "the bit of practical teaching (I) afterwards reviewed with the most curiosity was the course in Chemistry, which taught (me) a number of theories that befogged (my) mind for a lifetime." On reading Gibbs, Adams wrote, "I flounder like a sculpin in the mud" [1]. Regarding thermodynamics, Adams was at his best when he nominated Maxwell's demon for the Presidency on the strength of its ability to create order from chaos [18].

Opportunity Lost

Rukeyser, inspired by Adams's "most daring use of Gibbs that had yet been attempted" [1], goes even further. In a concluding chapter, she tried to compare the work of Gibbs to that of other creative individuals working during the turbulent period following the Civil War, especially Herman Melville (1819–1891). This disappointing chapter has been called "a climax of absurdity" as we are asked to "see the relationship between the Second Law of Thermodynamics and the White Whale" [19]. It is one thing to say that Gibbs and Melville were "deep and powerful expressions of their time" [1], and it is another thing to show how.

There is no record of Gibbs's views on war, slavery, or the role of science in society. He had an "innate tendency to avoid an expression of opinion on anything where he had not thoroughly explored all the implications or where for any reason he was not entirely sure of his ground" [6]. This must have been a source of frustration for Rukeyser, whose poetry crackles with social commentary. By equating Gibbs with Melville, a writer with whom she had a great affinity because he constantly dealt with the influences of war, Rukeyser put him within the context of the great changes through which he lived. Melville stuffed all of America in mid-century on Captain Ahab's ship; Gibbs was formed in mid-century. This seems to have been enough for Rukeyser to risk permitting Melville, as she permitted Adams, to speak for Gibbs. She admitted that Adams had laid himself "open to all the charges, and he was at once called a crazy old man and a charlatan, by the specialists" [1]. She undoubtedly recognized that she put herself in a similar position. The narrative might have closed round had Rukeyser offered Melville's views on slavery and science in the tradition of John Quincy Adams and Josiah Gibbs. She did not.

Rukeyser discussed practically all of Melville's major works, though conspicuously absent is any reference to *Benito Cereno* (1855), a novella about an uprising aboard a Spanish slaver told through the eyes of a confused American sailor that is squarely based on the Amistad incident [20]. Teasing meaning from *Benito Cereno* is a challenge. Melville is not explicit but "the perceptive reader should hear the undertone…as [he] forewarns the United States of a tragic fate to result from its tragic flaw" [21]. On the other hand, in "The Portent" [22], Melville could not have been clearer in his conviction that civil war was writ in slave rebellions:

Hanging from the beam,
Slowly swaying (such the law)…
(Weird John Brown)
The meteor of war.

Melville, a citizen of the world having experienced a form of bondage as a sailor, was responding to the increasing volume of the debate over slavery. Given Rukeyser's "relentlessly holistic vision" [23], evidenced by her simultaneous evaluation of the connections between energy and entropy, poetry and science, fathers and sons, as well as war and peace, it is certain that had she known of the plot of *Benito Cereno*, she would have chosen to end her biography by returning to the Amistad incident. In so doing, she could have shown that Melville, Gibbs's only contemporary equal in imagination, was compelled to deal directly with the great American sin, the crime whose resolution in the Civil War was, according to her thesis, requisite for the full realization of American culture, of which Gibbs is the highest example.

The bibliography in *Willard Gibbs* offers insight into Rukeyser's omission. The Melville anthology that she recommended does not contain *Benito Cereno* [24]. The cited biography by Lewis Mumford [25] contains analyses of Melville's important works, but Mumford intentionally avoided revealing the plot of *Benito Cereno* because it is an exercise in suspense and he "will not spoil [it] for those who have not read it by revealing its mystery." Matthiessen is cited for *American Renaissance* [26], but he missed Melville's purpose, called *Benito Cereno* "comparatively superficial," and thereby deflected attention from it. It seems likely that Rukeyser had not read *Benito Cereno* and missed the chance to come full circle in a comparison of Gibbs and Melville through the prism of the Amistad case.

Science Before Gibbs

The Amistad case, Rukeyser believed, was truly emblematic of the battle between good and evil, foreshadowing the greater struggle to come that would finally determine whether America was a country with or without an official policy to put profit before liberty, and whose outcome would determine whether the national soil would be free of the poison of enforced labor and thereby suitable for growing a Gibbs. If we want to evaluate whether the resolution of the slavery question was requisite for the development of American science as evidenced by the achievement of Willard Gibbs, we must know what science was like before Gibbs, especially in New Haven at the time of the Amistad incident.

As a graduate student, Gibbs studied with Benjamin Silliman, Jr., son of Benjamin Silliman (1779–1864), America's foremost scientist and colleague of Josiah Gibbs. The elder Silliman established Yale as a center for chemistry, geology, and mineralogy but had difficulty distinguishing legitimate sciences from pseudosciences that captured the American imagination, especially phrenology, which Davies [27] calls "an example of the almost incredible narrowness and prejudice which prevailed among men of science at the very time they were making such splendid advances in other fields of thought and discovery." Phrenology, the idea that protrusions or indentations of specific places of the head revealed an abundance or deficiency of traits assigned to those places, was brought to America by Johann Spurzheim. Within one week of landing in New York in 1833, he was off to Yale, where he stayed in Silliman's home and attended commencement. Spurzheim's phrenology lectures, and those of his disciple George Combe, were attended by nearly all of the Yale faculty and noted "for numbers and respectability, such as rarely falls to the lot of a public lecturer" in New Haven. Silliman delivered congratulatory speeches following the lectures. Swept up in the phrenology fever, Cinque's first lawyer, Roger Sherman Baldwin, deposited his phrenological analysis with the Yale Library.

As a personality test, phrenology was a harmless amusement—John Quincy Adams wondered how two phrenologists could meet without bursting into laughter—until, according to Davies [27], it began its dissolution into peculiarly American social philosophies aimed at proving with science the inferiority of non-European races. The "self-evident truth" that "all men are created equal" was a challenge to slaveholders, who therefore sought phrenological evidence of racial differences that could be used to justify the subservience of one race to another, all the while blaming these differences and their social consequences on the Creator. This evidence was apparently provided in Samuel Morton's and Combe's *Crania Americana* (1839) [28], a bigoted, phrenological analysis of Native American skulls in the context of a sweeping comparison of cranial capacity among the various races. Silliman characterized *Crania Americana* "as the most extensive and valuable contribution to the natural history of man, which has yet appeared on the American continent" and "an honor to the country" that places Morton "in the first rank among natural philosophers" [29].

Even though Silliman was a believer in phrenology, misused by others for racial purposes, he was always an outspoken opponent of slavery (though his education was in part financed by his mother's sale of Africans). At his 1796 Yale commencement, Silliman lamented in verse: "Not for himself; He toils, nor for himself he lives; His life, His labour, are another's wealth" [30]. Later, Silliman endorsed plans for the voluntary return of slaves to Africa and, with Josiah Gibbs, protested the establishment of slavery in the Kansas Territory [31]. It is hard to reconcile Silliman's lifelong abolitionist sentiments with his enthusiasm for Morton and Combe. Perhaps, like the contemporary naturalist Louis Agassiz, Silliman believed that blacks were intellectually inferior but, notwithstanding, deserved to be treated equally before the law [32].

Of course, the nonsense that was phrenology could be used for good or ill, depending upon who applied its principles. The abolitionist and illustrator John Barber (Fig. 3) appealed to phrenology [33], especially in *A History of the Amistad Captives* [34], the first of many such histories to follow. On the other hand, Nott and Gliddon's popular *Types of Mankind* [35], a predetermined ranking of the "social scale Providence has assigned to each type of man," was uglier and more explicit than *Crania Americana*. Karcher showed how Melville, in writing *Benito Cereno*, was reacting to this racist tract with the understanding that moderate antislavery sentiments would prove ineffectual once "white supremacists abandoned the bludgeon of the proslavery argument for the scalpel of science" [36]. Melville maintained a deep respect for science and was contemptuous of its misuses. He was an expert cetologist; for him "a whale-ship was my Yale College." In *Moby Dick* he takes naturalists to task for their inaccurate depictions of whales that look like "amputated sows" or "squash" (Fig. 4). However, his harshest satire is reserved not for the blunders of honest scientists but for the pseudoscientists, especially phrenologists [37]:

> If you unload [Moby Dick's] skull of its spermy heaps and then take a rear view of its rear end, which is the high end, you will be struck by its resemblance to the human skull, beheld in the same situation, and from the same point of view. Indeed, place this reversed skull (scaled down to the human magnitude) among a plate of men's skulls, and you would involuntarily confound it with them; and remarking the depressions on one part of its summit, in phrenological phrase you would say—This man had no self-esteem and no veneration.

Melville as champion of social equality among the races and scientist committed to the faithful representation of nature: this is the image that would have brought *Willard Gibbs* home. It is unlikely that such an ending would have won over Rukeyser's critics, especially those scientists repelled by a poet writing outside of her preassigned domain. Nevertheless, her defense of *Willard Gibbs* is today being mythologized in a new generation of Gibbs poetry: "Rukeyser, who fought for his biography. Self-Appointed. Against her, cohorts of colleagues and family" [38].

Hopefully, a new audience of scientists will not be so bothered by a poet presuming to characterize a thermodynamicist, as we are encouraged to court public largesse and welcome the imperfect admiration of our constituents. Whether or not Rukeyser wins a new audience, the Amistad case has been the focus of an explosive reinvestigation; no less than nine new books, three television documentaries,

Fig. 3 Barber's lithograph of New Haven Green, 1840 [40].

Fig. 1.   *Fig. 2.*   *Fig. 3.*

Fig. 4   Of this figure Melville said, "There is some sort of mistake in the drawing of Fig. 2. The tail part is wretchedly cropped and dwarfed, & looks altogether unnatural. The head is good" [41].

and an opera told Cinque's story in 1997–98. Yet, despite this Amistad frenzy, the tale was never better told than by Rukeyser, who wrote about the world into which Gibbs was born through the eyes of the Africans paraded around New Haven in autumn [1]:

> This color that they saw, these flickering delicate elms, the wide sweep of the Green [Fig. 3], the profound sky—nothing in the tropics, nothing on the sea, could have predicted this! But there was more; for past the avenues of feathers gleamed a whiteness never seen before, in soft round pillars rising as marble never seen before, a new and enchanting whiteness, fluted intricately, and rising to support great shapes that floated like white reefs over these pale and columned porches, whose steps rose up to them in the whiteness of astounding sand; and beyond this, a warm red never seen before, warm walls taller than they had dreamed, with shining squares, the gleaming windows in the warm brick. More feathers, feathery trees in double and triple arches, fell into green shadows, green brilliance, wherever they looked…Even the grass was softer here, the leaves cut and curled into softness. The smells of farm-wagons, the fruit, the early fall vegetables, the oyster-booths at the corner of the Green, mixed with the grass-smells; and the rich shadows fell among this light more softly, more graciously, than shadows ever fell. But, as far as the Africans were concerned, this was their prison.

References

1. Rukeyser, M. *Willard Gibbs;* Ox-Bow Press: Woodbridge, Connecticut, 1988.
2. *Adams, J .Q. Argument before the Supreme Court of the United States. The Case of the United States, Appellants, Vs. Cinque, and Others, Africans Captured in the Schooner Amistad*; S. Benedict: New York, 1841.
3. Gibbs, J.W. *The Collected Works of J. Willard Gibbs*. Vol. I; Yale University Press: New Haven, 1948.
4. Gibbs, J. W. *Am. J. Sci.* **1839,** *38,* 41, 43, 45, 255.
5. Gibbs, J. W. *The Liberator,* November 15, 1839, p 6.
6. Wheeler, L.P. *Josiah Willard Gibbs: The History of a Great Mind;* Yale University Press: New Haven, 1951.
7. Rukeyser, M. *A Turning Wind;* Viking: New York, 1939.
8. *The Craft of Poetry;* Packard, W., Ed.: Doubleday: Garden City, New York, 1974.
9. Rukeyser, M. *The Life of Poetry;* Paris Press: Ashfield, Massachusetts, 1997.
10. Weaver, E.C. *J. Chem. Educ.* **1943,** *20,* 259.
11. Samuelson, P.A. In *Proceedings of the Gibbs Symposium;* Caldi, D. G., Mostow, G.D., Eds.; American Mathematical Society and American Institute of Physics, 1990.
12. Kertesz, L. *The Poetic Vision of Muriel Rukeyser;* Louisiana University Press: Baton Rouge, 1981.
13. Kraus, C.A. *J. Am. Chem. Soc.* **1943,** *65,* 2476.
14. Färber, E. *ISIS* **1943,** *34,* 414.
15. Kazin, A. *The New Republic*, December 7, 1942, p 752.
16. Adams, B.; Adams, H. *Degradation of the Democratic Dogma;* The MacMillan Company: New York, 1920.
17. Jordy, W.H. *Henry Adams: Scientific Historian;* Yale University Press: New Haven, 1952.
18. Burich. J.R. *J. History Ideas* **1987,** 467.
19. Krutch, J.W. *Nation,* January 16, 1943, p 97.
20. Karcher, C. L. In *Critical Essays on Herman Melville's Benito Cereno;* Burkholder, R.E., Ed.; G.K. Hall and Co.: New York, 1992.
21. Adler, J.S. *War in Melville's Imagination;* New York University Press: New York, 1981.
22. Melville, H. *Battle-Pieces,* Kaplan, S., Ed.; Scholars' Facsimiles and Reprints: Gainesville, Florida, 1960.
23. Barber, D. S. *CLIO* **1982,** *12,* 1.
24. *Herman Melville: Representative Selections*; Thorp, W., Ed.; American Book Company: New York, 1938.
25. Mumford, L. *Herman Melville*; The Literary Guild of America: New York, 1929.
26. Matthlessen, F.O. *American Renaissance*; Oxford University Press: London, 1941.
27. Davies, J.D. *Notes on Phrenology, Fad and Science*; Archon Books: Hamden, Connecticut 1971.
28. Morton, S.G., Combe, G. *Crania Americana*; J. Dobson: Philadelphia, 1839. See also: Gould, S.J. *The Mismeasure of Man;* W. W. Norton: New York, 1996.
29. Silliman, B. *Am. J. Sci. Arts* **1833,** *23,* 356; **1839–40,** *38,* 375; **1840,** *39,* 87.

30. Brown, C.M. *Benjamin Silliman*; *A Life in the Young Republic,* Princeton University Press: Princeton, New Jersey, 1989.

31. Fulton, J.F.; Thomson, E.H. *Benjamin Silliman 1779–1864: Pathfinder in American Science;* Henry Schuman: New York, 1947.

32. Stanton, W. *The Leopard's Spots: Scientific Attitudes toward Race in America, 1815–59*; The University of Chicago Press: Chicago, 1960.

33. Fowler, L.N. The Liberator. January 24, 1840, p 16.

34. Barber, J.W. *A History of the Amistad Captives;* E.L. Barber and J.W. Barber: New Haven, 1840.

35. Nott, J.C.; Gliddon, G.R. *Types of Mankind;* J.B. Lippincott, Gram-bo & Co.: Philadelphia, 1854.

36. Karcher, C.L. *Shadow over the Promised Land: Slavery, Race, and Violence in Melville's America;* Louisiana State University Press: Baton Rouge, 1980. See also: Edwards, M.K.B. http://amistad. mysticsea-port.org/discovery/themes/bercaw.benito.cereno.html.

37. Hillway, T. *Modem Language Note* **1949,** 64, 145.

38. Strickland, S. *Kenyon Review* **1995,** 17, 53.

39. Wislon. E.B. *Scientific Monthly* **1931,** 32, 210.

40. Osterweis, R.G. *The New Haven Green and the American Bicentennial;* Archon Books: Hamden, Connecticut, 1976.

41. Vincent, H.P. *The Trying-Out of Moby Dick;* Houghton Mifflin Company: Boston, 1949.

Bart Kahr was born in New York City in 1961. He attended Middlebury College. His graduate studies with Kurt Mislow at Princeton University were followed by postdoctoral research at Yale in the lab of J. M. McBride, who encouraged him to read more Melville. He joined the faculty of Purdue University in 1990, but New Haven was his favorite city until he moved to Seattle with his wife Ann Kurth and son Aden. He is currently associate professor of chemistry at the University of Washington. His research group is studying the growth, structure, and physical properties of crystalline materials, with generous support from the U.S. National Science Foundation.

Proteins Versus Polymers: History Needs Revising[a]

Charles Tanford[b,c] and Jacqueline Reynolds[b]

Recognition of the existence of macromolecules and understanding of their structures and related functions must rate high among the success stories of chemistry in the twentieth century. Synthetic macromolecules, such as nylon, were newly created. Natural macromolecules such as proteins, DNA, and rubber, were extracted from biological sources and studied intensely. This is all common knowledge now; past achievements and their origins are entering the historical record of chemistry as a whole. We suggest here that this historical record, as reported so far, may be misleading, correct as far as it goes, but sadly incomplete.

The likely reason for this is that the bulk of macromolecular research was for many years divided between two separate communities: (1) the "polymer chemists," as the term is commonly understood today, whose roots are in organic chemistry and whose explicit orientation is often toward chemical industry, e.g., synthetic fibers or novel plastics and (2) the protein chemists, who until recently often started out as physiologists or physiological chemists and learned their rigorous chemistry as they went along. There were, of course, polysaccharide chemists and others who do not fall into either category, but they were few in number and not by themselves professionally unified. The bulk of all published research, the most obvious sources of dramatic progress, came from people associated with one or the other of the communities we have named. And the point to be made is that the two communities were separate, hardly aware of each other's existence. Even as late in history as our own student days (around 50 years ago), protein chemistry (often formally a part of "biochemistry") was looked down upon by respectable academic chemistry departments, with their conventional branches of organic, inorganic, physical and analytical chemistry meeting strict rules for curriculum and course content, set (in the United States) by the American Chemical Society.

The problem with current historical accounts of macromolecular chemistry is that they have all been written by authors who are "polymer chemists" or are oriented toward "polymer chemistry." This is a younger science than protein chemistry, and its adherents are perhaps more enthusiastic because their history is more recent and thus more readily accessible. The old protein literature tends to be neglected, which is a pity, because, as we shall show, protein chemistry was usually a step ahead in the "pure" aspects of macromolecular chemistry, in the development of concepts, methods, and even institutional organizations.

The Macromolecular Concept

Consider, for example, the very concept of macromolecules as such, the notion that molecules of huge molecular weight actually exist, linked by the same kinds of bonds that link the atoms of small molecules. The conventional story [1, 2] casts Hermann Staudinger as the first effective proponent (in 1920) of the idea, fighting heroically against intransigent opposition from organic chemists and from colloid chemists, who believed that large particles in solution are invariably secondary aggregates of much smaller true (covalently bonded) molecular entities. Staudinger is said to have finally convinced the rest of the world in a dramatic confrontation at a meeting in Germany in 1926—and Herman Mark, one of the father figures of polymer chemistry, may not have been quite convinced even then. All of this is historically accurate, but only if we limit our perspective to the early days of the polymer industry and to the organic chemists who were its pioneers.

The history from the protein point of view is quite different from the very beginning, not because protein chemists had exceptional genius or prescience, but simply as a result of the constitution of the protein molecules themselves, as compared to the molecules of other polymeric materials. Almost all proteins contain a small amount of sulfur, derived from the amino acids cysteine and methionine, and a small amount of any elemental constituent in a seemingly pure substance inevitably dictates a high formula weight. Other natural molecules—for example, cellulose, starch, and rubber—are composed of C, H, and O alone, at comparable atomic levels, and analysis (allowing for reasonable experimental

[a]*Chemical Intelligencer* 1999(3), 24–27.

[b]Emeritus Faculty, Duke University

Tarlswood, Back Lane, Easingwold, York Y06 13BG, UK

[c]Deceased

B. Hargittai and I. Hargittai (eds.), *Culture of Chemistry: The Best Articles on the Human Side of 20th-Century Chemistry from the Archives of the Chemical Intelligencer*, DOI 10.1007/978-1-4899-7565-2_49, © Springer Science+Business Media New York 2015

error) will inevitably yield plausible minimal molecular weights in the range of what we now know to be the monomer in a polymeric chain. Proteins are unique in yielding high molecular weights by this route.

S.P.L. Sørensen and his research group in 1909. (Photograph by C. Freslew & Co., Copenhagen. Carlsberg Foundation Picture Archives, Copenhagen).

The tiny, but reproducible iron content of hemoglobin, coupled with the crystallinity of that protein and evident purity by all available criteria, was even more convincing than sulfur analysis. By 1872 it had become textbook material; for example, Thudichum's *Manual of Chemical Physiology* [3] may be cited:

> Persons who have not studied this branch of chemistry and... do not read of atomic weights rising above 500, may wonder at the high atomic weight here assigned [to hemoglobin]. But this body can now be obtained pure...and always contains 0.4% iron.

The term "atomic weight" was still being used to designate the sum of the weights of atoms in a molecule. The assigned high number, based on a single atom of iron and therefore *minimal*, was 16,700.

As criteria for protein purity got better over the years and analytical methods in general improved, the sulfur limitation remained virtually unchallengeable support for the macromolecular concept. An extensive paper in 1902 by the Connecticut plant chemist Thomas Osborne, famous for the meticulous accuracy of his work, included results for animal as well as vegetable proteins, 24 altogether, all yielding minimal molecular weights of about 15,000 [4]. In another paper, he exploited the new ability to differentiate more than one form of nitrogen and recognized that the diversity of amino acids within a given protein inevitably leads to similarly high molecular weight values, if integral stoichiometry is to apply [5]. Osborne is articulate and quite unambiguous about use of the term "molecular weight" exactly as we understand it today. He simply ignored the colloid association theory that

entered a brief period of prominence around 1910 and that some chemists enthusiastically supported [6].

Physical methods for measuring molecular weight began to replace data based on chemical analysis and confirmed what everyone had supposed all along. For example, S. P. L. Sørensen, the famous biochemist and inventor of the pH scale, measured the molecular weight of crystalline egg albumin (in solution) by osmotic pressure and, in the course of this work, demonstrated that the laws of thermodynamics for a *pure* solute were obeyed; colloidal aggregates were expected to be heterogeneous, and some colloidologists even believed that thermodynamics would be altogether inapplicable. Unlike Osborne, Sørensen didn't just ignore the colloid chemists, but condemned them. "Colloidal chemistry has, in my opinion, not contributed to further progress, but rather the reverse," he said in 1915, in the first of a series of papers reporting on this work [7]. A little later, T. Svedberg's ultracentrifuge provided the definitive tool for measuring protein molecular weights [8]. He actually had begun his career as a colloid chemist but, when given the Nobel Prize in 1926, made no mention in his award address of any raging controversy regarding the macromolecular concept *per se*.

Valid questions might have remained. How are amino acids joined together to build macromolecules? Can we emulate nature and build them in the laboratory? (One must keep in mind that no organic chemist worth his salt believes in a chemical structure unless he can synthesize it with his own hands!) In fact, both questions had been answered already. Emil Fischer and Franz Hofmeister in 1902 independently demonstrated that peptide bonds provided the necessary link; Emil Fischer went on to synthesize polypeptides up to a length of 18 amino acid residues. A good historical account is given by Fruton [9].

Electrochemistry

Proteins are charged, multipolar electrolytes. This became known almost as soon as free charges in aqueous solution were first accepted at all, following publication of the theory of electrolyte dissociation by Arrhenius in 1887. G. Bredig, a student of Wilhelm Ostwald at the Physical Chemistry Institute in Leipzig, pointed the way in 1894. He measured dissociation constants of a huge number of organic bases, including tetramethylammonium derivatives, which lack a dissociable H^+ and thus do not possess a nitrogen atom that can act as a true base. Betaine was one of the substances he studied, and he realized that it must be a dipolar ion, what he called an "internal salt," with the formula $N(CH_3)_3^+ - CH_2 - COO^-$ [10]. Ordinary amino acids were soon found to be dipolar ions, too, on the acid side of the titration range of their amino groups; i.e., glycine is $NH_3^+ - CH_2 - COO^-$ and not $NH_2 - CH_2 - COOH$.

This was quickly followed by recognition of the amphoteric nature of proteins—like betaine, but multiplied 50-fold or more—with lysine, arginine, and histidine residues contributing positive charges near neutral pH and glutamic and aspartic acid contributing negative charges. (And, perhaps, one free amino group and one free carboxyl group from the ends of each polypeptide chain). Proteins in solution were seen to move in opposite directions in an electric field, as cations at high pH and anions at low pH, consistent with their amphoteric state. Electromotive force cells were used (as early as 1898) to determine free H^+ or OH^- that remained in solution after addition of measured amounts of acid or base, yielding (as difference) a quantitative measure of how much was bound to the protein. The binding was shown to be a proper equilibrium process, with a saturation limit that was equivalent to a count of acidic and basic groups per molecule. Saturation levels corresponding to around 50 bound H^+ or OH^- per protein molecule were typical [11].

And on the theoretical side, the famous Debye-Hückel theory for the distribution of free ions about any small central ion (published in 1923) was adapted within a year to a multicharged central ion by Kai Linderstrøm-Lang [12], the protein chemist/physiologist who followed in Sørensen's footsteps at the Carlsberg laboratory in Copenhagen. Linderstrøm-Lang's model allowed for changes in net charge as a function of pH but (because of the mathematical difficulties involved) could not pinpoint the actual coordinates of individual charges on the protein molecule. This remaining problem was solved by the theoretical physical chemist J. G. Kirkwood in 1934.

In contrast, the first synthetic organic polyelectrolyte that was given serious study was poly(acrylic acid)—and that only happened in the 1930s. Hermann Staudinger was enthusiastic about synthetic polymers as models for natural macromolecules, and he thought that there was a specially urgent need for proteins, the structure of which, to him at least, remained a mystery. To quote directly [13], "there are few natural products of high molecular weight for which work on a more easily analyzable model substance appeared as necessary as for proteins." It was proposed that work should be begun with a study of poly(acrylic acid), before more complex amphoteric polymers were tackled.

X-Ray Diffraction

The history is, of course, not entirely one-sided. Polymer chemists were the first to exploit X-ray diffraction from oriented fibers as a tool for structure analysis, a not surprising priority in view of the link between polymer chemistry and industrial fibers. The pioneers of the method were specialists, belonging to neither the protein nor the polymer group. But they were first welcomed by the polymer group and, for a few years (the 1920s), actually exerted a potent influence by their support for the colloid aggregation theory: the X-ray data gave them unit-cell dimensions, repeat distances along the fiber, which were invariably small, just a few angstroms. The simplest interpretation was that the molecule could not be larger than the unit cell and that the very idea of a macromolecule was absurd. This topic is well discussed by Morawetz [1].

Ultimately, of course, X-ray diffraction served industrial and biological science equally well. The double-helical structure of DNA is based on fiber analysis, and protein structures in their ultimate detail are based on diffraction from three-dimensional crystals.

Conclusion

Our thesis has been that protein chemists formed the vanguard of the forces that established the existence of the macromolecular state, years ahead of polymer chemists in general and Hermann Staudinger in particular. Any reader still unconvinced might consult a penetrating review by Edwin J. Cohn, entitled "The Physical Chemistry of the Proteins" and published in 1925, that is, at the very time when Staudinger's battles were supposedly taking place. Cohn reviewed the existing data for molecular weights, and electrochemical properties and for viscosity as well (which we have not covered); each named protein is clearly a unique individual. "Large organized molecules" he called them, to distinguish them from colloidal aggregates [14].

Cohn, whose training included postdoctoral stints with Osborne and Sørensen, went on to create a grand institute for protein research at Harvard University, partnered after 1926 by John T. Edsall. One of the fruits of that collaboration was a coauthored treatise, *Proteins, Amino Acids and Peptides as Ions and Dipolar Ions*, which became a veritable bible for protein research in mid-century [15]. More than 20 years had to elapse before polymer chemistry acquired an institute of comparable reputation in the United States, founded by Herman Mark at Brooklyn Polytechnic University in 1946.

As a final comment, it is worthwhile to recall that the words "protein" and "polymer" were both coined by the same person: one of the fathers of chemistry, Jöns Jacob Berzelius (1779–1848). The word "protein," suggested by Berzelius in 1839 [16], is derived from the Greek and means "to be in the lead, out in front"—a remarkably prophetic choice, we would say.

References

1. Morawetz, H. *Polymers: The Origins and Growth of a Science;* John Wiley & Sons: New York, 1985, reprinted by Dover Publications: New York, 1995.

2. Staudinger, H. *Arbeitserrinerungen;* Hüthig: Heidelberg, 1961: English translation entitled *From Organic Chemistry to Macromolecules;* John Wiley & Sons: New York, 1970.

3. Thudichum, J. L. W. *A Manual of Chemical Physiology;* Longmans: London, 1872.

4. Osborne, T. B., *J. Am. Chem. Soc.* **1902,** *24,*140–167.

5. Osborne, T. B.; Harris. I. F. *J. Am. Chem. Soc.* **1903,** *25,* 323–353.

6. A detailed account of reactions to the colloid association theory is being published elsewhere: Tanford, C.; Reynolds, J. *Ambix* **1999,** *46,* 33–51.

7. Sørensen, S. P. L. et al. *C. R. Trav. Lab. Carlsberg* **1915–1917,** *12,* 1–372.

8. Svedberg, T.; Pederson, K. O. *The Ultracentrifuge;* Clarendon Press: Oxford, 1940.

9. Fruton, J. S. *Proc. Am. Phil. Soc.* **1985,** *129,* 313–370.

10. Bredig, G. Z. *Phys. Chem.* **1894,** *13,* 289–326, esp. footnote, p 323.

11. Bugarszky, S.; Liebermann, L. *Pflüger's Arch.* **1898,** *72,* 51–74.

12. Linderstrøm-Lang, K. *C. R. Trav. Lab. Carlsberg* **1924,** *15,* 1–29.

13. Trommsdorf, E.; Staudinger, H. In *Die hochmolekularen organischen Verbindungen, Kautschuk und Cellulose*; Springer: Berlin, 1932; p 333.

14. Cohn, E. J. *Physiol. Rev.* **1925,** *5,* 349–437.

15. Cohn, E. J; Edsall, J. T. *Proteins, Amino Acids and Peplides as Ions and Dipolar Ions*; Reinhold: New York, 1943.

16. Victory, H. B. *Yale J. Biol. Med.* **1950,** *22,* 387–393.

Charles Tanford received his Ph.D. in physical chemistry from Princeton University and went on to postdoctoral work in the Cohn and Edsall group. He has held positions at the University of Iowa and Duke University.

Jacqueline Reynolds has a Ph.D. degree in physical chemistry from the University of Washington and did postdoctoral work with J. Steinhardt, himself an earlier investigator with the Cohn and Edsall group. She also became a member of the faculty of Duke University. Both authors retired from academia in 1988 and have lived in England since, engaged jointly in historical research and writing. Their best-known book, *The Scientific Traveller*, was published by John Wiley & Sons. New York, in 1992.

PHOTOGRAPH: Photograph of the authors at a meeting of the Humboldt Foundation in Germany in 1985.

Ubiquitous Cyclodextrins[a]

József Szejtli[b]

For some of the newly developed areas of chemistry, it has been observed that the first papers are scattered in various journals and are published by various authors. Only a few authors stick to the topic for decades and become the dominant personalities in that specific field. There are a lot of advantages in having researchers in different laboratories, and with different backgrounds, interests, and experimental facilities, working in the same field. Following a period of slow growth, the number of publications suddenly begins an exponential climb. It is in this phase, after the first patents are filed and their utilization becomes routine, that the number of papers reaches an inflection point and eventually the curve reaches a plateau.

Financial rewards can be expected for those who start their own projects before the inflection point is reached. Those who join in after the inflection point have a diminishing probability of producing something really original. The above characterization finds a spectacular example in the cyclodextrin story.

The Beginnings

In 1891 a French scientist, Villiers, published his observation of a crystalline substance formed in rotting potatoes [1]. He thought it was a new type of cellulose and called it "cellulosine." The next paper appeared in 1904 and was authored by Schardinger [2], a microbiologist in Vienna. He studied the microorganisms isolated from rotting potatoes, called *Bacillus macerans.* He observed two different crystalline substances, but he did not consider them to be a form of cellulose. They were easy to hydrolyze by mineral acids. They were cyclodextrins, but he did not know that; even the structure of starch was not yet known.

It was only in the 1930s that the chemical structure of cyclodextrins was elucidated. In the 1950s, French [3] at the University of Iowa and Cramer [4] in Heidelberg developed methods for the preparation of cyclodextrins and discovered their ability to form inclusion complexes. It was also at this

time that the cyclodextrins were ascribed high toxicity, based—as it turned out eventually—on erroneous experiments.

In the 1970s the number of papers on cyclodextrins began to increase. It had taken 80 years to publish the first thousand papers on cyclodextrins. The second thousand papers were published within four years, and two years sufficed for the third thousand. Since the beginning of the 1990s, the annual production has been more than a thousand papers and patents. Although many of these papers merely reproduce previous results or speculate without solid experimental background, their sheer amount alone indicates a tremendous impact. As of 1997, the growth of the number of publications in the cyclodextrin field has not yet reached an inflection point!

What Are They?

In natural photosynthesis, the two substances produced in the largest amounts are cellulose and starch. They are both macromolecules and built of thousands of D-glucopyranose units. The difference between them is in the glucosidic linkage, as shown in Fig. 1. The cellulose macromolecule has an extended zigzag-type structure, because the direction of the glucosidic linkages alternates. In starch, however, these linkages are oriented in the same direction and, because of the tetrahedral bond angle, these molecules form a helical structure. Degrading starch to smaller fragments by the use of acids, enzymes, or heat, or even mechanically, produces the dextrins. They are present in the crust on bread (during baking, the starch is degraded by the heat) and in beer (upon malting of barley, the starch is degraded by enzymes), and even the sticker on the beer bottle is glued by dextrins to the glass surface. The dextrins are more or less water-soluble, hygroscopic, heterogeneous substances, and their aqueous solutions are viscous and sticky.

When starch is degraded by an enzyme produced by *Bacillus macerans,* the product is a mixture of cyclic and acyclic dextrins (Fig. 2). Linear dextrins containing six, seven, or eight glucose units close to form α-, β-, and γ-cyclodextrins, respectively (Fig. 3). The cyclodextrins look like crystalline sugar, are not hygroscopic, and cannot be used as adhesives. In most of their physical properties, they differ fundamentally from the heterogeneous linear dextrins. Their peculiar property is their doughnut, or cylinder like,

[a] *Chemical Intelligencer* 1999(3), 38–45.

[b] CYCLOLAB, Cyclodextrin Research & Development Laboratory Ltd, P.O. Box 435, H-1525 Budapest, Hungary (deceased)

B. Hargittai and I. Hargittai (eds.), *Culture of Chemistry: The Best Articles on the Human Side of 20th-Century Chemistry from the Archives of the Chemical Intelligencer,* DOI 10.1007/978-1-4899-7565-2_50, © Springer Science+Business Media New York 2015

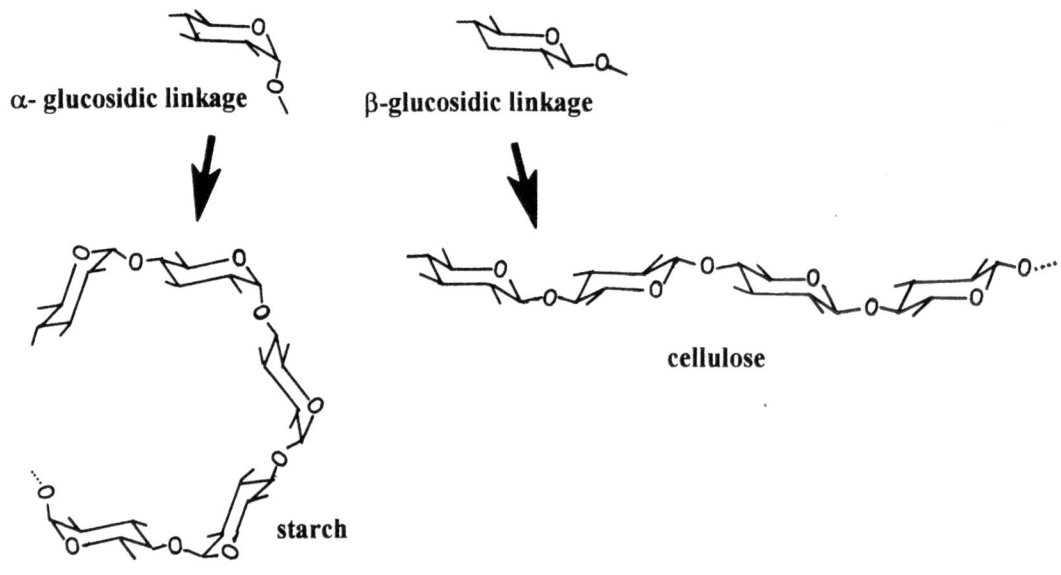

Fig. 1 Structural differences between starch and cellulose.

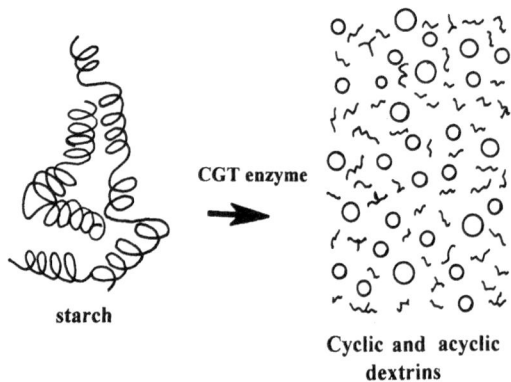

Fig. 2 Degradation of starch with cyclodextrin glycosyltransferase (CGT) enzyme. A mixture of cyclic and acyclic dextrins is formed.

shape. The planes of the glucopyranose units are aligned parallel to a six-, seven- or eightfold symmetry axis. All secondary hydroxyls of the glucose units are located on one rim of the cylinder, and all primary hydroxyl groups on the other. The internal wall of these cylinders is formed by hydrogen atoms and glucosidic oxygen atoms. This wall is apolar, while the secondary hydroxyl rim and the outside wall are polar (Fig. 4).

The water solubility of the "parent" cyclodextrins (Fig. 3) varies: at room temperature, it is 14, 1.8, and 22 g/100 ml for α-, β-, and γ-cyclodextrin, respectively. The unexpectedly low solubility of β-cyclodextrin may be attributed to its high symmetry. The planes of all glucopyranose units are aligned parallel to the central axis of the cylinder, forming seven hydrogen bonds between the C2- and C3-hydroxyl groups of the neighboring glucopyranose units. In α-cyclodextrin, two glucopyranose units have a distorted conformation. Their

planes are not parallel to the axis of the cylinder, and there are only four hydrogen bonds in the structure. γ-Cyclodextrin is not perfectly planar, because eight glucopyranose units cannot be arranged in one plane while maintaining the tetrahedral bond angles at the glucosidic bridge oxygens. One of the glucopyranose units is about 0.5 Å above the plane of the ring. β-Cyclodextrin readily forms dimers even in aqueous solution, hence having the highest tendency to crystallize and the lowest solubility.

When a cyclodextrin dissolves in water, its cavity remains apolar. Water molecules may enter this cavity but encounter repulsive forces. Thus, the internal wall of the cyclodextrin structure is not "wetted" by water. When apolar molecules are added to aqueous cyclodextrin solutions, and their sizes are compatible with the dimensions of the cyclodextrin cylinder, "inclusion complexes" are formed (Fig. 5). Upon removal of the water, the solid products are easily isolable in microcrystalline or amorphous form. They are molecular capsules; the "host" cyclodextrin molecules are filled with the "guest" molecules. No chemical bond is established between the host and the guest, yet the physical properties of the guest are strongly modified. Under appropriate conditions, these inclusion complexes are easily redissociated and the guest molecules recover their original physicochemical properties.

At the beginning of the 1970s, the cyclodextrins seemed to me to be interesting and promising substances, yet they were considered highly toxic, and thus unsuitable for human consumption, and they were very expensive.

The alleged toxicity had been published in only one paper, but it had a tremendous impact! There were, however, no further details available, such as experimental conditions,

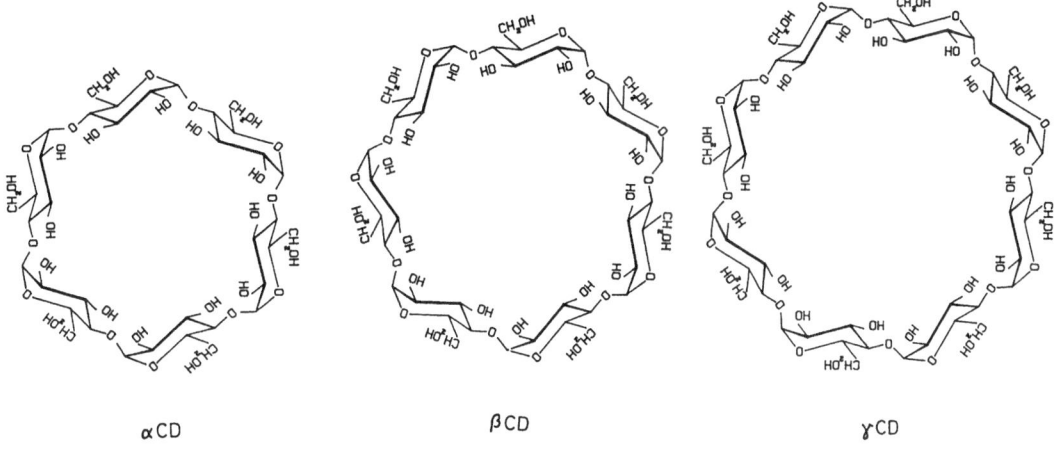

Fig. 3 Chemical structure of the three "parent" cyclodextrins (CDs).

secondary hydroxyl side

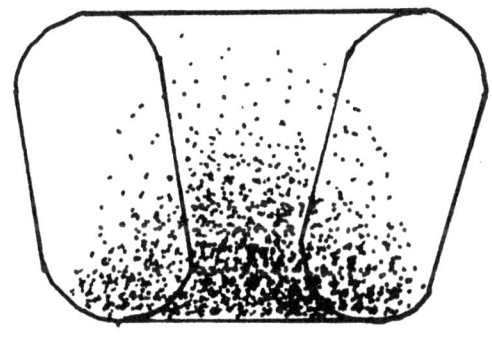

primary hydroxyl side

Fig. 4 Axial cross section of a cyclodextrin molecule. The darker areas represent the more hydrophobic regions, and the light areas are hydrophilic.

number of animals treated, analytical data for the consumed cyclodextrins, etc. On the other hand, the raw material for cyclodextrin production, starch, is available in very large amounts at low prices. It was primarily the work of Hungarian and Japanese laboratories in the early 1970s that finally showed that the notion about the toxicity of cyclodextrins was erroneous, giving a tremendous push to research on and applications of cyclodextrins.

The utilization of cyclodextrins on an industrial scale relied on two industries: the starch industry and the drug industry. The starch industry produces large amounts of starch and starch-derived products and seemed capable of developing the appropriate technologies for the production of α- and β-cyclodextrin. The pharmaceutical industry, in view of the exponentially growing costs of drug research, was looking for new drug-formulating technologies to improve the stability and bioavailability of drugs.

Entering the Field

In the early seventies, after having worked for 15 years in polysaccharide chemistry, mainly on the chemistry and application of starch, I found myself in the research division of a big pharmaceutical company in Budapest. Faced with the everyday problems of drug formulation, which in most cases uses carbohydrates, such as starch, lactose, and cellulose derivatives, it seemed quite obvious to me to try the cyclodextrins for such purposes. However, in 1971 β-cyclodextrin was still very expensive and available only as a fine chemical, although heavily contaminated with organic solvents. At the same time, Professor K. H. Frömming [5] at the Pharmaceutical Institute of the Free University of Berlin, in cooperation with the Schering Company, Berlin, launched a CD-drug-formulation project and published very interesting results about the very first study in humans of bioavailability enhancement of a drug by this technique. Their project was severely hindered though by the lack of availability of cyclodextrins in kilogram amounts. At this point we decided to screen a lot of *Bacillus* strains and boost their enzyme-producing capacity. We found ways of producing α- and β-cyclodextrin on an industrial scale. While the world production of cyclodextrins in 1970 did not exceed several tens of kilograms (almost all in Japan), in 1981 it was estimated to be about 1 ton, and presently it is several thousand tons. I anticipate that this figure will multiply in the next decade (Fig. 6).

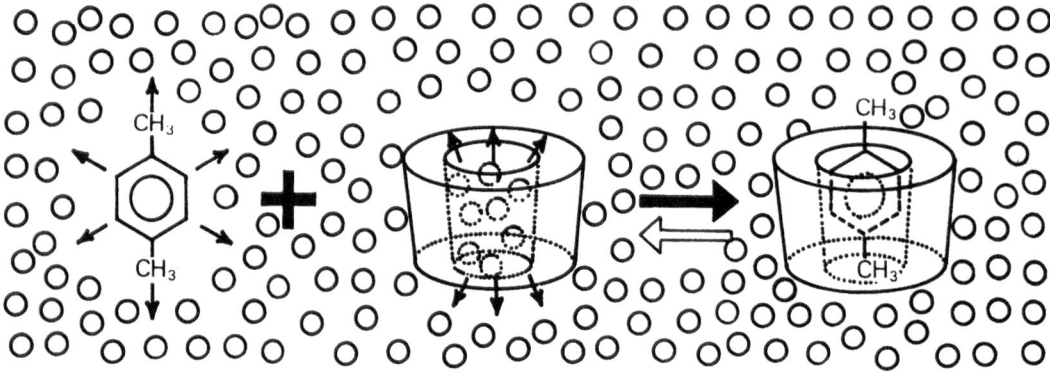

Fig. 5 Schematic representation of the formation of an inclusion complex. Small circles represent water molecules, which are repelled both by the hydrophobic (potential guest) p-xylene molecules and the hydrophobic cavity of the truncated cyclodextrin cylinder. The driving force for the inclusion is mainly the exchange of the polar/apolar interactions (e.g., between apolar CD cavity and polar water or apolar potential guest p-xylene and water) for apolar/apolar interactions between guest and CD cavity.

Applications

When substances are complexed by cyclodextrins, many of their physicochemical properties are altered, but their original properties reappear upon dissociation of the complex. For example, complexed volatile fluids, essential oils, or even gases are not volatile even under vacuum or at elevated temperatures, until the complexes are put in contact with water. Encapsulation protects easily oxidizable substances against atmospheric oxygen (Fig. 7). Complexed explosives, such as nitroglycerin, do not explode. Water-repelling, water-insoluble lipid like substances complexed with cyclodextrins are wettable and water-soluble. Similar modifications of the properties of a wide variety of substances offer enormous possibilities for diverse industries.

A further possibility is to substitute the hydroxyl groups of the cyclodextrins, converting them to extremely water-soluble or water-insoluble cyclodextrin derivatives. One can prepare acidic or basic cyclodextrin derivatives, or one can immobilize the cyclodextrins on solid surfaces.

Adequate toxicological studies proved that β-cyclodextrin administered orally is not toxic. Because the cyclodextrin rings have no end groups, β-amylases cannot split the ring, and even α-amylolytic enzymes hydrolyze cyclodextrins much more slowly than starch. Humans have no appropriate enzyme to metabolize the cyclodextrins. Upon oral administration, cyclodextrins are absorbed from the intestinal tract only in very small amounts. However, the colon microflora can easily metabolize the cyclodextrins. Upon opening of the ring, the metabolism of the linear dextrins progresses like the metabolism of starch or sugar, to carbon dioxide and water. When administered parenterally, only γ-cyclodextrin is excreted without interacting with various components of mammalian organisms. Thus, γ-cyclodextrin may be injected even in large amounts without resulting in any toxic effect.

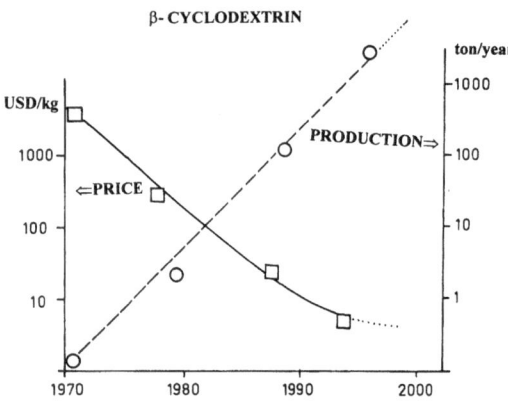

Fig. 6 Increase of β-cyclodextrin production and drop in its price during the last 25 years.

This is not the case for β-cyclodextrin, because it forms insoluble complexes with cholesterol, which is then accumulated in the kidneys. Modifying β-cyclodextrin, for example, by substituting several hydroxyl groups with hydroxypropyl groups or sulfobutyl groups, eliminates this noxious effect. Thus, these highly soluble β-cyclodextrin derivatives can be applied as parenteral drug carriers for injectable drug solutions.

Released When Needed

The elimination of the volatility and sensitivity to oxygen of essential oils by cyclodextrin complexation is used in solid flavor preparations. These preparations do not lose their flavoring capacity even after years and can be stored without protection. For example, flavored teas can be prepared by complexing the fragrance and flavor of jasmine, mango, banana, apple, etc., with cyclodextrins and mixing the gran-

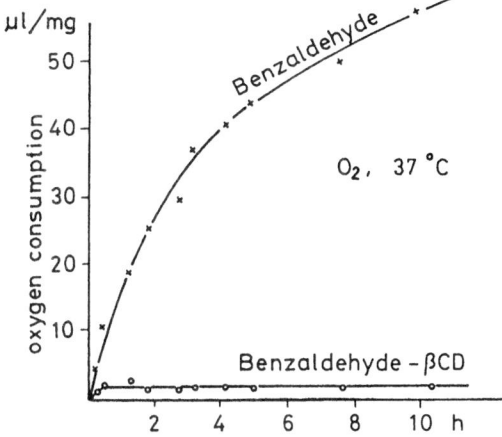

Fig. 7 Effect of complexation with β-cyclodextrin on the rate of oxidation of benzaldehyde. Benzaldehyde is oxidized rapidly to benzoic acid in an oxygen atmosphere; however, when it is complexed with β-cyclodextrin, it is well protected against oxidation.

Helping Drug Action

The most spectacular and economically rewarding applications of cyclodextrins are found in the pharmaceutical industry. Many promising drugs have never reached the market because they lack stability and/or bioavailability. Even marketed drugs often have very low bioavailability. In some patients, the absorption may be complete whereas in other patients it is insignificant. Even in an individual patient, the bioavailability may be very different at different times, depending on the state of the gastrointestinal tract and other physiological conditions. There are drugs that cannot be stored even for weeks without considerable loss of the active ingredient. Such drugs cannot be marketed, because the minimum requirement is that at least 95% of the active ingredient remain detectable after one-year storage at ambient temperature. Complexation with cyclodextrins may extend their stability.

In 1976 we prepared a garlic oil/β-cyclodextrin complex, which was left, inadvertently, on a laboratory shelf until 1987, when the laboratory was repainted. Reanalyzing it after 11 years, we found that the gas chromatogram was up to 90 % identical with the 11-year-old gas chromatogram. Our garlic oil/β-cyclodextrin complex soon became one of the numerous and popular preparations on the DM 100 million/year German garlic product market. Recently, it has also become available in the United States. These preparations suppress high blood pressure and reduce the blood cholesterol level.

Many drugs dissolve too slowly and only reach a rather low concentration in aqueous systems. When complexed to cyclodextrin, most such poorly soluble compounds are dissolved in water within minutes and are absorbed very effectively.

The pharmacokinetics of many drugs can be improved via complexation with cyclodextrins. For example, piroxicam, a nonsteroidal anti-inflammatory and analgetic agent, which is used widely to treat headache, dismenorrheal problems, rheumatic pains, etc., has its maximum pain-relieving effect about two hours after oral administration. The same amount complexed with cyclodextrin produces a similar or even better pain-relieving effect within half an hour. In addition, the complexed drug has a reduced incidence of stomach-irritating side effects. For many drugs, the doses can be reduced when they are complexed with cyclodextrin, owing to the enhanced absorption of the complexes.

Extremely stable inclusion complexes can be prepared by interlinking two cyclodextrin rings with appropriate spacers. Such an approach increases the complex association constants by a factor of 1000 to 100,000. By connecting appropriate "antennas" to duplex cyclodextrins, drug targeting becomes possible. An "antenna" is a short-chain oligosac-

ulated complexes with tea in filter bags; no tin-box packaging is necessary for storage of these teas, even for years. When these tea bags are dropped into a cup of hot water, the complex rapidly dissolves, releasing the flavor and fragrance substances. Essential oil components of various spices, like onion, marjoram, dill, and caraway, complexed with cyclodextrin and mixed with salt can be used in the preparation of foodstuffs. Under the combined effect of heat and humidity, the entrapped flavors and fragrances are released immediately. Lemon-peel oil is very sensitive to atmospheric oxygen and is marketed as a cyclodextrin complex, mixed with powdered sugar. The lemon-peel oil is released from the complex during baking or when dissolved in water. Nonwoven textiles on which fragrances complexed with cyclodextrins are immobilized are used as fabric-softener sheets in washer-dryers. During the drying cycle, the fragrances are released and adsorbed onto the washed fabrics.

In the cosmetics industry, cyclodextrins are used to eliminate some inevitable odor components in suntan preparations or to protect oxidation-sensitive components like squalene.

As mentioned above, β-cyclodextrin, when injected in mammalian organisms, forms insoluble complexes with cholesterol, which damages the kidneys. This complexation has, however, been taken advantage of in a technology for the production of low-cholesterol butter. When β-cyclodextrin is mixed with molten butter and a small amount of water, the cholesterol/β-cyclodextrin complex can be easily removed and the cholesterol-free (or very low cholesterol containing) butter can be isolated. Cholesterol can also be removed from egg yolk in a similar manner. Such low-cholesterol eggs have been launched on the market in the United States.

charide that reacts only with the specific receptors on cell surfaces of the target organ. By complexing, for example, light-sensitive porphyrinoid structures with such antenna-bearing duplex cyclodextrins, the injected light-sensitive porphyrinoid guests can be transported specifically to the target organ and concentrated on the surface of its cells. Upon irradiation of this organ with strong light of appropriate wavelength, the porphyrinoids will be converted to highly toxic photoconversion products, and their effect will be focused exclusively on the target organ. This is the essence of photodynamic tumor therapy.

Molecular Devices

One particular example of cyclodextrin complexes is the water-soluble γCD complex of C_{60} buckminsterfullerene. This complex makes it possible to study fullerene reactions in aqueous solutions.

Cyclodextrins are among the most appropriate rotaxane-forming molecules. A long slim guest molecule is threaded through the cyclodextrin cavity, and then both ends are terminated by bulky groups or these terminal groups are ionized, so that the threaded molecule cannot slip out of the cavity. Under various environmental effects (pH, irradiation, electric field, etc.), this threaded molecule may rotate around its axis, but its mobility is otherwise restricted. Similarly, the mobility of the cyclodextrin ring is also restricted, and it can only move along its axis.

By threading a long slim guest through a number of cyclodextrin rings, a "molecular necklace" can be obtained (Fig. 8). "Molecular tubes" can be prepared by complexing polyethylene glycol-bis-amine with α-cyclodextrin. Then the polyrotaxane is reacted with 2,4-dinitrofluorobenzene. In this way, both ends of the long-chain guest are terminated by bulky groups (Fig. 9). When this polyrotaxane is reacted with epichlorohydrin the vicinal cyclodextrin rings will be interconnected through glyceryl bridges between the primary and secondary sides of the cyclodextrins. Finally, under the effect of a strong alkali agent, the dinitrofluorobenzene groups will split off and the long polymer chain will slip out from the polymeric tube (Fig. 10).

Helping or Fighting Microorganisms

The application of microorganisms has made possible the conversion of some steroids into reaction intermediates or end products in industrial processes. However, steroids are poorly soluble in water, but only the dissolved substrates can be converted to the desired products on an industrially feasible

time scale. The presence of organic solvents is not tolerated above a very low level by the microorganisms converting the steroid. When β-cyclodextrin is added to a reactor converting hydrocortisone to prednisolone, hydrocortisone is fully dissolved and converted completely, and no hydrocortisone + prednisolone mixed crystals are formed. The microbial conversion is also faster than without cyclodextrin. This technology was developed in the mid-1980s and was the first industrial example of the application of cyclodextrins in biotechnology.

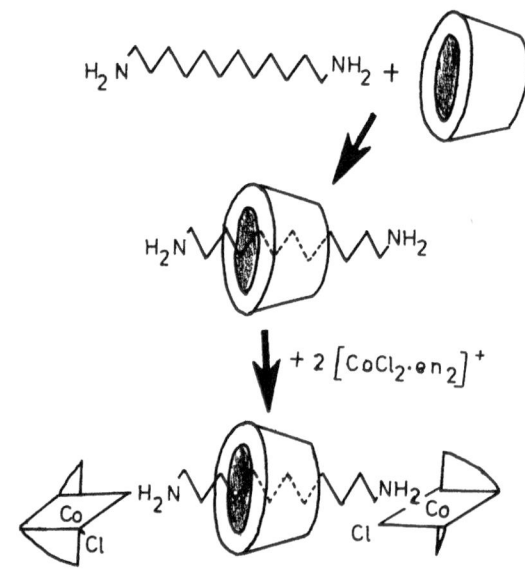

Fig. 8 Rotaxane: An α,ω-diaminoalkane is threaded through a CD ring, and then both terminal amino groups are converted to bulky groups [e.g., by reacting them with dichlorobis(ethylenediamine) cobalt ion]. The "axis" molecule cannot slip out from the CD ring but can freely rotate within it.

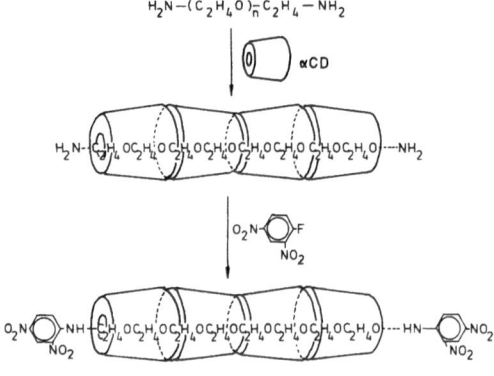

Fig. 9 "Molecular necklace" = polyrotaxane: A long slim polyethylene oxide chain can be threaded through a series of CD rings; by terminating both ends of the chain with bulky substituents (e.g., by reacting them with 2,4-dinitrofluorobenzene), the structure is stabilized without any covalent links.

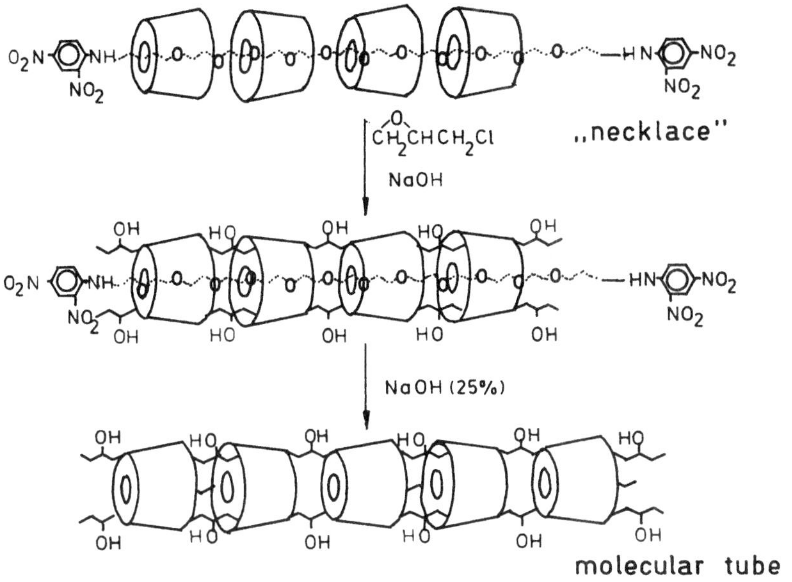

Fig. 10 "Molecular tube": The CD rings in the "molecular necklace" can be interconnected for example, by epichlorohydrin alkaline solution. After the terminating bulky groups are removed (by strong alkali), the "axis" molecule will slip out from the "tube".

According to all microbiological textbooks, the *leprae bacillus (Mycobacterium leprae)* cannot be cultivated in vitro. Professor Kátó, at the Université de Montréal, dedicated 30 years to the problem of cultivating *leprae bacillus* under in vitro conditions. In vitro cultivation of any pathogenic microorganism is necessary for screening potential drugs. Kátó had studied many *leprae bacillus* strains, collected in the jungle leprosy hospitals of Africa and South America, and came to the conclusion that in vitro the *leprae bacillus* cannot take up its primary energy source, palmitic or stearic acid, because of its thick hydrophilic shell. When palmitic acid was complexed with dimethyl-β-cyclodextrin, it became water-soluble and could easily penetrate through the hydrophilic shell of *leprae bacillus,* making it possible to cultivate the microorganism in vitro.

The vaccine against whooping cough is produced by *Bordatella pertussis*, which is a slow-growing microorganism. This microorganism produces the hemagglutinin factor only on solid medium. In submersed culture, the vaccine production is inhibited by a fatty-acid-like inhibitor, produced by the microorganism itself. The production of the vaccine on solid surfaces is laborious, slow, and expensive. However, with the addition of dimethyl-β-cyclodextrin to the submersed culture in amounts as low as about 2 mg/ml, the vaccine production is enhanced up to 100-fold. Dimethyl-β-cyclodextrin forms a soluble complex with the cell surface lipids and by removing them enables the cells to produce the vaccine continuously.

I anticipate broad application of the cyclodextrins in the pesticide industry. For example, in the cultivation of tea, the tea leaf roller insects cannot be killed by pyrethroids because the insects are inside the rolled leaves, and sunshine degrades the pyrethroids within hours. Complexing the pyrethroids with cyclodextrins extends their insect-killing effect for days, until the insect chews through the rolled leaves and becomes intoxicated. Honeybees do not become intoxicated because they do not eat the leaves, and simple contact with the pyrethroid/cyclodextrin complex does not result in absorption of the insecticide. On the other hand, when the leaf-eating insects eat the complex, together with the leaf, the complex dissociates, and thus the toxic effect of the insecticide is manifested.

Cyclodextrins are among the most widely used chiral separating agents in chromatography. Cyclodextrin derivatives are used in either the stationary phase, dissolved or bound to the solid support, or in the mobile phase in both high-performance liquid chromatography (HPLC) and capillary zone electrophoresis. The electrically charged cyclodextrin derivatives form complexes with neutral guest molecules and make possible their separation. More than 50% of the chiral chromatographic separations published during the past few years employ cyclodextrins either in the mobile or in the stationary phase.

Covalently bound cyclodextrins are also used for controlled release of entrapped guest molecules. For example, when monochlorotriazinyl β-cyclodextrin is reacted with textiles, the cyclodextrin is bound covalently to the textile fibers. Thereafter, these cyclodextrin cylinders can be filled with a perfume, a fungicide, or an insecticide, and the non-bound guests can then be fully removed. Thus, for example, a perfumed, dry T-shirt in which the cotton fibers contain about 6% bonded cyclodextrin has no smell, but when a person wears it, body heat and especially perspiration result in a slow release of the perfume. Simultaneously, hydrophobic components, such as bad-smelling short-chain fatty acids present in perspiration, will be complexed by the immobilized cyclodextrins. Not only the surface of hydrophilic cotton fibers, but also the surface of the less hydrophilic synthetic fibers can be modified with immobilized cyclodextrins, opening up new possibilities in the textile industry.

Cyclodextrins and their complexes have been at the forefront of supramolecular chemistry. The first cyclodextrin inclusion complex—although its structure was unknown—was described as early as 1911. In 1953, the first patent of a drug/cyclodextrin complex disclosed almost all the fundamental aspects of the use of cyclodextrins in pharmaceutical formulations. As we enter the era of applied supramolecular chemistry, only our imagination limits the potential uses of the cyclodextrins.

References

1. Villiers, A. *Compt. Rend.* **1891**, *112*, 536.
2. Schardinger, F. *Wien. Klin. Wochschr.* **1904**, *17*, 207.
3. French, D. *Adv. Carbohydr. Chem.* **1957**, *12*, 189.
4. Cramer, F. *Einschlussverbindungen.* Springer-Verlag, Berlin, 1954.
5. Frömming, K. H.; Weyermann, I. *Arzneim.-Forsch.* **1973**, *23*, 424.

József Szejtli (b. 1933 in Nagykanizsa, Hungary) is Director of CYCLOLAB, the Cyclodextrin Research & Development Laboratory Ltd. in Budapest. He graduated as a chemical engineer in 1957 and got his Ph.D. at the Budapest Technical University in 1961. He then did postdoctoral work at the Norwegian Technical University Trondheim (1963–64), was a Research Fellow at the Institute of Nutrition in Potsdam, Germany (1965–66), and worked as Professor of Chemistry at the University of Havana, Cuba (1967–70). After his return to Hungary, he headed the Biochemistry Research Laboratory of the CHINOIN Pharmaceutical and Chemical Works from 1971 through 1988. Since 1989 he has directed CYCLOLAB, an independent research organization, basing its operations on contracts worldwide and concerned exclusively with the utilization of cyclodextrins in the pharmaceutical, food, petroleum, paper, cosmetics, and other industries. Dr. Szejtli has also been active in academia. He received the D.Sc. from the Hungarian Academy of Sciences in 1976, has been a Professor at the Kossuth Lajos University of Debrecen since 1980, received the Award of the Hungarian Academy of Sciences in 1986, the Gold Medal of the Incheba (Bratislava, Slovakia) in 1988, and the Moët-Hennessy Prize (Paris) in 1991. He is Editor of *Cyclodextrin News* and Member of the International Organizing Committee of the Cyclodextrin Symposia and the Symposia of Molecular Recognition. Dr. Szejtli has authored or coauthored more than 450 research papers, 6 books, and over 90 patents. As an author, his latest book is *Cyclodextrin Technology* (Kluwer, 1988), and his latest coedited book (together with Professor Osa of Japan) is Volume 3 on cyclodextrins in the series *Comprehensive Supramolecular Chemistry* (1996).

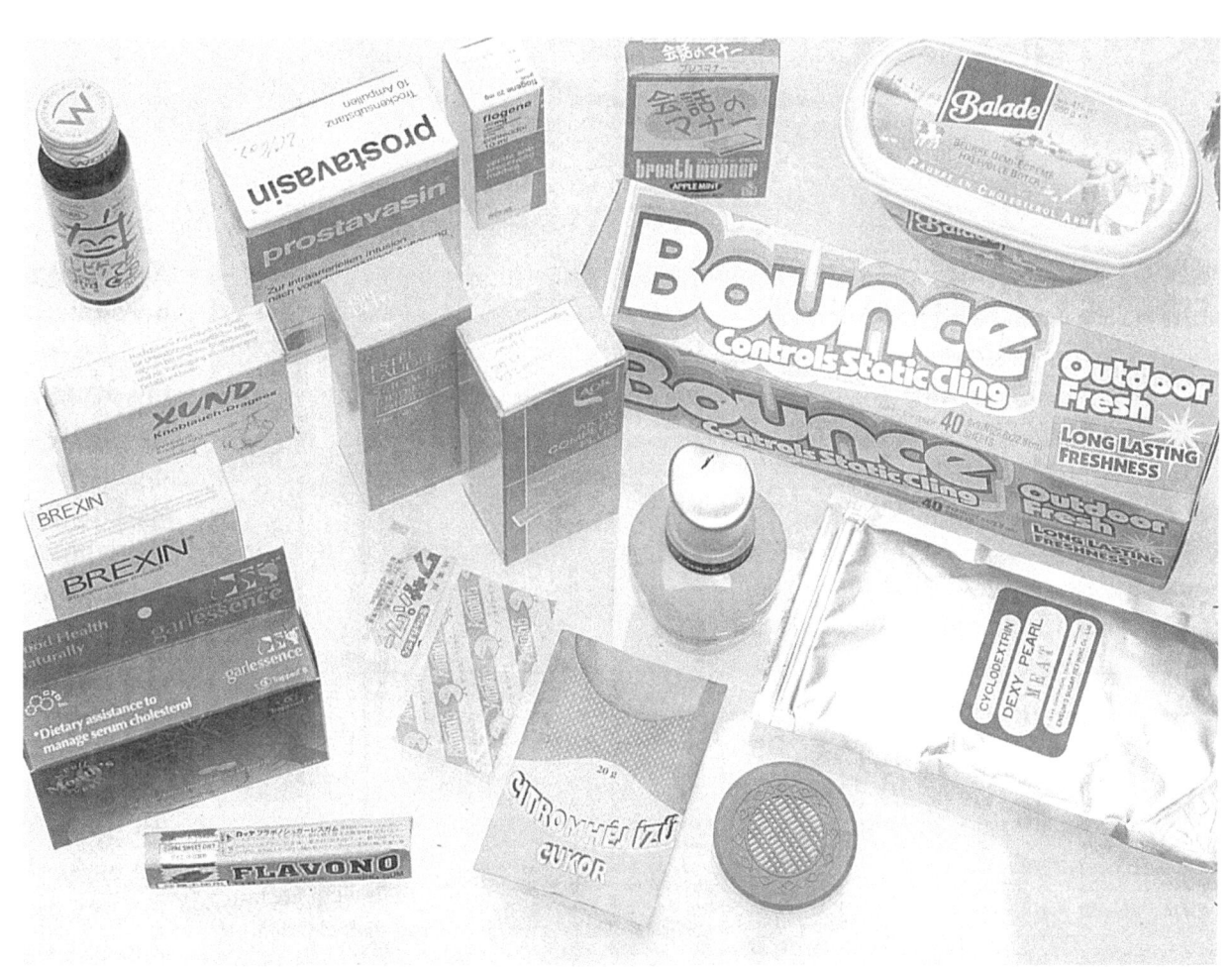

Sampler of cyclodextrin-related products:

Vitamin concentrate solution: cyclodextrin is solubilizer and stabilizer (Japan)
Prostavasin: Prostaglandin E_1/aCD complex, vasodilator, for intraarterial infusion (Germany)
Flogene: Piroxican/bCD complex analgetic, pediatric solution (Brasil)
Breath-manner: applemint/CD complex, chewing-gum (Japan)
Balade: low cholesterol butter (the original cholesterol content of butter is reduced to about $1/10^{th}$ by using bCD) (Belgium)
Xund: Garlic oil/bCD: reduces blood pressure and blood lipid level (Germany)
Estee-Lauder: Sun tan lotion, contains cyclodextrin to solubilize the active ingredients
AOK active complex plus: contains cyclodextrin to reduce the unwanted odour of one vital component
Bounce: perfume/b-cyclodextrin containing non-woven tissue, to be used in washer-dryer for perfuming of washed textiles
Brexin: Piroxican/b-cyclodextrin analgetic tablets (as well sachets, suppositories) (Italy)
Washabi: horse radish extract (allylisothiocianate/b-cyclodextrin) humidity activated indoor disinfectant (Japan)
Autoki-5: Anti-bad-breath b-cyclodextrin chewing tablet (Japan)
Vivace: perfume/hydroxypropyl-b-cyclodextrin long lasting perfumed composition (Japan)
Meatflavour/b-cyclodextrin: for food seasoning (Japan)
Garlessence: garlic oil/b-cyclodextrin complex (USA)
Flavono: Spearmint/b-cyclodextrin complex containing chewing-gum (Japan)
Limon-peel oil/b-cyclodextrin: (diluted with sugar) for household use e.g. in cakes (Hungary)
Plastic grill: for telephone-speaker, contains perfume/b-cyclodextrin complex (Japan)

Of Sandwiches and Nobel Prizes:
Robert Burns Woodward[a]

Thomas M. Zydowsky[b]

"The notice in The Times *of London (October 24, p. 5) of the award of this year's Nobel Prize in Chemistry leaves me no choice but to let you know, most respectfully, that you have—inadvertently, I am sure—committed a grave injustice."*
—LETTER FROM R.B. WOODWARD TO THE NOBEL COMMITTEE FOR CHEMISTRY, DATED OCTOBER 26. 1973.

R.B. Woodward lecturing. (Photo by Ormond V. Brody).

[a]*Chemical Intelligencer* 2000(1), 29–34.

[b]25 Harley Drive, #6, Worcester, MA 01606, USA
e-mail: tmzinc@aol.com

Ernst O. Fischer and Geoffrey Wilkinson received the 1973 Nobel Prize in chemistry for their pioneering work, performed independently, on the chemistry of the organometallic sandwich compounds [1], The decision to award the Nobel Prize to Fischer and Wilkinson was hardly questioned, since it was a fitting tribute to their extensive, groundbreaking efforts over the preceding two decades. However, the decision *not* to award a share of the Nobel Prize to Robert Burns Woodward *was* questioned, and even after 25 years, it continues to be a sensitive and emotional issue in some circles [2].

Perhaps Woodward himself provided the most emotional and historically significant response to the 1973 Nobel Prize. His public response varied, but in many situations he said little, if anything, about the prize [3]. His recently discovered private response, which he mailed to the Nobel Committee for Chemistry two days after the winners of the 1973 Nobel Prize were announced, reflected his intense desire to receive credit for his seminal contributions to organometallic sandwich chemistry [4].

We must examine events from 1952 to understand Woodward's reaction to the 1973 Nobel Prize in chemistry. In late 1951 and early 1952, two independent research groups published papers describing the synthesis of an unusually stable iron-containing compound: Kealy and Pauson from Duquesne University published a paper entitled "A New Type of Organo-lron Compound" [5], and Miller, Tebboth, and Tremaine from The British Oxygen Company published a paper entitled "Dicyclopentadienyliron" [6]. Kealy and Pauson's paper was submitted to *Nature* on August 7, 1951, published in England on December 15, 1951, and arrived in the United States about one month later. Miller, Tebboth, and Tremaine's paper was submitted to the *Journal of the Chemical Society* on July 11, 1951, published in England on March 24, 1952, and arrived in the United States about four to six weeks later.

B. Hargittai and I. Hargittai (eds.), *Culture of Chemistry: The Best Articles on the Human Side of 20th-Century Chemistry from the Archives of the Chemical Intelligencer,* DOI 10.1007/978-1-4899-7565-2_51, © Springer Science+Business Media New York 2015

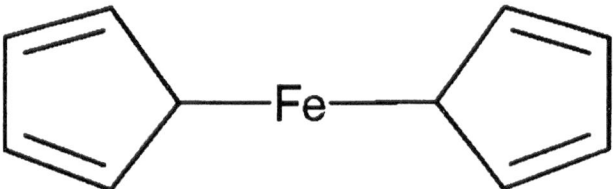

Fig. 1 Linear structure for dicyclopentadienyliron.

The two papers described the serendipitous synthesis, preliminary chemical characterization, and tentative structure assignment for dicyclopentadienyliron (see Fig. 1). The Duquesne group discovered their synthesis while trying to prepare dihydrofulvalene from ferric chloride and cyclopentadienylmagnesium bromide, whereas the London group uncovered their route during attempts to synthesize amines by reacting nitrogen and cyclopentadiene over iron filings. Both groups assigned the linear structure shown in Fig. 1 to their unexpected product. In doing so, they promptly attracted a contingent of chemists who questioned the veracity of the linear structure.

R.B. Woodward discussing ferrocene.

Harvard colleagues Geoffrey Wilkinson and Robert Burns Woodward were part of that contingent. In 1952 Wilkinson was a first-year assistant professor of inorganic chemistry, and Woodward was a full professor of organic chemistry. Wilkinson (1921–1997) had received his Ph.D. in nuclear chemistry from Imperial College of Science and Technology in London in 1946, and before his appointment at Harvard, he had held postdoctoral fellowships at the University of California at Berkeley and at the Massachusetts Institute of Technology (MIT). While at MIT, Wilkinson switched from nuclear chemistry to inorganic chemistry. Woodward (1917–1979) was already an established star on the international chemistry scene in 1952. He had been a child prodigy and had received his Ph.D. from MIT at age 20. By 1952, he had

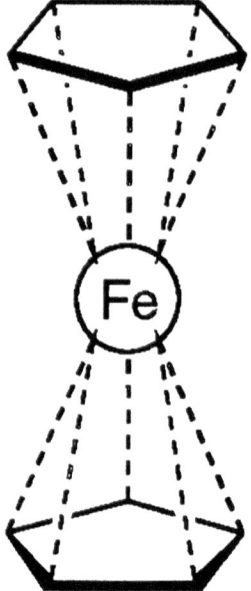

Fig. 2 Sandwich structure tor dicyclopentadienyliron.

already begun to publish some of the outstanding work in organic synthesis, structure elucidation, and theory that would subsequently earn him numerous honors, including the 1965 Nobel Prize in chemistry.

Myron Rosenblum was a graduate student in Woodward's group in 1952. He recalled that Woodward came into his lab one day in early January 1952 and began to discuss Kealy and Pauson's *Nature* paper [7]. According to Rosenblum, Woodward drew the linear structure for dicyclopentadienyliron on a blackboard and said that he thought that it was wrong. Woodward then carefully drew the now familiar sandwich structure for dicyclopentadienyliron (see Fig. 2) on the same blackboard. Without offering any insight into his reasoning, he told Rosenblum: "I think that this is the right structure. Why don't you take a few days off from your work and make the compound and let's look at it." Rosenblum repeated Kealy and Pauson's synthesis, and on January 21, 1952, he had crystals of the bright orange compound ready for testing.

At around the same lime, Wilkinson had also come across Kealy and Pauson's *Nature* paper, and he independently thought up the sandwich structure for dicyclopentadienyliron [8]. Through a subsequent conversation with Rosenblum, Wilkinson learned of Woodward's plan to investigate the novel compound. After discussing their mutual interest in the problem, Wilkinson and Woodward agreed on a series of experiments that would be used to verify their structure proposal.

On April 2, 1952, less than four months after Kealy and Pauson's paper appeared, Wilkinson, Rosenblum, postdoctoral fellow Mark Whiting, and Woodward (order of authors on the paper) published a one-page communication in the *Journal of the American Chemical Society* describing the results of two experiments that ruled out the linear structure for dicyclopentadienyliron [9]. The Harvard group reported that the dipole moment of dicyclopentadienyliron was effectively zero and that the infrared spectrum showed only one type of C—H bond. In place of the linear structure, the Harvard chemists proposed their new structure in which the iron atom was sandwiched between two cyclopentadienyl groups, hence the name sandwich compounds [10]. The dipole moment and infrared data supported the sandwich structure; it is important to emphasize, however, that Wilkinson and Woodward dreamed up the sandwich structure for dicyclopentadienyliron before any synthetic work or physical characterization had even begun.

Wilkinson and Woodward were not the only chemists to challenge the linear structure proposed for dicyclopentadienyliron. William E. Doering at Columbia not only questioned it, but in September 1951, he actually suggested the sandwich structure to Peter Pauson [11, 12], and, somewhat later, John R. Johnson at Cornell also suggested the sandwich structure [13]. W. C. Fernelius and E. O. Brimm had their doubts, and at their suggestion, Penn State College physicists Ray Pepinsky and Philip Eiland determined the molecular structure of dicyclopentadienyliron by using X-ray methods [14]. Meanwhile, over in Germany, E. O. Fischer and W. Pfab also used X-ray methods to solve the structure. For Fischer, it was his first step on the way to the 1973 Nobel Prize in chemistry [15].

Jack D. Dunitz and Leslie E. Orgel.

Nothing like the sandwich structure had ever been seen before. In 1952, Marshall Gates was the assistant editor of the *Journal of the American Chemical Society,* and he handled Woodward's manuscript submissions. In a letter to Woodward dated March 28, 1952, Gates wrote: "We have dispatched your communication to the printers but I cannot help feeling that you have been at the hashish again. 'Remarkable' seems a pallid word with which to describe this substance" [16].

Wilkinson and Woodward's "remarkable" structure enticed yet another team of chemists to work on organometallic sandwich compounds. Jack Dunitz and Leslie Orgel were Research Fellows in England in 1952, and Dunitz's account of their decision to work on dicyclopentadienyliron once again underscores the novelty and lure of the sandwich structure. In a 1992 paper celebrating the 40th anniversary of the discovery of ferrocene (dicyclopentadienyliron) [17], Dunitz said, "I think it is difficult today to appreciate just how surprising, unorthodox, even revolutionary, this structure must have appeared to chemists forty years ago. At any rate, I have to confess that my first reaction was one of extreme skepticism, if not plain disbelief." Dunitz came across Wilkinson and Woodward's paper shortly after it appeared, and according to Dunitz: "I opened the library copy of the JACS and came across this astonishing Harvard proposal: two parallel cyclopentadienyl rings with an iron atom sandwiched between them. I thought: what nerve these Harvard chemists have! To publicly put forward such a structure on such scanty evidence."

On his way out of the library, Dunitz ran into Orgel, and together they scrutinized Wilkinson and Woodward's paper. Orgel was as skeptical as Dunitz, so they decided to investigate the new compound. According to Dunitz, "We found that the compound was easy to prepare in crystalline form. We decided to make it and, by determining its crystal structure, demonstrate the incorrectness of the Harvard proposal."

Dunitz and Orgel soon learned that Wilkinson and Woodward's sandwich structure was indeed correct [18]. Their work also provided a novel explanation for the stability of this remarkable structure in terms of molecular orbital theory.

Woodward also predicted that dicyclopentadienyliron was aromatic and that it would have properties characteristic of typical aromatic compounds such as benzene. Later in 1952, a follow-up paper from Woodward's group (Woodward, Rosenblum, and Whiting) confirmed the predicted aromatic properties of the new compound, and in that paper they also proposed the name ferrocene for dicyclopentadienyliron [19]. That second communication was Woodward's penultimate paper in the ferrocene series, although his group continued to work on sandwich compounds of other transition metals for at least two more years.

Wilkinson was an assistant professor in search of research topics on which to build an independent career. He was undoubtedly aware of the significance of the new field that he had helped to create, and he recognized the longterm research potential of the sandwich compounds. Working independently of Woodward, Wilkinson published four ferrocene-related papers in 1952, and many more throughout his career. He subsequently became one of the world authorities on the chemistry of organometallic sandwich compounds and earned numerous awards for his work in that field, including the biggest prize of all—the Nobel Prize.

During his Nobel Prize award address in Stockholm, Wilkinson described the two factors which, in 1952, had led him to propose the sandwich structure for ferrocene [20]. The first factor was the well-known (to him) instability of transition-metal alkyls and aryls, and the second factor was his gut feeling, at that time unproved, concerning the binding scheme of several unrelated organometallic compounds. In 1951 Wilkinson was already thinking about transition-metal complexes of unsaturated ligands (cyclopentadiene-like), so he was clearly a "prepared mind" waiting for the right chance (ferrocene) to come along [21].

Wilkinson recounted the events leading up to his independent proposal of the sandwich structure in a 1975 paper [8]. He described his thinking when he came across Kealy and Pauson's *Nature* paper during his weekly visit to the departmental library' in this way:

> On seeing the structure I, which was also the one Miller, Tebboth, and Tremaine had drawn in their paper which appeared later, I can remember immediately saying to myself "Jesus Christ it can't be that!" Now I don't know why it was not the Sedgwick view quoted above that first occurred to me but the chelate diene structure, but I remember scribbling out on a piece of paper the structure II in which both double bonds were coordinated, and almost immediately III, as the significance of the resonance structures (I had been much impressed by Pauling) dawned and the equivalence of the carbons became obvious, "It's a sandwich." The thing that really excited me was the thought that if iron did this, the other transition metals must also form sandwich compounds.

Geoffrey Wilkinson speaking at an international conference in Munich, 1958.

Wilkinson went on to say that he and Woodward independently, and for different chemical reasons, proposed the sandwich structure for dicyclopentadienyliron, and, over lunch at the Harvard Faculty Club one afternoon, they agreed

to carry out the experiments needed to verify their structure proposal. He also acknowledged that Woodward suggested that ferrocene would behave like a typical aromatic compound and that he (Wilkinson) had not considered that possibility.

Woodward was on sabbatical leave in England when the Nobel Committee announced the winners of the 1973 Nobel Prize in chemistry. In an unpublished letter to the Chairman of the Nobel Committee for Chemistry dated October 26, 1973, Woodward reacted to the press release for the 1973 Nobel Prize in chemistry in this way [4]:

> The notice in *The Times* of London (October 24, p. 5) of the award of this year's Nobel Prize in Chemistry leaves me no choice but to let you know, most respectfully, that you have—inadvertently, I am sure—committed a grave injustice.

Woodward went on to quote several newspaper articles that had described Fischer and Wilkinson's contributions to organometallic sandwich chemistry, especially their role in the structure elucidation of ferrocene. The articles stressed the novelty and significance of the exciting new sandwich compounds but never once mentioned Woodward's contributions to the ferrocene story.

Woodward then gave his account of the events leading up to the proposal of the correct structure for ferrocene:

> The problem is that there were two seminal ideas in this field—first the proposal of the unusual and hitherto unknown sandwich structure, and second, the prediction that such structures would display unusual, "aromatic" characteristics. *Both* of these concepts were simply, completely, and entirely mine, and mine alone. Indeed, when I, as a gesture to a friend and junior colleague interested in organo-metallic chemistry, invited Professor Wilkinson to join me and my colleagues in the simple experiments which verified my structure proposal, his initial reaction to my views was close to derision.... But in the event, he had second thoughts about his initial scoffing view of my structural proposal and its consequences, and all together we published the initial seminal communication that was written by me. The decision to place my name last in the roster of authors was made, by me alone, again as a courtesy to a junior staff colleague of independent status.

Wilkinson and Woodward gave vastly different accounts of their early contributions to organometallic sandwich chemistry. According to Wilkinson's 1975 account, he thought up the sandwich structure for ferrocene while reading Kealy and Pauson's *Nature* article, several days prior to his conversation with Woodward at the Harvard Faculty Club. He regarded himself as a well-trained independent investigator who had spent considerable time thinking about the bonding in transition-metal complexes and naturally claimed co-inventorship for the sandwich structure. He also felt that from the beginning, he and Woodward agreed on the new structure, and that theirs was a collaborative effort in which both parties contributed to the scientific ideas.

On the other hand, Woodward claimed sole inventorship for both ideas (sandwich structure and aromaticity). He recalled that Wilkinson initially derided his (Woodward's) sandwich structure proposal but eventually embraced the structure and its consequences. Woodward also stated that he did Wilkinson a favor by letting him participate in the experiments that verified the structure proposal and by putting Wilkinson as first author. Wilkinson thought he and Woodward were peers, whereas Woodward saw himself as the mentor and Wilkinson as his protégé.

Woodward closed his letter to the Nobel Committee by saying that he had not seen the actual award citation issued by the Swedish Academy of Sciences or the official press release:

> Regrettably the precise citation issued by The Academy in connection with the award is not available to me here in England, nor have I been able to find a complete account of the ancillary material released to the press. Quite possibly the former does not signalise the special importance of the unique structural proposal and the demonstration of its correctness, and the latter well make a clear acknowledgment—ignored by the press—of my definitive contributions in those respects. Should these things be true—though in all candor I have to say that the actual press reports here provide no basis for supposing that they are—the problem is much minimized. But, I am sure that you will understand that I cannot read with equanimity such distorted and historically incorrect statements as those quoted above.

In fact, neither the award citation nor the ancillary material released to the press mentioned Woodward by name. In a reply to Woodward's letter, Arne Fredga, then Chairman of the Nobel Committee for Chemistry, wrote [22]:

Geoffrey Wilkinson discussing azulene complexes, Ethyl Corporation, around 1960.

Your letter of 26th October was received. It contains information not evident from the publications, but of great interest for the history of science. … The committee does not make available to the press information about a newly elected Nobel Laureate.… it is customary not to mention co-workers

and co-authors who are not sharing the prize, and this rule has been followed also in the present case.

Woodward's letter apparently induced at least one member of the Nobel Committee to overlook the rule. Acting either on his own or with the approval of his colleagues, Professor Ingvar Lindqvist acknowledged Woodward's contributions on two separate occasions during his introduction of Fischer and Wilkinson at the 1973 Nobel Prize ceremony [1]. Lindqvist said:

> The facts were available for all to see. Once the correct hypothesis was arrived at, by fantasy or intuition, it readily lent itself to simple process of logical deduction. I am of course referring to the way in which they, together with the former Nobel Laureate Woodward, reached the conclusion that certain compounds could not be understood without the introduction of a new concept, namely that of the sandwich compounds. … This they did by the successful synthesis of a large number of compounds which were analogous to the initially discovered ferrocene (named by Woodward in analogy to benzene), but with other metals than iron.…

The fact remains that Fischer and Wilkinson received the 1973 Nobel Prize in chemistry for their extensive investigations on the chemistry of organometallic sandwich compounds, not the discovery. Despite having a clear understanding of the importance of the field that he helped to establish, Woodward subsequently directed his efforts to other areas of organic chemistry. As a result, he missed out on a share of the 1973 Nobel Prize—a share that Woodward felt he deserved.

Woodward's longtime friend and fellow Nobelist Sir Derek Barton summarized Woodward's feelings in this way [23]:

> And when Geoff got a Nobel Prize for his work on ferrocene and its congeners, which he shared with E. O. Fischer, Bob Woodward said to me that it was rather strange, that he deserved to have that Nobel Prize. He didn't object to Geoff having one, too. But he certainly objected to the fact that he was not on that Prize. And they could have done that quite easily, because there was room for another person.

Roald Hoffmann of Cornell University thinks that Woodward made a strategic error by not expanding his organometallic research efforts to include other transition metals and instead concentrating on the aromatic properties of ferrocene [24]. Myron Rosenblum of Brandeis University feels that Woodward left the field because "perhaps he was more interested in the art and intellectual drama of organic synthesis" [25].

In retrospect, it is hard to second-guess Woodward's decision to concentrate on organic synthesis, structure elucidation, and theory. While Fischer and Wilkinson conducted their Nobel Prize-winning research in organometallic chemistry, Woodward received the 1965 Nobel Prize in chemistry for his contributions to the "art of organic synthesis," developed the Woodward-Hoffmann rules for the conservation of orbital symmetry with Roald Hoffmann, and together with

Albert Eschenmoser, led the team of chemists that completed the 100-step total synthesis of vitamin B_{12}.

The Nobel Foundation's official record regarding the 1973 Nobel Prize in chemistry is closed to the public until 2023; however, a letter from then Nobel Chemistry Committee Chairman Holger Erdtman to Woodward's good friend and fellow Nobel laureate Lord Todd offers an unofficial explanation for the Nobel Committee's decision *not* to include Woodward in the 1973 Nobel Prize [26]. This letter, dated December 13, 1973, states:

> Thank you for your confidential letter of Nov. 30, from which I understand that Bob was distinctly upset—and perhaps not unreasonably—by the press reports of the award. However, I feel that the name Woodward has come a little out-of-the-way (if you understand that dictionary expression!). In the final declaration to the Academy it is said that Woodward made a point contribution of value (of certain importance).

Geoffrey Wilkinson and Robert Burns Woodward left a rich chemical legacy upon which future generations of chemists will continue to build. They also left a story, albeit an incomplete and unresolvable one, which speaks to the emotions of the people behind the scientific advances and discoveries.

Acknowledgments I want to thank Ed Atkinson, Derek Barton, Michael Becker, Mary Ellen Bowden, F. A. Cotton, Jack Dunitz, Dick Hill, Roald Hoffmann, Gail McMeekin, Myron Rosenblum, Leslie Orgel, Linda Simon, Leo Slater, Arnold Thackray, Lise Wilkinson, Crystal Woodward, Eudoxia Woodward, Marcia Yudkin, and members of the Harvard University Archives staff for their help and support during various phases of this work.

References and Notes

1. Lindqvist, I. In *Nobel Lectures in Chemistry 1977–1980*; Frangsmyr, T., and Forsen. S., Eds.; World Scientific: Singapore, 1993; pp 99–100.
2. This is my personal observation based on interviews with some of Woodward's former co-workers.
3. One unsubstantiated exception involves Woodward and a prominent English chemist.
4. HUG(FP) 68.10, Box 25. Nobel Prize II (folder 2). "By permission of the Harvard University Archives."
5. Kealy, T. J.; Pauson, P. L. *Nature* 1951, *168*, 1039–1040.
6. Miller, S. A.; Tebboth, J. A.; Tremaine, J. F. *J. Chem. Soc. (London)* **1952,** 632–635.
7. Professor Myron Rosenblum, tape-recorded interview with Tom Zydowsky, Waltham, MA, August 28, 1997.
8. Wilkinson, G. *J. Organometal. Chem.* **1975,** *100,* 273–278.
9. Wilkinson, G.; Rosenblum, M.; Whiting, M. C.; Woodward, R. B. *J. Am. Chem. Soc.* **1952,** *74,* 2123–2124.
10. Geoffrey Wilkinson used the term "sandwich" in a 1952 paper (Wilkinson, G. *J. Am. Chem. Soc.* **1952,** *74.* 6148–49). Jack Dunitz and Leslie Orgel used the term "molecular sandwich" in their 1953 *Nature* paper (see Ref. 18, below) that they submitted three weeks after Wilkinson's.
11. See footnote 41 in: Pauson, P. L. *Quart. Rev.* **1955,** 391–414.
12. I want to thank Professor Roald Hoffmann for bringing this reference to my attention.
13. HUG(FP) 68.10, Box 13, Correspondence-Personal, 1950–1953, "By permission of the Harvard University Archives."
14. Eiland, P. F.; Pepinsky, R. *J. Am. Chem. Soc.* **1952,** *74,*4971.
15. Fischer, E. O.; Pfab, W. *1. Naturforsch.* **1952,** *7b,* 377–379.
16. HUG(FP) 68.8, Box 13, Ferrocene (folder 1). "By permission of the Harvard University Archives."
17. Dunitz, J. In *Organic Chemistry: Its Language and Its State of the Art;* Kisakürek, M. V., Ed.; Verlag Helvetica Chimica Acta: Basel; VCH: Weinheim, New York, 1993; pp 9–23.
18. Dunitz, J. D.; Orgel. L. E. *Nature* **1953,** *171,* 121–124.
19. Woodward, R. B.; Rosenblum, M.; Whiting, M. C. *J. Am. Chem. Soc.* **1952,** *74,* 3458–3459.
20. Wilkinson, G. In *Nobel Lectures in Chemistry 1971–1980;* Frangsmyr, T., and Forsen, S., Eds.; World Scientific: Singapore, 1993; pp 137–154.
21. Seyferth, D.; Davison, A. *Science* **1973,** *168,* 699–701.
22. HUG(FP) 68.10, Box 25. Nobel Prize II (folder 2), "By permission of the Harvard University Archives."
23. Professor Derek Barton, tape-recorded interview with Tom Zydowsky, College Station, TX, November 22, 1997.
24. Professor Roald Hoffmann, tape-recorded interview with Tom Zydowsky, Ithaca, NY, May 14, 1998.
25. Professor Myron Rosenblum, personal communication to Tom Zydowsky, December 20, 1998.
26. HUG(FP) 68.10, Box 25, Nobel Prize II (folder 2), "By permission of the Harvard University Archives."

Thomas M. Zydowsky is an organic chemist and writer living in Worcester, Massachusetts. He received his BS and MS in chemistry with David M. Piatak at Northern Illinois University, his Ph.D. In organic chemistry with Richard K. Hill at The University of Georgia, and had postdocs with Heinz G. Floss and Leo A. Paquette at The Ohio State University. He is writing a series of articles on Woodward that will form the basis for a full biography. Anyone interested in contributing to the biography may contact Tom via e-mail or regular mail.

Aleksandr N. Nesmeyanov[a]

"There are instances when an outstanding scientist is also a brilliant organizer of collective research work," Nobel laureate Peter L. Kapitsa said. Aleksandr N. Nesmeyanov was such a scientist.

Emiliya G. Perevalova[b]

Aleksandr Nikolayevich Nesmeyanov (1899–1980), full member of the Science Academy, was a Russian or, rather, Soviet scientist who occupied a commanding position in Soviet science for a long period of time. Nesmeyanov contributed significantly to the development of organometallic and organoelement chemistry. He coined the term "organoelement" to designate the organic derivatives of all elements except those belonging traditionally to organic chemistry, that is, H, O, N, S, and the halogens.

Nesmeyanov created a great school of exceptional productivity, which published over 1000 papers and monographs. His pupils have become leading scientists in Russia and in foreign countries.

Nesmeyanov studied under the great organic chemist Nikolai D. Zelinskii at Moscow University but established his own research direction early on. In 1929, he discovered a new way of making metal-organic compounds via aryldiazonium salts. This became known as the Nesmeyanov reaction. This technique was then extended to the synthesis of broad classes of organic derivatives of non-transition elements, both metals and nonmetals. This marked a transformation from metal-organic to element-organic chemistry· He worked a great deal with transition-metal organic compounds, such as ferrocene and its derivatives, aryl complexes, metal carbonyls and their organometallic derivatives, and many other classes of organometallic complexes. He was also interested in such fundamental phenomena as tautomerism, conjugation, the stereochemistry of electrophilic and homolytic substitution, and so on. He was keenly interested in the production of foodstuffs from nontraditional sources, in part because he was a vegetarian who considered the killing of animals barbaric. He had exceptional chemical intuition and a rich imagination that prompted him to look for new substances of unusual structure.

As a result of Nesmeyanov's organizational skills, several new research institutes and science centers were created, including the Institute of Element-Organic Compounds, which was named after him in 1980. He made an important contribution to the creation of the new campus of Moscow University.

Nesmeyanov's concern for the new generations of scientists was demonstrated by his efforts to send young researchers abroad. In connection with this, he corresponded with Alexander Todd of Cambridge University in 1955. That this endeavor was far from trivial is shown by the fact that it necessitated the intervention of then deputy Prime Minister A. N. Kosygin, who had met Todd on a visit to Cambridge. Eventually, an agreement was reached under the terms of which two Soviet researchers were to be received in Cambridge. Todd insisted that their selection should be based on scientific merit alone. The first two participants in this program, N. K. Kochetkov and E. A. Mistryukov, arrived in Cambridge in the fall of 1956. Kochetkov later became the director of the Institute of Organic Chemistry of the Academy of Sciences. Todd later reminisced about this [1]: "For me this brought the friendship of these two young Soviet colleagues and strengthened my friendship with Aleksandr Nesmeyanov."

For a long period of time (1935–80), Nesmeyanov was in charge of two large laboratories, one in his institute and the other at the university. He was Rector of Moscow University (1948–51) and President of the Soviet Academy of Sciences (1951–61). He was Head of the Department of Organic Chemistry of Moscow University (1944–78) and Director of the Institute of Organic Chemistry and of the Institute of Element-Organic Chemistry of the Science Academy (1954–80). He held various other positions such as Chairman of the State (formerly, Stalin) Prizes of the Soviet Union (1947–61) and was a member of the Supreme Soviet (the Soviet Parliament) (1950–62). He was highly decorated in the Soviet Union and was elected to 17 foreign science academies and was made an honorary doctor and professor of numerous foreign universities.

[a]*Chemical Intelligencer* 2000(2), 32–36.

[b]Leninskii Prospect 13/97, 117071 Moscow, Russia

B. Hargittai and I. Hargittai (eds.), *Culture of Chemistry: The Best Articles on the Human Side of 20th-Century Chemistry from the Archives of the Chemical Intelligencer,* DOI 10.1007/978-1-4899-7565-2_52, © Springer Science+Business Media New York 2015

S. S. Nametkin, N. D. Zelinskii, and A. N. Nesmeyanov in Moscow in the late 1940s. Sergei Semenovich Nametkin (1876–1950) was a hydrocarbon chemist. The Nametkin transformation described by him and L. Ya. Bryusova concerns camphenes and other terpenes. Nikolai Dmitrievich Zelinskii (1861–1953) worked on the chemistry of alicycic compounds, heterocycles, and organic catalysis, and, in particular, on the applications of platinum and palladium catalysts. He discovered what he called irreversible catalysis. He prepared numerous organic substances and used them to model gasoline and its fractions. All three scientists have appeared on Soviet stamps, Nametkin in 1976, Zelinskii in 1961, and Nesmeyanov in 1980. (see below).

I worked under Nesmeyanov in his university laboratory for 35 years (1945–80), first as a graduate student and later, for 20 years, as a professor. In 1980, I was put in charge of this laboratory and held this position for the next eight years. We worked a lot, and there was a good and creative atmosphere. Nesmeyanov came twice a week (later once a week). He gave his lecture to the students and then talked with his graduate students and associates. His lectures had an unhurried style as if he were thinking things through while lecturing.

Nesmeyanov was a well-educated and polite man who never raised his voice. He was not happy when my group shifted from ferrocene to organogold compounds, but he did not prevent this move and continued to show interest in our work. He had such great authority that his recommendations and advice were followed without the need for him to formulate them as orders. It is quite possible that the infamous discussions of the "reactionary" theory of resonance could have led to heavier sacrifices, similar to what had happened in biology, had it not been for Nesmeyanov's restraining influence.

Although he was often ill during his last years, he remained in charge of his two laboratories to the end. By then, his public functions had been greatly reduced, but he also stayed on as Director of the Institute of Element-Organic Compounds until his death in 1980.

Memoirs about Nesmeyanov have been published [1–4] in which he is depicted not only as a scientist but also as a family man, storyteller, poet, tourist, collector of mushrooms, and truly a renaissance man.

I complete this brief review with a virtual interview with Nesmeyanov in which the answers to the virtual questions are excerpted from his own reminiscences [3].

Q. What impact did your teacher, Professor Zelinskii, have on you?
A. Nikolai D. Zelinskii, along with many others, left the University in 1911 in protest against the actions of the Minister of Education. He returned to the University in 1917 and became Head of the Department of Organic and Analytical Chemistry. He was at the peak of his creative abilities at that time. I have often asked myself about his influence on me. We had no joint publications, he did not give me any ideas, he did not even teach me how to use the literature. Nonetheless, as time goes, to my own surprise, I appreciate his role in my life more and more. If I was different from the rest of his students, it may only have been because I sought independence from the very beginning and expected nothing from him except the possibility of working in his laboratory. I must be infinitely grateful to him for not having fired this independent graduate student, later assistant (1922–28), and let him do what he wanted to, and, on top of that, he even looked after his material wellbeing.

Q. What aspect of your research is closest to your heart?
A. It is difficult to distinguish whether I like some more than others. I like what we did on the chemistry of ferrocene, my diazo technique, stereochemistry, and my latest work, which was also my interest in the very beginning, on synthetic foodstuffs. I was known for the diazo technique for the synthesis of organomercury compounds for many years, and in a way it makes me feel like Conan Doyle must have felt when he said, "But I have written more than just Sherlock Holmes."

Q. How did your interest in chemistry begin?
A. It happened when I was 13, after I had had some experience in various branches of biology. I was spending the summer with my maternal grandmother, and there I found an old textbook of inorganic chemistry that caught my fancy more than the books of Jules Verne and H. G. Wells. A year later, I acquired an organic chemistry text and started a home lab. I fell in love with materials, their colors, smells, and shapes.

Q. What impressions do you retain from your childhood?
A. I was not well built physically. I was shy and introverted in the extreme. I was even too shy to go to the store. When I was 12, this was no longer a problem, but I still found it difficult to enter a room if there were people in it. Throughout my university years, through graduation, I avoided speaking in public. I valued honesty above all, and for me honesty was equivalent to becoming a scientist. I wanted to become a professor and accomplish something really big in science. I yearned for immortality in my deeds, in people's memory. I did not know that science was not the best way to do something immortal. Until the age of about 10 or 11, I was religious. Later, as I was trying to evaluate myself, I felt more inclined toward the arts than science. I lacked my dad's computer-like logic; I was visual rather than abstract, and abstract thinking is so characteristic of modern science.

Q. How did you become a vegetarian?
A. I was 9 or 10 when I declared to my parents that I would no longer eat meat. It depressed me to see the hopeless situation of animals selected for slaughter. Then, beginning in 1913, I no longer ate fish either. My mother and others stated that "the animal world is organized in such a way that some animals are the foodstuff of others and that this was the law of nature." To this I responded, "Man has science to establish his own laws and order in nature rather than follow it blindly. According to the law of nature, man does not fly, yet man utilized other laws of nature and enabled himself to fly. The goal of mankind is to overcome this law, soaked in blood, according to which some creatures are consumed by others and first of all by man."

It was not easy to stay vegetarian during the famine of 1919–21 when fish was so essential. When I say that it was not easy, I mean because of hunger, not because of any lessening of my will. I would have died rather than eat meat. This is how fanaticism and sects are born. I was aware of such dangers and tried to avoid them, tried to avoid placing myself against the rest of the people. I did not turn declining to eat meat into some sort of protest.

Q. Were you a member of the Soviet Communist party?
A. From the beginning of the war, I felt this to be a moral necessity and applied for membership in 1943. First, I was accepted as a candidate. Then, a year later, I became a member of the party.

Q. You were the Rector of Moscow University. What was your role in creating the new campus of the university?
A. As soon as I became Rector, I initiated discussions about a new campus. Yu. A. Zhdanov, who was then in charge of the Science Division of the Central Committee of the Communist Party, told me that he would give me a signal when the time came. Then one day he told me that it had been decided to build several high-rises in Moscow and I could request one of them for the University. At once, we wrote a letter to Stalin, and the decision came back quickly. A difficult period followed, which was full of tension.

Q. In 1951, you were elected President of the Academy of Sciences.
After the unexpected death of the previous President, S.I. Vavilov, I first heard about my possible candidacy from our drivers, who were usually well informed. Indeed, G.M. Malenkov, who was a member of the Politburo of the Central Committee of the party, invited me for a talk, and it became clear to me that my presidency had already been decided.

When I became President of the Academy of Sciences, I got acquainted first with its research institutes. Most of the greatest physicists were busy with nuclear energy and with the production of the atomic bomb, and later of the hydrogen bomb. This was under the jurisdiction of the corresponding ministry and not of the Academy.

Thinking about the frontiers of science, I determined that we should pay special attention to the interfacing of its various branches. This included the interaction of biology with other fields, but, first of all, biology had to be freed from its suffocating pseudoscience. One of our measures was the creation of a new Institute of Biophysics. Another new institute was the Institute of Element-Organic Compounds, as an interface between organic and inorganic chemistry. Finally, we created a new Institute of Scientific Information.

A.N. Nesmeyanov (on the left) examines the model of the new university building. The chief architect, Lev Rudnev, is on the right.

Q. In your work on the Committee of the Stalin Prize, did you have meetings with Stalin?
A. Our committee only made recommendations. The decisions were made in the Politburo of the Central Committee. One of the members of the Politburo participated in our activities. First, this was A.A. Zhdanov, and later G.M. Malenkov. My report served as the basis for a critical evaluation. Then came "judgment" day. I was notified that later in the day I would be called to the Kremlin. The session started at 10 or 11 P.M. and lasted till 2 A.M. These meetings took place in Stalin's office in the building of the Council of Ministers in the Kremlin. I knew that I was supposed to make the presentation, but nobody gave me a signal to start. Finally, I asked whether I should begin. Stalin, with slight irritation, said, "We are waiting for you." I tried to be as concise and clear as possible. Sometimes A. A. Zhdanov, who was at the time our instructor from the Politburo, would interject some clarification. Sometimes Stalin would ask a question. When we shifted from science to inventions and constructions, and these were mostly of a military nature, Stalin was in his element. He knew every airplane, tank, and piece of war machinery, he knew their merits and their problems, and it sufficed just to name the item without any detailed discussion.

I was observing Stalin with interest. He was dressed in a gray coat with large stars, emblematic of his rank of Marshall, and he was walking back and forth along the long desk, smoking cigarettes (rather than a pipe), thinking hard. Sometimes he would stop and speak. It happened sometimes that he would come directly to me when he wanted to look at my papers, and I could see his large hands dotted with

mottles. His hands were as close to me as my own. Although these night sessions were full of tension, sometimes I could detach myself from the discussion and think about Stalin. I longed to understand this man.

I chaired the award committee from 1947 through 1961 so I attended about four or five such Politburo meetings, up to the end of Stalin's life in 1953. Subsequently, the process was simplified, and the committee was given the right to make decisions rather than making recommendations only.

Q. What kind of interactions did you have with Khrushchev?
A. Once I. V. Kurchatov and I initiated a conversation about the impossible situation in biology, which was being suppressed by pseudoscience. We decided to ask to be received by Khrushchev and talk with him about it. The meeting did not start in the best way. Kurchatov told Khrushchev about the gains that the United States had derived from hybrid corn and how we were losing out a lot by lacking modern genetics in our science. Khrushchev became agitated and he withdrew a couple of long ears of corn from his desk. He started waving them in our direction and telling us that this was our corn and that we understood nothing about agriculture. He advised us to stay with our physics and chemistry and keep out of biology. After that, he became visibly bored while we were telling him about the poor state of biology in our country and about Lysenko's mistakes.

A.N. Nesmeyanov lecturing with his assistant (now full member of the Academy) N.K. Kochetkov.

Professors of Organic Chemistry of Moscow University in the lab:
Emiliya G. Perevalova and A. N. Nesmeyanov, around 1960.

On my return from the meeting, I got a phone call from Khrushchev. He told me, "Comrade Nesmeyanov, hands off Lysenko or else heads will roll." This was the end of this story, and I was busy with other things. I kept attending the meetings of the Council of Ministers, and there were more interactions than before, and more unpleasant situations as well. Sometimes, they may have been unintentional but, in other cases, it was unmistakable that Khrushchev meant to interfere in the affairs of the Academy, under the guise of giving instructions for the improvement of our activities. It

was becoming more and more clear that he was applying the saying, "For the watch to go, you have to shake it." This "shaking" was Khrushchev's only means of interfering in our affairs though, and he applied it with increasing frequency. Then an incident happened at the end of 1960. Khrushchev hinted at the unsatisfactory performance of the Science Academy, and he said that the reason was that the Academy dealt with little flies. I stood up at that point and, to the horror of the taciturn Politburo members, I declared that it was important to investigate these little flies too. It was unheard of and unprecedented to say anything that contradicted Khrushchev's viewpoints, and I added: "It is possible to replace the President of the Academy by someone better suited for this position, M. V. Kel'dish, for example." "I think so too," snapped back Khrushchev. The meeting then continued. For me, all that was left to do was just "wait."

References

1. *Aleksandr Nikolayevich Nesmeyanov: Scientist and Human Being. A Collection of Memoirs* (in Russian); Nauka: Moscow, 1988.
2. Chatt, J.; Rybinskaya, M. I. *Biog. Mem. Fellows R. Soc.* **1983,** *29,* November.
3. Nesmeyanov, A. N. *On the Swings of the 20th Century* (in Russian); Nauka: Moscow, 1999.
4. Nesmeyanova, M. A. *Light of Love. Reminiscences about Aleksandr Nikolayevich Nesmeyanov* (in Russian); Nauka: Moscow, 1999.

Moses Gomberg and the Nobel Prize[a]

Lennart Eberson[b]

Almost every textbook of organic chemistry tells the reader that Moses Gomberg discovered the first free radical, triphenylmethyl, in the year 1900, by allowing triphenylmethyl chloride to react with zinc metal (Fig. 1, middle) [1]. In retrospect, we can see that this was one of the most important discoveries in organic chemistry in the twentieth century: it reinstated the old, supposedly closed discussion of the existence of free radicals and paved the way for radical chemistry as we know it today, pervading many areas of theoretical, experimental, and technological chemistry [2].

The Beginnings of the Nobel Institution

In the same period, the stipulations of Alfred Nobel's will were institutionalized and the first Nobel prizes awarded, beginning in 1901. For chemistry, it was ruled that "the most important discovery or improvement" should be awarded and that prizes should be given "to those persons who shall have contributed most materially to benefit mankind during the year immediately preceding." The time restriction, almost impossible to uphold in view of the strong but healthy skepticism of the research community when confronted with new ideas and discoveries, had been modified in § 2 of the Code of Statutes of the Nobel Foundation, laid down by King Oscar II in 1900: "The proviso in the Will to the effect that for the prize-competition only such works or inventions shall be eligible as have appeared 'during the preceding year,' is to be so understood, that a work or an invention for which a reward under the terms of the Will is contemplated, shall set forth the most modern research of work being done in that of the departments, as defined in the Will, to which it belongs; works or inventions of older standing to be taken into consideration only in case their importance have not previously been demonstrated." The most important reason for this more flexible rule is found in § 5: "No work shall have a prize awarded to it unless it have been proved by the test of experience or by the examination of experts to possess the preeminent excellence that is manifestly signified by the terms of the will."

Another important regulation concerned the nomination procedure. To be eligible for a Nobel Prize in a given year, a candidate must be nominated by a person entrusted with this task. Such persons were (and still are) (i) members of the Royal Academy of Sciences in Stockholm, (ii) members of the Nobel committee for Chemistry, (iii) previous Nobel Prize awardees, (iv) professors of the chemical sciences at universities in Scandinavia, (v) holders of similar chairs at six or more other universities, selected each year by the Academy to provide a fair representation of countries and sites of learning, and (vi) other scientists whom the Academy may see fit to select in any given year. Rules (v) and (vi) ensured an influx of new nominators each time nominations were invited.

Triphenylmethyl Problem "Solved" and Ripe for Nomination After 1911

The beginnings of the Nobel Institution and Gomberg's discovery nearly coincided, but not until 1915 was Gomberg nominated for the first time. The difficult task of testing the discovery of stable free radicals had occupied many prominent members of the chemical community in the period between 1900 and 1911. This story has been told elsewhere and need not be repeated here [3]. The problem was finally "solved" by W. Schlenk's preparation of compounds that existed nearly entirely as free radicals, such as tribiphenylylmethyl [4] (Fig. 1, background). By 1911, it was beginning to be accepted that triphenylmethyl was a free radical, existing in low concentration (2–3 %) in equilibrium with its dimer, hexaphenylethane, and thus Gomberg's discovery was ripe for nomination [5].

He was nominated for the first time in 1915 by L. Chugaev from Petersburg, Russia, in a letter dated January 12, 1915. This would seem to be early enough for the letter to reach the committee before the deadline of January 31. However, the Russian calendar in this period was 13 days behind the Gregorian one, and thus the actual date was January 25. World War I was raging, and the censorship exercised by the *Ochrana,* the Czar's secret police, delayed all mail to and from Russia. Thus, the letter arrived in Stockholm too late, and, in accordance with the statutes, the nomination was disallowed but kept aside until the next year. However, in 1916

[a]*Chemical Intelligencer* 2000(3), 44–49, 57.

[b]Deceased

B. Hargittai and I. Hargittai (eds.), *Culture of Chemistry: The Best Articles on the Human Side of 20th-Century Chemistry from the Archives of the Chemical Intelligencer,* DOI 10.1007/978-1-4899-7565-2_53, © Springer Science+Business Media New York 2015

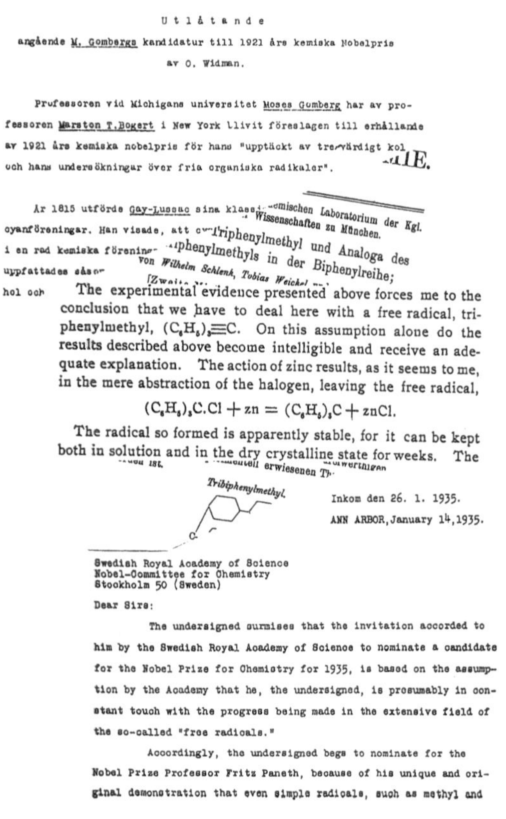

Utlåtande

angående M. Gombergs kandidatur till 1921 års kemiska Nobelpris

av O. Widman.

Professoren vid Michigans universitet Moses Gomberg har av professoren Marston T. Bogert i New York blivit föreslagen till erhållande av 1921 års kemiska nobelpris för hans "upptäckt av trevärdigt kol och hans undersökningar över fria organiska radikaler".

År 1815 utförde Gay-Lussac sina klass...

cyanföreningar. Han visade, att c...

i en rad kemiska förenin...

uppfattades såso...

hol och

...mischen Laboratorium der Kgl. ... Wissenschaften zu München.

Triphenylmethyl und Analoga des ...phenylmethyls in der Biphenylreihe;

von Wilhelm Schlenk, Tobias Weichel ...

The experimental evidence presented above forces me to the conclusion that we have to deal here with a free radical, triphenylmethyl, $(C_6H_5)_3\equiv C$. On this assumption alone do the results described above become intelligible and receive an adequate explanation. The action of zinc results, as it seems to me, in the mere abstraction of the halogen, leaving the free radical,

$$(C_6H_5)_3C.Cl + zn = (C_6H_5)_3C + znCl.$$

The radical so formed is apparently stable, for it can be kept both in solution and in the dry crystalline state for weeks. The

Tribiphenylmethyl.

Inkom den 26. 1. 1935.

ANN ARBOR, January 14, 1935.

Swedish Royal Academy of Science
Nobel-Committee for Chemistry
Stockholm 50 (Sweden)

Dear Sirs:

The undersigned surmises that the invitation accorded to him by the Swedish Royal Academy of Science to nominate a candidate for the Nobel Prize for Chemistry for 1935, is based on the assumption by the Academy that he, the undersigned, is presumably in constant touch with the progress being made in the extensive field of the so-called "free radicals."

Accordingly, the undersigned begs to nominate for the Nobel Prize Professor Fritz Paneth, because of his unique and original demonstration that even simple radicals, such as methyl and

Fig. 1 Sections of text from references quoted. Middle: Gomberg's original claim regarding the free-radical nature of triphenylmethyl [1]; top: beginning of Widman's report in 1921 [12]; bottom: beginning of Gomberg's letter of nomination in 1935; background: Schlenk's paper on biphenyl analogs of triphenylmethyl [4].

it was again disallowed because Chugaev did not have the right to nominate that year! After this unlucky start, allowed nominations of Gomberg were received in 1921, 1922, 1924, 1927, 1928, 1929, 1938, and 1940, from a total of 16 people.

The Nobel Committee for Chemistry

The task of the selection of Nobel prizewinners was entrusted to Nobel committees, consisting of five members. Each year, the committee delivered a report to the chemistry class of the Academy, recommending that the prize should be given to one or several persons or, sometimes, that no prize should be awarded because no prizeworthy candidate had been identified. After a decision by the chemistry class, the final vote was taken by the whole Academy. The Academy report consisted of a summary, based on special reports regarding each nominee, written in the same year or previously. The setting up of this system and its operation in the first 15 years of the Nobel Institution have been described in detail by E. Crawford in her book *The Beginnings of the Nobel Institution. The Science Prizes, 1901–1915* [6]. This book gives an excellent account of the early history of the science Nobel prizes and has chapters describing the work of the committees and their interpretation of the Code of Statutes. Of particular importance in this context is the observation that the recency rule had been interpreted in a way that made it possible to consider work carried out within "the last two decades."

The Code of Statutes gave most of the power over Nobel prize decisions to the committee. The role of the committee was further strengthened by the fact that all special reports were commissioned from members of the committee. The only "outside" reporter was Svante Arrhenius in his capacity as the Director of the Nobel Institute of Physical Chemistry. Another important factor was the small number of nominees (Fig. 2), which made it possible to evaluate the work of a candidate on the first occasion a nomination appeared. Thus, many candidates were evaluated early in their careers, which might have had negative consequences for them in comparison with those evaluated as established scientists. Considering the very long mandates of the committee members, it is fair to conclude that the Nobel prizes of the first 25–30 years were controlled by the committee as it was constituted in 1915 (Table 1). This was the year of Gomberg's first nomination and is thus a suitable starting point for examining how the committee handled the discovery of free radicals. The committee of 1935 (Table 1) exercised similar power until about 1950, the present limit of the period open for research in the Nobel Archives.

The committee of 1915 had three members whose doctoral training was in organic chemistry, but it was primarily O. Widman who was assigned the task of writing special reports about nominees from this branch of chemistry. Widman was born in 1852 and defended his D.Phil. thesis on chloronaphthalenes at Uppsala University in 1877, having carried out his studies with one of Berzelius's pupils, L. Svanberg, until 1874 and then with P. Cleve (the first chairman of the Nobel Committee for Chemistry; †1905). Widman traveled extensively in Europe during the period 1878–80, culminating in a stay in Adolf von Baeyer's "Laboratorium der Akademie der Wissenschaften" in Munich, where he aquired techniques and theories of "classical" organic chemistry, then at its height in Germany and strongly represented in Munich. During the time that he was active as a professor in Uppsala (1882–1917), he was considered to be the most prominent organic chemist in Sweden. Outside science, he was active politically in the City Council of Uppsala and related organizations for many years.

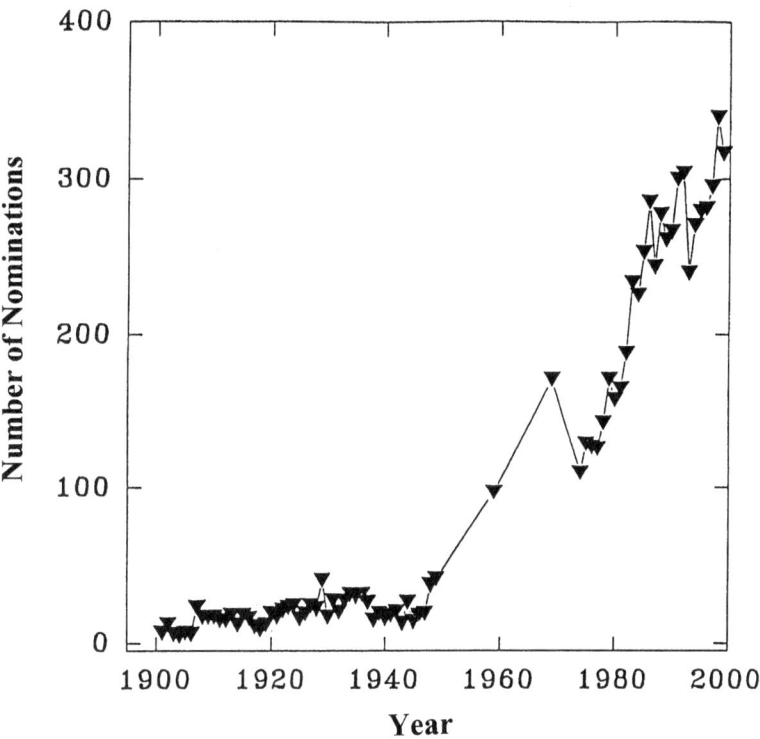

Fig. 2 Number of nominations for the Nobel Prize in chemistry throughout the history of the prize.

Table 1 Members of the Nobel Committee for Chemistry in 1915 and 1935

NAME (YEAR OF BIRTH)	PERIOD OF MEMBERSHIP	OFFICIAL POSITION	SCIENTIFIC TRAINING
In 1915			
Å.G. Ekstrand (1846)	1913–24	Government service, Stockholm	Organic chemistry
O. Hammarsten[a] (1841)	1905–26	Professor of medicinal and physiological chemistry, Uppsala University	Medical doctor, physiology
P. Klason (1848)	1900–25	Professor of chemistry and chemical technology, Royal Institute of Technology, Stockholm	Organic chemistry
H. Söderbaum (1862)	1900–33	Professor of agricultural chemistry at the Academy of Agriculture, Uppsala	Inorganic chemistry[b]
O. Widman (1852)	1900–28	Professor of organic chemistry, Uppsala University	Organic chemistry
In 1935			
H. von Euler-Chelpin[c] (1873)	1929–46	Professor of general and organic chemistry, Stockholms Högskola	Physical chemistry
B. Holmberg (1881)	1934–53	Professor of organic chemistry, Royal Institute of Technology, Stockholm	Organic chemistry
W. Palmaer[d] (1868)	1926–42	Professor of theoretical chemistry and electrochemistry, Royal Institute of Technology, Stockholm	Inorganic chemistry
L. Ramberg (1874)	1927–40	Professor of chemistry, Uppsala University	Organic and analytical chemistry
The Svedberg[e] (1884)	1925–64	Professor of physical chemistry, Uppsala University	Physical chemistry

[a] Chairman, 1910–26. [b] Also active in the history of chemistry. [c] von Euler-Chelpin was mainly active as a biochemist and received the Nobel Prize in chemistry in 1929. [d] Chairman, 1934–39. [e] The Svedberg received the Nobel Prize in chemistry in 1926.

Handling the Nominations of Gomberg and Schlenk

The first time Gomberg's discovery was mentioned in a special report was in 1918, a year in which he was not himself nominated. This was in a report on the works of W. Schlenk, who was nominated for the first time in 1918 for "his work on organometallic compounds and the valence problem of nitrogen" (and subsequently nominated in 1920, 1924, 1925, and 1929). Widman wrote an 11-page report [7] which covered all aspects of Schlenk's work, and only two pages were devoted to the triphenylmethyl problem. The first paragraph of this section immediately introduced the image of a hesitant Gomberg, even retracting his own idea, an impression that would become part of most later judgments of his work.

After Gomberg had discovered triphenylmethyl in 1900, this body has been the subject of great interest. Gomberg already from the beginning stated the view that this was a compound which contained a trivalent carbon atom, i.e., a "free radical." This immediately raised objections, and Gomberg found himself forced to give up his idea, if only for a while.

While this statement may be correct (see below), it should be noted that the only documentation for it in Gomberg's entire scientific production consists of two sentences in an account of preliminary work from 1906 [8]. His retraction, if there ever was one, lasted only about six months [9].

After listing the problems that occupied Gomberg and other researchers in the period 1900–10 [10], Widman concluded that the decisive proof of the existence of triphenylmethyl-type radicals was provided by Schlenk's isolation of a number of species with nearly 100% radical character, for example, tribiphenylmethyl (Fig. 1, background). Schlenk was also mentioned as being the first to verify experimentally the existence of the hexaphenylethane $\rightleftarrows$ triphenylmethyl equilibrium by ebullioscopy in benzene at ca. 80 °C. Here we must note that Gomberg had earlier obtained similar indications from cryoscopy in naphthalene, but the high temperature, ca. 80 °C, made him careful in his interpretation since he could not exclude decomposition. As is also well known today, Schlenk developed new methods and apparatus to deal with air- and water-sensitive compounds, which earned him great praise from Widman. These inventions were used mostly in dealing with organometallic compounds, and the triarylmethyl work was performed in an apparatus described by another free-radical chemist, J. Schmidlin [11]. Widman concluded that Schlenk's work on triarylmethyl radicals, even if it definitively proved the existence of free radicals, was not new and original enough. Gomberg was the one who discovered triphenylmethyl, and Schmidlin suggested the equilibrium hypothesis.

On a different note, Schlenk's work on alkylmetals, e.g., alkyllithiums, was deemed interesting, but these reagents were judged not likely to become of any greater use in the service of organic synthetic chemistry because of the extreme difficulty in handling them.

In the 1918 report to the Academy, the committee summarized Widman's special report, citing Schlenk's rare experimental skill in handling air- and moisture-sensitive compounds, but pointed out that it was Gomberg who had made the discovery of free radicals. The committee also endorsed the statement about the bleak future of alkylmetals. That year the Nobel Prize was not awarded; it was awarded to Fritz Haber the following year.

In 1921, Gomberg was properly nominated for the first time, and his work was the subject of a five-page review [12] by Widman (Fig. 1, top) in accordance with the practice of the committee. After referring to the long discussion about the possible existence of free radicals in the period 1815–65 and the ensuing acceptance of the dogma of tetravalent carbon, Widman described the nature and impact of Gomberg's discovery. He pointed out the problems that Gomberg had encountered in his further studies on the structure of the dimer of triphenylmethyl (definitely proven only in 1968 [13]), the electrical conductivity of triphenylmethyl solutions in liquid sulfur dioxide, the color of solutions containing triarylmethyl radicals, and the molecular weight determinations aimed at showing the monomeric nature of triphenylmethyl. He also cited parts of the two conclusions upon which the contention that Gomberg had retracted his free-radical hypothesis was based: "This hydrocarbon can hardly possess the simple formula $(C_6H_5)_3C$, however satisfactorily this symbol describes all other properties of this strongly unsaturated compound" and "The fact suggests in all probability that a conversion into quinoid compounds of some kind has taken place." These were quoted from studies on halogenated triphenylmethyls [8, 9], the results of which were not easily understandable at that time because no adequate theory existed. From this, Widman concluded again that Gomberg had found himself forced to give up his original view of the trivalency of carbon in triphenylmethyl, even if only for a short period.

The full text of Gomberg's conclusions [8] is given below, since there was and still is a problem of nomenclature in understanding them properly:

2. The constitution of the body formed by removal of the "carbinol-chlorine" from the halotriphenylmethyl chlorides can hardly be expressed by the formula $(C_6H_4Hlg)_3C$. Such a formula would indicate a similar function of the three phenyl groups which in fact does not exist. However, the same conclusion can now be drawn regarding triphenylmethyl itself: also this hydrocarbon can hardly possess the simple formula $(C_6H_5)_3C$, however satisfactorily this symbol describes all other properties of this strongly unsaturated compound;

3. The fact that the removal of the "carbinol-chlorine" causes one of the three phenyl groups (or one of the six groups of the dimolecular triphenylmethyl) to assume a function different from the two others, suggests in all probability that a conversion into quinoid compounds of some kind has taken place. None of the so far suggested formulas is however in full agreement with the findings reported in this paper.

The main difficulty with the interpretation of these sentences lies in the meaning of the word "triphenylmethyl," which at that time could be taken to mean the monomeric radical, a dimer or a mixture of dimers, or a mixture of the monomer and the dimer(s). In most cases, the exact meaning must be deduced from the context in which the word is used in documents of the period.

After pointing out the contributions of Schmidlin and H. Wieland, who suggested that an equilibrium between a dimeric species (hexaphenylethane and/or a quinol, a second possible dimer of somewhat higher complexity) and the monomeric free radical would explain the experimental observations, Widman stated that as of 1910, the problem still had not been settled. Still, a majority of chemists considered triphenylmethyl to be either a labile hexaphenylethane or a quinol.

The next sentence of Widman's report introduced Schlenk's contributions. One cannot avoid noticing the admiration for German organic chemistry implicit in the following sentence: "In this year W. Schlenk started to publish his masterly studies, emanating from the famous Munich Laboratory." Then Widman described briefly Schlenk's work on monomeric triarylmethyls [4], referring to his 1918 special report. He also mentioned that Pummerer and Frankfurter in 1914 had prepared another type of compounds having a trivalent carbon, the α-ketomethyls, and drew attention to Wieland's discovery in 1911 that tetraphenylhydrazine can dissociate into free diphenylamino radicals, in principle, the same phenomenon as hexaphenylethane dissociation. Thus, the discussion of the constitution of the triarylmethyls had been concluded around 1911, and Widman went on to his final assessment of Gomberg's discovery:

> As seen from the above, the observation made by Gomberg 21 years ago has led to exceedingly important theoretical results. However, the credit for these does not belong to Gomberg alone but to a very significant degree Schlenk, whose work in this and related areas (see my report on Schlenk's work from 1918) must in themselves be regarded as more prominent than Gomberg's. Even if one disregards the fact that Gomberg's discovery presumably is too old now to be awarded by the Nobel prize, it would not be fair to award him with exclusion of Schlenk. Anyway, the question of a possible sharing of the prize between the both is presently not pertinent, since Schlenk has not been nominated for the Nobel prize this year.

Here two statutory rules were cited. I have already touched upon them above, but some additional comment is required. A strictly upheld rule is that a person has to be nominated in a given year in order to be eligible for the Nobel Prize in that year—for the obvious reason that the committee would otherwise find itself occupied with a steadily accumulating and unmanageable list of candidates. The rule about the recency of discoveries had been used flexibly, to the extent that work done during the past two decades could be considered, as concluded by Crawford [6]. Thus, Widman's conclusion presumably reflected an unwritten rule of the committee that 20 years was about the time limit for the age of a discovery. The facts that the rule of § 5 had prohibited any award to Gomberg before 1910–12 and that World War I had interrupted the awarding of Nobel prizes for two years were not taken into account; if they had been, the corroborated and generally accepted discovery of free radicals might well have been said to be only 7–9 years old in 1921.

The committee quoted from Widman's report almost verbatim in 1921, and in 1922 the nomination was dealt with negatively by a short reference to the report of 1921. In the critical year of 1924, both Gomberg and Schlenk were nominated; the former received two nominations, one of which was together with G. N. Lewis. The committee relied on the previous special reports from 1918 and 1921, respectively, for its one-page statement on the candidacies of Gomberg and Schlenk. In summary, it was noted that Gomberg discovered a compound in 1900 which he denoted as triphenylmethyl, containing a trivalent carbon atom and thus being a free radical. On the basis of his own work and criticism from other researchers, he had to retract his view, if only for a short time. After 10 years of scientific discussion, Schlenk was finally able to prove the existence of triarylmethyls and solve this theoretically interesting valence problem. Therefore, the Nobel Prize could not be awarded to Gomberg alone, especially since Schlenk's work must be considered to be more prominent. On the other hand, Gomberg made the first discovery and Schlenk's work, even if it unambiguously confirmed Gomberg's suggestion, was based on results by others, not only Gomberg but also Schmidlin. Thus, neither of the candidates could justly be awarded the Nobel Prize at the exclusion of the other.

The possibility of a shared prize was briefly introduced but met a difficulty in relation to the recency rule: "Gomberg made his discovery 24 years ago and its importance was clearly established in 1910, i.e., 14 years ago. To award Gomberg now would according to the views of the committee not be in good agreement with this statute and by its consequences actually be equal to putting this rule out of force."

Thus, the candidacies of Gomberg and Schlenk were ruled out in 1924 by the recency statute. No Nobel Prize was awarded in that year because of the lack of suitable candidates, which is somewhat surprising in view of the fact that the prizewinners of the three upcoming years (R. Zsigmondy 1925, The Svedberg 1926, H. Wieland 1927) were all nominated in 1924. The 1924 prize was reserved for the next year and, in the end, forever. So, there remains a question which will be addressed tentatively below: Was the committee really so deeply concerned about the recency statute, or did it simply not want to award the discovery of free radicals? Gomberg was to be nominated several times in the years to come but the committee, now with largely different members (Table 1), always dealt with these nominations by reference to the Academy report of 1924 and Widman's special reports of 1921 and 1918.

As a curiosity, it can be mentioned that Gomberg was asked to nominate for the Nobel Prize in chemistry for 1935. The beginning of his one-page letter in Fig. 1 (bottom) shows that he nominated F. Paneth for his demonstration that simple alkyl radicals exist. Can one detect the slightest hint of irony in the first paragraph?

Conclusion

The events related above permit a tentative conclusion to be drawn as to why no Nobel Prize was awarded for the discovery of the first free radical. The formal reason in the critical year, 1924, was based on the recency rule. However, it is difficult to imagine that a determined champion of a free-radical award on the committee would not have been able to circumvent this argument and convince the other members about the prizeworthiness of Gomberg's and Schlenk's work. No member, and particularly not Widman, who had been the reporter on free-radical chemistry, wanted to play this role in the critical year of 1924. Thus, the opportune moment was lost. The arguments used were based on Widman's special report on Gomberg in 1921, which reviewed Gomberg's work with some emphasis on negative aspects. In particular, the quotation of Gomberg's retraction of his idea, also mentioned in the special report on Schlenk in 1918, which represented two sentences out of a published output of then more than 400 pages, appears odd and somewhat out of context. In contrast, the praise of Schlenk for his construction of new devices to handle air- and water-sensitive compounds does not have any solid basis in this context, because for his work with triarylmethyls Schlenk used an apparatus developed by Schmidlin. Besides, Gomberg was the first to construct a special apparatus for this purpose in 1904 [14] (Fig. 3). In short, it seems that the committee did not consider the discovery of free radicals important enough for a Nobel Prize. That this was an absolute verdict is shown by the fact that the Nobel Prize of 1924 was reserved forever. Later, when the ramifications of free-radical chemistry had started to pervade organic chemistry, the recency rule became indeed valid.

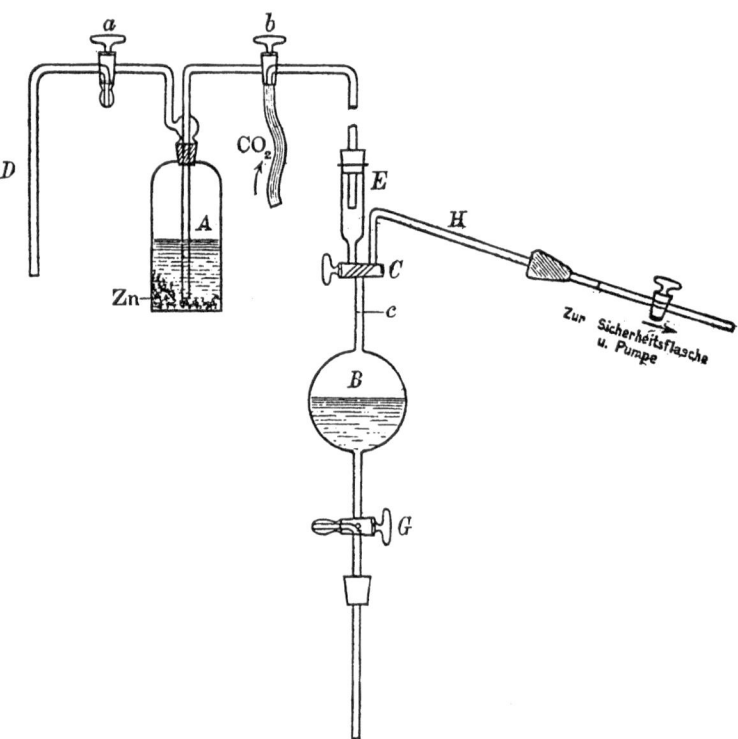

Fig. 3 Apparatus for handling triarylmethyls under oxygen-free conditions, constructed by Gomberg and Cone [14].

A similar opinion on stable free radicals was expressed later by C. Walling in his book *Free Radicals in Solution* [15], published in 1957: "However, because their structural requirements for existence are possessed by only rather complicated molecules, they have remained a rather esoteric branch of organic chemistry." The persistence of Walling's opinion about stable free radicals is shown by the following quotation from his autobiography [16], published in 1995: "For the field I was inadvertently entering [Walling had asked Kharasch if he could join his group], 1937 was a landmark year. Free radicals of course first entered organic chemistry in 1900 with Gomberg's preparation and identification of triphenylmethyl, but the chemistry and properties of such 'stable' or 'persistent' species had remained largely a chemical curiosity." If this statement could be made so much later, how could the Nobel Committee of 1924 have had a different view?

Acknowledgments I gratefully acknowledge the permission of the Royal Swedish Academy of Sciences, Stockholm, to carry out research in its Nobel Archive. I also thank Anders Lundgren, Karl Grandin, Salo Gronowitz, and Tom Tidwell.

Editor's Note: I am grateful to (Mrs.) Ann-Marie Eberson for recovering this manuscript from among the late Professor Eberson's papers under sad and strenuous circumstances and to Professor Torvard Laurent for his manifold assistance in bringing out this paper.

References

1. Gomberg, M. *J. Am. Chem. Soc.* **1900**, *22,* 757; also published in German: Gomberg, M. *Ber. Dtsch. Chem. Ges.* **1900**, *33,* 3150.
2. Tidwell, T. T. *Adv. Phys. Org. Chem.,* in press; *The Chemical Intelligencer,* **2000**, *6(3),* 33–38.
3. McBride, J. M. *Tetrahedron* **1974**, *30,* 2009.
4. Schlenk, W.; Weickel, T.; Herzenstein, A. *Justus Liebigs Ann. Chem.* **1910**, *372,* 1.
5. For reviews, see Gomberg, M. *J. Am. Chem. Soc.* **1914**, *36,* 1144; Walden, P. *Chemie der freien Radikale;* S. Hirzel: Leipzig, 1924.
6. Crawford. E. *The Beginnings of the Nobel Institution. The Science Prizes, 1901–1915;* Cambridge University Press: Cambridge 1984.
7. Widman, O. Special Report on W. Schlenk, 1918.
8. Gomberg, M.; Cone, L. H. *Ber. Dtsch. Chem. Ges.* **1906**, *39,* 3274.
9. Gomberg, M. *Ber. Dtsch. Chem. Ges.* **1907**, *40,* 1847.
10. Eberson, L. *Adv. Phys. Org. Chem.,* in press.
11. Schmidlin, J. *Ber. Dtsch. Chem. Ges.* **1908**, *41,* 423.
12. Widman, O. Special Report on M. Gomberg, 1921.
13. Lankamp, H.; Nauta, W. Th.; MacLean, C. *Tetrahedron Lett.* **1968**, 249.
14. Gomberg, M.; Cone, L H. *Ber. Dtsch. Chem. Ges.* **1904**, *37,* 2033.
15. Walling, C. *Free Radicals in Solution;* Wiley: New York, 1957; p 3.
16. Walling, C. *Fifty Years of Free Radicals;* American Chemical Society: Washington. D.C., 1995; p 11.
17. Schoepfle, C. S.; Bachman, W. E. *J. Am. Chem. Soc.* **1948**, *69,* 2921; *Great Chemists;* E. Farber, Ed.; Interscience: New York, 1961; Chapter 85.

Molecular Biology and Peterhouse[a]

John Meurig Thomas[b,c]

Depending upon one's perspective, definition, and preferences, one may locate the origins of molecular biology in several distinct places or periods. The late P. B. Medawar used to argue that it was W. H. Bragg and W. T. Astbury at the Davy Faraday Laboratory in London who began it all in the early 1920s when they investigated, by X-ray crystallography, the structure of materials such as silk, wool, and hair. Others say that J. D. Bernal is the progenitor of the subject. The term molecular biology was coined by Warren Weaver, a mathematician who headed the Natural Sciences Section of the Rockefeller Foundation, in his report to the Foundation's President in 1938. One of the hotbeds of the subject was Peterhouse, a constituent College of the University of Cambridge.

About the time that the Mongol warrior Kublai Khan overthrew the Sung Dynasty and conquered China—in 1284 to be precise—the Bishop of Ely, the small town in East Anglia with a magnificent cathedral, decided to found Peterhouse, the first of the colleges of the University of Cambridge. Although the University itself dates from 1209, when migrants from the University of Oxford had begun to congregate in Cambridge, it was the establishment of Merton College in Oxford (in 1274) that provided a precedent for a group of scholars sharing premises under a set of governing statutes to allow them a more coordinated existence in the service of scholarship. Peterhouse was granted a royal charter in 1285, thereby establishing it with an independent identity, similar to that of Merton, and bestowing upon its small community of scholars the key concepts of lodging, library, chapel, and dining room. These concepts are still much in evidence more than seven centuries later.

By 1400, 7 of the 51 present-day colleges of the University of Cambridge existed in some form, including Trinity, which did not adopt its modern identity until 1546. The number had risen to 17 by 1800; and in the nineteenth century the University underwent a rapid expansion, partly in response to the rise of science in a fast changing world. Nevertheless, 70 or so years ago, Peterhouse was still a small college with but 100 undergraduates, 3 graduate students, 9 Fellows (and 6 Honorary Fellows), and, as Master, the formidable Sir

Adolphus Ward (founder of the Cambridge Historical Series). Although it has grown significantly in the last 60 years—it now has 250 undergraduates, 90 graduate students, 32 Fellows on its Governing Body and a total of 37 Honorary, Emeritus, and Junior Research Fellows, 5 Bye Fellows, and a Master—it remains the smallest of the ancient colleges.

For anyone unfamiliar with the universities of Oxford or Cambridge, the mode of operation and system of governance of the individual colleges within the single-site ancient federal universities are rather confusing. Students from all the colleges attend lectures or go to laboratories, which are not, however, controlled by the colleges, but by the noncollegiate public university via subject-specific units of administration, the faculties or departments. Typically, a student applies to a college for a place and, after acceptance, will matriculate (enroll) in the University of Cambridge as an undergraduate to study a particular subject (such as mathematics, classics, history, engineering, or natural sciences) at the age of 18 or 19. That student's life in Cambridge will then be based in his or her college (e.g., King's, Christ's, Corpus Christi, or Churchill) but will involve regular attendance at noncollegiate sites determined by the subject he or she is reading (e.g., Department of Chemistry or Faculty of Engineering or Faculty of Law). Postgraduate students are also members of a college, but many of them spend less time there than the undergraduates. They may even be primarily based at other academic centers, such as the Laboratory of Molecular Biology (primarily funded by the Medical Research Council of the U.K.) or a Cancer Research Fund unit within university departments in the biological or medical sciences.

Students in the humanities, especially those reading history, English, or law, receive most of their core tuition at the individual colleges, where Fellows—who may or may not also hold a University appointment—give so-called supervision that entails detailed analysis and criticism of written work or exercises to very small groups (two or three students at a time, but sometimes just one) of undergraduates. These exercises are assessed by college Fellows or other senior members (supervisors) who may be from other Cambridge colleges. In each major subject area—medicine or modern and medieval languages, English, etc.—there is a Director of Studies, whose task is to ensure that each student under his or her aegis follows a carefully selected set of supervisions drawn both from within and from outside the college.

[a]*Chemical Intelligencer* 2000(4), 25–33.

[b]The Master's Lodge, Peterhouse, Cambridge CB2 1QY, UK

[c]Davy Faraday Research Laboratory, Royal Institution of Great Britain, 21 Albemarle Street, London W1X 4BS, UK

B. Hargittai and I. Hargittai (eds.), *Culture of Chemistry: The Best Articles on the Human Side of 20th-Century Chemistry from the Archives of the Chemical Intelligencer*, DOI 10.1007/978-1-4899-7565-2_54, © Springer Science+Business Media New York 2015

An undergraduate course of study in most cases lasts for three years—although, of late, the move to four-year courses in the sciences has grown inexorably—and involves an examination known as a Tripos [1]. Tripos examinations take place in about 30 subject areas; and by far the most populous Tripos course is natural sciences (accommodating some 2000 undergraduates in all, in such subjects as anatomy, biochemistry, physics, pathology, and zoology).

Some famous students to emerge from the Cambridge system include Dirac, Cockcroft (who was the first to split the atom), Mott and Abdus Salam (all of St. John's College); Isaac Newton, Lord Rayleigh, J. J. Thomson, Lord Rutherford, Bertrand Russell, and John Pople (Trinity); William Harvey (who discovered the circulation of the blood), and James Chadwick (discoverer of the neutron) (Gonville and Caius); the economist Maynard Keynes (King's); John Milton (author of *Paradise Lost*), Charles Darwin, and C. P. Snow (Christ's); and, from Peterhouse, Henry Cavendish (who was the first to establish that water was a compound and the first to determine the density of the Earth), William Thomson (Lord Kelvin, the founder of much of thermodynamics), James Clerk Maxwell (one of the greatest theoretical physicists of all time, who provided the mathematical interpretation of Faraday's electromagnetic field), Charles Babbage (father of the modern computer), James Dewar (inventor of the thermos flask), Frank Whittle (pioneer of the turbojet), and Christopher Cockerell (inventor of the hovercraft).

J.D. Bernal and His Influence

In 1936, there arrived in Cambridge from Vienna, his hometown, a 22-year-old graduate whose intention was "to seek the Great Sage" [2], who taught him that the riddle of life was hidden in the structure of proteins and that X-ray crystallography was the only method capable of solving it. The young man in question, Max Perutz, son of a prosperous, Anglophile textile manufacturer, was admitted to Peterhouse as a graduate student on the first of October that year, having started his research work for a Ph.D. in the Cavendish Laboratory some days earlier.

Max Perutz had entered Vienna University in 1932, and for five semesters he felt that he was wasting his time in an exacting course of inorganic chemistry. His curiosity was, however, aroused by organic chemistry and especially by a course of organic biochemistry in which Sir Gowland Hopkins's exciting work (on vitamins and enzymes) at Cambridge was outlined. With financial help from his father, he arrived in Cambridge with the intention of solving a great problem in biochemistry. Guided a little later by a cousin, who lived in Prague, he soon convinced himself that an appropriate target for his ambitions was to solve the structure of hemoglobin, the respiration protein of the red blood cells.

In 1936, the colorful, brilliant 35-year-old J. D. Bernal was at the height of his powers. In Max Perutz's words, Bernal had a wild inane of fair hair, sparkling eyes, and lively, expressive features, and "he was a bohemian, a flamboyant Don Juan, and a restless genius always searching for something more important to do than the work of the moment." Perutz to this day says that Bernal was the most brilliant conversationalist he has ever met.

Bernal had matriculated at Emmanuel College, Cambridge, in 1919 and left, after graduation in 1923, to work with Sir William (W. H.) Bragg at the Royal Institution of Great Britain (R. I.) in London, where, among other things, he solved the structure of the layered mineral graphite by X-ray crystallography. He returned to Cambridge in 1927 as a Demonstrator in crystal physics, later becoming Assistant Director of Research at the Cavendish Laboratory. In his group, prior to the arrival of Max Perutz, he had working alongside him Dorothy Crowfoot (Hodgkin), who had joined him from Oxford in 1932, and Isidore Fankuchen (from Brooklyn Polytechnic).

In the context of molecular biology and the modern study of enzymes and other biological catalysts, perhaps the most crucial of all the breakthroughs that Bernal and his associates achieved was their realization of the significance of the sharp spots in the X-ray diffraction photographs of crystalline specimens of the enzyme pepsin. This was explicitly stated in the paper that Bernal and Dorothy Crowfoot published in *Nature* in 1934 [3]. Before this work, many scientists felt that biological macromolecules had no well-defined molecular structure, certainly not in solution. Many believed that large, biologically significant molecules had (on the microscopic scale) the appearance of spaghetti—intertwined strands of variable length, bent, folded, and so forth in a manner that was difficult physically to disentangle and structurally to describe. But Bernal and Crowfoot noted that "from the intensity of the more distant (X-ray diffraction) spots, it can be inferred that the arrangement of atoms inside the protein molecules is also of a perfectly definite kind."

Of that period in the early 1930s when Dorothy Crowfoot worked in Bernal's group, she later wrote [4]:

> Every day, one of the group would go and buy fresh bread from Fitzbillies [5], fruit and cheese from the market, while another made coffee on the gas ring in the corner of the bench. One day there was talk about anaerobic bacteria at the bottom of a lake in Russia and the origin of life, another, about Romanesque architecture in French villages or Leonardo da Vinci's engines of war, about poetry or painting. We never knew to what enchanted land we would next be taken. More serious scientific discussions took place in the "Space Groups," an informal series of colloquia in crystallography.

Bernal's effulgence, enthusiasms, generosity of spirit, and abilities had a life-enhancing influence upon those who came within his ambit. The Scandinavian biologist Professor Lindstrøm-Lang once said: "When Bernal comes to see me,

I feel that my research is worthwhile." And John Kendrew (1917–1997), who first came into contact with Bernal when they rubbed shoulders on World War II service duties in Ceylon, said of him [6]:

> He had an infectious delight in new ideas, whether his own or another's; the question of credit did not arise, for all that mattered was that the idea was exciting and that it had to be pursued. Other people's results gave him as much pleasure as his own. He had an immensely stimulating influence on scientists of his own and younger generations, which was far beyond, and possibly more important than, his own personal contributions.

Brilliant as Bernal undoubtedly was as a creative scientist, his organizational skills were sometimes inadequate. Max Perutz was disappointed that Bernal had no major problem in structural biochemistry to challenge him on his arrival from Vienna. Instead, he was sent to learn the principles of X-ray crystallography in the Cambridge Department of Mineralogy and Petrography, where, as he once put it, he was given a fragment of "a nasty crystalline flake which someone in Mineralogy had picked off a slag heap," known as rhodonite [7], to investigate. This training did, however, prove valuable, for Perutz became adept in X-ray crystallography.

Bernal was later to influence, as we note below, John Kendrew, Rosalind Franklin, Aaron Klug, John Finch, and Ken Holmes, all of whom were early pioneers in the burgeoning field of molecular biology. In 1938, he, Perutz, and Fankuchen reported [8] important results on their X-ray crystallographic work on chymotrypsin (a proteolytic enzyme) and hemoglobin, where they realized afresh that these profoundly important biological molecules had well-defined structures down to the atomic scale.

Bernal left Cambridge to take up the chair of Physics at Birkbeck College, University of London, in 1938. With the advent of World War II, he did work of supreme importance on the intelligence front, work that won him the admiration of Earl Mountbatten of Burma and others. Perutz also undertook work of considerable national importance, but only after suffering the trauma of being rounded up, in May 1940, along with hundreds of other German and Austrian refugee scholars, mostly Jewish and all anti-Nazi, and packed off to an internment camp in Quebec [9].

Max Perutz, John Kendrew, and the Founding of the MRC Unit for Molecular Biology

In 1938, Lawrence Bragg was appointed Rutherford's successor as Cavendish Professor of Experimental Physics in Cambridge. Soon he grew to appreciate the enormous task that Max Perutz had set himself: the determination of the structure of hemoglobin. After the cessation of hostilities of World War II, Max Perutz had resumed his investigations, and Bragg helped Perutz to obtain an Imperial Chemical Industries Fellowship, which was to run until 1947. On the Cambridge scene, two years earlier, there reappeared John Kendrew. After graduating in chemistry in 1939 (and doing some physical organic chemistry with E. A. Moelwyn Hughes), he had left for work first on radar and later in operational research and ended up with the honorary rank of Wing Commander, being demobbed while in the Far East.

Kendrew, having already been aroused by Bernal's zeal for the structural determination of biologically significant molecules, decided to return to England via California, where he took the opportunity of pursuing such thoughts with Linus Pauling at Caltech. Among the varied provinces of Pauling's protean genius, his penetrating insight into the structural elucidation of (small) biological molecules was particularly exciting. Pauling's stimulus greatly influenced Kendrew, so that, when he returned to Cambridge, he had already decided to commence work on the structure of proteins.

When John Kendrew joined Max Perutz in the autumn of 1945, most experts regarded their prospects of success as bleak. At the Memorial Meeting [10] (November 1997) for John Kendrew, Aaron Klug said: "John had joined Max Perutz on a voyage of discovery (building the ship as they went along) where the land sought was clear—the three-dimensional structures of proteins—but with no route through uncharted waters:. . . the conventional wisdom was that the goal was unreachable." But 12 years after he started work with Perutz, during which time he had become a teaching Fellow of Peterhouse (1947) and its Director of Studies in Natural Science, its Librarian, Keeper of Portraits, Steward, and Wine Steward, Kendrew saw something no one had seen before—a three-dimensional (3D) picture of a protein molecule, myoglobin. The picture was a crude one; but two years later, in 1959, using the linear diffractometer devised and built by Arndt and Phillips at the Davy Faraday Laboratory of the Royal Institution (where Perutz and Kendrew were Honorary Readers, 1954–68), a much sharper picture of myoglobin, in all its glorious complexity, was obtained, with the identities of the amino acid residues clearly discernible.

At the same time, Perutz, having earlier convinced himself and some of his skeptical contemporaries that the heavy-atom substitution method could work for proteins, solved the structure of hemoglobin, which contains four times as many atoms (10,000 in all) as myoglobin. Here were two quite independent structural determinations of related proteins done by pure physics without any assumptions about the chemical nature of myoglobin and hemoglobin or the relationship between them. This exhilarating information revealed that, fundamentally, the intrinsic structures of the two proteins, replete with heme groups, numerous folds, and (Pauling's) α-helices, were essentially similar. The inescapable conclusion was that each had to be right.

At the Kendrew Memorial Meeting, Max Perutz recalled how, in the autumn of 1959,

> [John Kendrew] secluded in the vast, bleak, windowless room of the Cavendish Laboratory, which had housed its first cyclotron, could be seen building up the first atomic model of a protein molecule. He erected a towering forest of 1/8 steel rods on a wide wooden platform and marked the co-ordinates of the atoms derived from his X-ray crystallographic analysis on the rods with coloured "meccano" clips. ...he clamped about 1300 brass "atoms" ... until his model was complete. It became the Eighth Wonder of the World, and John was immensely proud of it. It is now on permanent exhibition in the Science Museum, London.

As Keeper of Portraits, John Kendrew served his college with distinction. Not only did he catalog all, and trace the provenance of many, of the pictorial possessions of the college, he also undertook to have several of them X-rayed and cleaned, and in this his love of renaissance art helped him greatly. His exceptional knowledge of classical music, and of the best recordings, made him popular with students and Fellows alike: some of them (now themselves retired) recall, as does Max Perutz, that it was in John Kendrew's room on C staircase in Old Court, Peterhouse, that they first heard "Hi-fi."

After solving the structure of myoglobin, John Kendrew lost his interest in personal research but not in science. In 1959, he had founded and edited (from his rooms on C staircase in Peterhouse) the *Journal of Molecular Biology*, the first, and for many years, the leading journal in the subject. He continued editing the journal up to the mid-1980s, by which time he was Director (its first) of the European Molecular Biology Laboratory (EMBL) at Heidelberg.

The *Journal of Molecular Biology* (JMB), from its inception, has chronicled many of the crucial steps in the growth and development of the twin streams of molecular genetics and structural biology. Statistics from the Institute of Scientific Information place JMB among the top 10 scientific journals (in all fields) in terms of impact measured over a 15-year period. John Kendrew, who recruited Sydney Brenner, Sir Andrew Huxley, James D. Watson, Maurice A. F. Wilkins, Matthew Meselson, and Paul Doty as members of his Editorial Board, and Max Perutz, Sir Peter Medawar, Melvin Calvin, Seymour Benzer, Francis Crick, Francois Jacob, Arthur Kornberg, Salvador Luria, Leslie Orgel, Alexander Rich, R. C. Williams, and others as his Advisory Board, had remarkable perspicacity when he decided to mastermind it all from his rooms in Old Court, Peterhouse. In the words of *Life Magazine* (1998), "JMB announced (in 1959) that a new discipline of study had been firmly established."

Perutz and Kendrew, long before their scientific goals were reached, had been the founding members of the Medical Research Council's Research Unit for the Study of the Molecular Structure of Biological Systems at the Cavendish Laboratory. Out of this unit grew the present Laboratory of Molecular Biology (LMB), now housing about 400 scientists. The LMB is arguably one of the most successful research laboratories in the world. (To date, its scientists have been awarded 10 Nobel prizes [11].)

Shortly after Perutz and Kendrew shared the Nobel Prize in chemistry in December 1962, they were the prime movers in the founding of the European Organisation for Molecular Biology (EMBO), with Kendrew as Secretary of its first Council. They had been concerned that, whereas American universities had quickly grasped the promise of molecular biology, most European ones ignored it. American postdoctoral workers could readily obtain fellowships to take them to Europe, but European ones had no funds to gain experience abroad. America had summer schools to spread the gospel, but Europe had none. EMBO fellowships and seminar schools are still going strong and have had a decisive impact on molecular biology in Europe. However, as Max Perutz has often said, from the very start of EMBO, John Kendrew's aim was the creation of the EMBL. This great laboratory was opened in Heidelberg in 1974 with Kendrew as its first director. It could never have come into existence but for John Kendrew's determination and brilliant organizational and diplomatic skills. He used these attributes to the fullest in his days as a teaching Fellow at Peterhouse, Research Scientist at the MRC Unit, Trustee of the British Museum, and President, in turn, of the British Association for the Advancement of Science and of the International Council of Scientific Unions, with headquarters in Paris.

The Laboratory That Perutz Created

When Max Perutz and John Kendrew realized that they needed a special building to pursue molecular biology, they saw that they were very short of expert biochemists. The only member of Perutz's team in this area was Vernon Ingram, who worked at the MRC Unit from 1952 to 1958. The MRC Unit for the Study of the Molecular Structure of Biological Systems was recreated in 1961 as the Laboratory of Molecular Biology in a new building on the future site of the University's Medical School. Fred Sanger, who had already won his first Nobel Prize, was invited to join the new Laboratory, of which Perutz was Chairman. Sanger, who had already been supported by MRC in the Department of Biochemistry at Cambridge—but where (as did several others) he found it difficult to get along with its Head of Department—maintained a large degree of independence in the new Laboratory.

Fred Sanger brought with him two bright stars, Brian Hartley and Ieuan Harris. Harris, in turn, had two outstanding colleagues of his own: John Walker, later a Nobel laureate in chemistry (1997), and Richard Perham, later Head of Biochemistry, University of Cambridge. At the same time

(1961), Hugh Huxley, of University College, London, and Aaron Klug, of Birkbeck College, London, came to join the Structural Studies Unit, of which Kendrew was Head.

Francis Crick, who started research at University College, London with E. N. da C. Andrade on measurement of the viscosity of water, switched to biology in 1947, and on the advice of the eminent biochemist and physiologist A. V. Hill, went to the Strangeways Laboratory, Cambridge, where he was assigned an impossibly difficult project on tissue culture. Having lost interest in his work, he sent a mathematician friend of his on a reconnaissance mission to ascertain the prospects of joining Perutz and Kendrew, who quickly agreed to accept him. So began his Ph.D. work on protein structure by X ray diffraction analysis. (To this day, Perutz delights in telling the tale that one of Crick's first acts was to demolish his supervisor's (Perutz's) then model of hemoglobin [12].) At that time, Kendrew had recruited Hugh Huxley as his first research student: on completing his Ph.D. in 1952, Huxley moved to London. A later research student, based in Peterhouse and working for John Kendrew, was Peter Pauling. In 1951, another young American student, James Watson, who had already completed a Ph.D. at the University of Indiana, joined the Perutz-Kendrew unit. Two other key American visitors were Richard Dickerson and Howard Dintzis.

With J. D. Bernal's retirement looming in Birkbeck College, London, Perutz, in 1958, invited Aaron Klug and Rosalind Franklin to join the new Laboratory when it was being planned and their NIH grant was about to run out. Klug brought Kenneth Holmes (later of EMBL) and John Finch with him. Alas, Rosalind Franklin passed away in 1958.

The record of molecular-biological achievement registered by these individuals in the Perutz's Laboratory is dazzling.

In 1955, after some 10 years' work, Sanger determined, by chemical methods, the complete amino acid sequence of the protein bovine insulin, the first of any protein. It established that the amino acids in protein chains are arranged in a definite sequence, which was not known before, and that this sequence is genetically determined. Sanger received the Nobel Prize in chemistry in 1958 for this fundamental work, which opened a new chapter in biochemistry [13]. It is acknowledged that, but for the technique of paper chromatography (invented by a former student and Honorary Fellow of Peterhouse, A. J. P. Martin, who shared the Nobel Prize in chemistry with R. L. Synge in 1952), Sanger's work on insulin would have been impossible. Martin's paper chromatography, aided by electrophoresis, was also instrumental in Ingram's beautiful work on sickle-cell anemia. In 1949, Linus Pauling and three of his students discovered a chemical difference between normal hemoglobin and the hemoglobin of those suffering an inherited blood disease, sickle-cell

anemia. Perutz got some sickle-cell blood from the United States, and he passed it on to Ingram, who split the hemoglobin into fragments (using the enzyme trypsin, which breaks the peptide bonds). Chromatography and electrophoresis then enabled Ingraim to show that sickle-cell hemoglobin differs from the normal one at just one pair of sites, where the amino acid valine replaces glutamic acid of the normal form. This experiment showed that replacement of only one pair of amino acids, of the 287 pairs that make up hemoglobin, produces a catastrophic effect: the link between structural chemistry and molecular biology was incontrovertibly established. Even more important, Ingram's experiment demonstrated for the first time that genetic mutations lead to the replacement of single amino acids in a protein. That discovery posed the question of the genetic code that specifies the sequence of amino acids in proteins.

Shortly after his arrival in the Perutz-Kendrew unit, Watson convinced Crick that the structure of DNA might be even more fundamental than the structure of proteins. Using prior chemical knowledge and the X-ray diffraction data of Maurice Wilkins and Rosalind Franklin, Crick and Watson built their famous double-helical model of the structure of DNA, arguably one of the most important structures ever revealed by X-ray crystallography [14]. This was to earn Wilkins, Crick, and Watson the 1962 Nobel Prize for physiology or medicine.

Perutz has recalled [15] how the chemical and physical data available at the time were apparently insufficient to build an atomic model of the structure of DNA and how a lucky coincidence played a key role in the solution of the problem. John Kendrew happened to be friendly with Erwin Chargaff, the Austrian-born biochemist at Columbia University, who was working on the chemistry of DNA. One day Chargaff visited Cambridge, and Kendrew invited him to dinner at the high table in his College, Peterhouse, together with Watson and Crick. There, Chargaff drew their attention to a paper he had recently published in the somewhat obscure Swiss journal *Experientia*, showing that in the DNA from several different sources the ratios of adenine (A) to thymine (T) and guanine (G) to cytosine (C) were always near unity. It was a vital clue. If DNA was made of two helices, Chargaff's result suggested that A in one chain is linked to a T in its opposite chain, and G in one chain is always linked to C in its opposite chain. It was a vital part of the jigsaw which Watson and Crick used with great ingenuity.

In the late 1950s, thanks to the contributions of Dickerson, Dintzis, and other young collaborators who succeeded in obtaining heavy-metal-substituted crystals of myoglobin, Kendrew achieved his goal of solving the three-dimensional structure of myoglobin, including the positions of most of the 2500 atoms.

Hemoglobin turned out to be a moving molecular mechanism, and Perutz spent many years trying to unravel its works by determining the detailed atomic structure of its two forms, one without and the other with oxygen bound. When atomic models of the two forms were finally ready, they revealed the molecular movements involved within a few days. It was of a kind that no one could have guessed, subtle and simple at the same time, infinitely rewarding by its intrinsic beauty, as a first model of a protein mechanism, and because it led to an understanding of several inherited diseases due to malfunctions of hemoglobin. Perutz gained new insights into molecular evolution and into the delicate (sometimes quite major) differences exhibited by hemoglobin in a wide range of living species. For example, the frogs of Lake Titicaca high in the mountains of Bolivia, unlike the frogs of Lake Michigan, which are at much lower level, have evolved a type of hemoglobin that is better able to absorb oxygen. In recent years, Perutz has attacked some of the possible causes of neurodegenerative diseases such as Huntington's chorea and found evidence for so-called "polar zippers" in the macromolecules of patients suffering from such diseases.

In communicating knowledge to other scientists, few individuals can rival Perutz's ability to give seminars that extend the frontiers of knowledge in so many different disciplines. His two published collections, full of warmth, humanity, insight, and broad culture, entitled *Is Science Necessary?* (1989) and *I Wish I'd Made You Angry Earlier* (1998), are gems. At Peterhouse, he contributes richly to the intellectual and cultural life of the College through his regular participation in the Kelvin Club (the students' science society).

Orgel, Klug and Their Successors

Another Fellow of Peterhouse and its Director of Studies in Natural Sciences from 1957 to 1964, Leslie Orgel, is now at the Salk Institute in California. Trained as a graduate student in Oxford in theoretical chemistry, he was largely responsible for developing ligand-field theory and authored a book on transition-metal chemistry. In 1958, he provided the first convincing explanation for the red color of rubies [16]. But a year earlier he had already begun to contribute to molecular biology when he wrote an ingenious mathematical attempt (with Crick and J. S. Griffith) to guess the nature of the genetic code [17]. It tackled the coding problem, which is how, in protein synthesis, a sequence of four things (the nucleotides A, T, G, and C) determines a sequence of many more things (amino acids, of which there are 20 naturally occurring ones commonly found in proteins). Orgel says of this paper that it is often quoted as "an example of a pretty theory that was totally wrong" [18]. Orgel's move away from theoretical inorganic chemistry and into molecular biology, which he consciously underwent [19], prompted further collaboration with Crick while Orgel was still at Peterhouse [20].

As outlined elsewhere [21], Orgel has made major contributions to the question of the origin of life. It was he who suggested that RNA came first, not just before DNA, but before proteins, in the profoundly complicated story of the origin of the organized complexity that is such a taxonomic feature of the processes of life [22]. Biochemists were quick to appreciate the significance of Orgel's view. If RNA could somehow catalyze its own replication, then life may have begun with a soup of RNA molecules acting both as genetic storehouses and, when folded into suitable three-dimensional shapes, as catalysts. Crick and Orgel, doubtless playing the role of devil's advocate proposed the idea, in explaining biogenesis, of "directed panspermia," according to which Earth was deliberately seeded with life by intelligent aliens [23]. It is interesting to note that one of the nineteenth century's leading scientists, engineers, and businessmen, Lord Kelvin, a lifelong Fellow of Peterhouse and a former student (who matriculated in 1841), had views not too dissimilar. The following is an excerpt from his address to the British Association for the Advancement of Science in 1871:

> Because we all confidently believe that there are at present, and have been from time immemorial, many worlds of life besides our own, we must regard it as probable in the highest degree that there are countless seed-bearing meteoric stones moving about through space. If at the present instant no life existed upon this earth, one such stone falling upon it might ... lead to its becoming covered with vegetation.

Even before Aaron Klug moved from Birkbeck College, London (where he, Rosalind Franklin, and John Finch were doing pioneering work on viruses), he had first come across John Kendrew one night in Cambridge in 1953, when he was a member of the Department of Colloid Science. Kendrew and Klug were fellow users of EDSAC, the first electronic, digital computer in the university. (With access to EDSAC, which stands for Electronic Delay Storage Automatic Calculator, one of the world's first fully operational computers, Kendrew had set out to replace the tedious calculations necessary in analyzing X-ray data, traditionally done by hand from tables or with the aid of primitive analog machines.) In 1962, when Klug joined the MRC Laboratory, Kendrew was instrumental in getting him elected a teaching Fellow (and, soon after, Director of Studies) in Natural Sciences, duties which he performed at Peterhouse until his retirement in 1993. (He took up the Presidency of the Royal Society in 1995.) Like Kendrew before him, Klug won the Nobel Prize (outright) in chemistry (in 1982) while he was a teaching Fellow at Peterhouse. All the while, however, he was establishing himself as a world authority in some five distinct subdisciplines of molecular biology.

Over a 30-year period, Klug and his associates, notably John Finch, Tony Crowther (both of Peterhouse), Donald Caspar, and David DeRosier, deployed electron microscopy to determine the internal structures of a range of virus particles, of organelles such as chromatin, muscle filaments, and bacterial flagella, and of enzymes such as hemocyanin, catalase, and purple membrane (bacteriorhodopsin). In particular, Klug and his associates established the technique of 3D image reconstruction from a set of 2D projections. (This principle later laid the basis for the X-ray CAT scanner.) Klug's breadth and depth of coverage of the molecular-biological landscape is exceptional and includes determination of the structure of nucleosomes (with Jo Butler, for many years a supervisor in biochemistry at Peterhouse, and Tim Richmond) and of chromatin (with Roger Kornberg and Jean Thomas). His early (1971) picture of the range of pH and ionic strength at which cylindrical rods of tobacco mosaic virus nucleate and grow was amplified in his Nobel Lecture [24]. He and his colleagues found that a crucial intermediate (the two-layer protein disk) is used to initiate assembly of the virus by recognizing a specific region on the RNA, it dislocates, and then the rod grows by adding more disks which break down and add to the tip.

One of Klug's most important discoveries (with J. Miller and A. D. McLachlan), and the most recent one, is that of zinc-containing proteins (known as "zinc fingers") that recognize specific DNA sequences, combine with them, and initiate the transcription of the neighboring genes [25]. About 1–2 % of the human genome codes for zinc fingers. The key to the success is that the protein is built on modular principles, out of a combination of "fingers" repeating in tandem, each recognizing a short stretch (some three base pairs of DNA).

In the hands of other members of the LMB, notably Richard Henderson, Nigel Unwin, and Tony Crowther (whom Aaron Klug "recruited" as a Fellow at Peterhouse in 1981, to supervise mathematics for natural scientists), enormous progress has been achieved in elucidating (respectively) the nature of the light-driven proton pump bacteriorhodopsin, the neurotransmitter-neuroreceptor link (that is, the synapse which is the chemical junction between two nerve cells), and the hepatitis B virus.

Other Fellows of Peterhouse who do not pursue their experimental work at LMB, but who contribute to molecular-biological research, have either been stimulated directly by problems drawn to their attention by Klug or have been cooperating with former members of the LMB. Belonging to the first of these categories is Chris Calladine, who is Professor of Structural Mechanics in the University of Cambridge, and, to the second, Sophie Jackson, who works in the University Chemical Laboratories, and Andrew Lever, Reader in Infectious Diseases at the Department of Medicine and Director of Studies in Medicine at Peterhouse.

Calladine, an authority on engineering plasticity and the theory of shell structures, became involved in discussions with Klug at Peterhouse after the latter (in the late 1960s) had become intrigued by corkscrew-like flagellar filaments by which bacteria swim. The problem was to explain how a helical structure could be constructed from a single kind of building block. Calladine proposed, in 1976, a convincing mechanism of the mode of action of the filaments (made of the protein flagellin) in terms of building blocks with "mechanical" bistable features. The same kind of structural engineering approach was used by Calladine to elucidate the well-known switching of DNA between A and B right-handed helical forms. Calladine and a visiting scientist, Horace Drew (a former student of Dickerson in the United States who came to Cambridge to work at the LMB, just as his research supervisor had done 20 years earlier), have ingeniously tackled the problem posed by the nonplanarity of the base pairs for their stacking in DNA. This they have done by combining the principles of mechanical engineering and molecular structure. Indeed, Dickerson, in his analysis of the sequence-dependent structure of DNA oligomers, sought correlations between the base-stacking arrangements and the degrees of twist, slide, and roll of the dinucleotide steps and described his general scheme in terms of "Calladine's rules."

During the last 30 years, the 3D structures of more than 10,000 proteins have been determined experimentally. Now, for the first time, one may inspect the universe of protein folds and examine how the chemical sequence of the amino acids governs the unique 3D structure and how, in turn, this folded structure directs the biological function of the protein. This general problem lies at the heart of understanding the deeper causes of Alzheimer's and Parkinson's diseases as well as Creutzfeldt-Jakob disease. Central to the whole issue is the mechanism of folding: how does the highly flexible "unfolded" state change into a unique, compact and active "folded" state? And how does this transformation take place so rapidly? It was once thought that, in order to fold rapidly, proteins must change along defined pathways, characterized by discrete intermediates. This view held sway until Sophie Jackson, as a Ph.D. student in Cambridge, showed that a protein did not need to populate intermediate states to undergo fast folding. Her work earned her a Research Fellowship at Peterhouse, where she is now an official teaching Fellow (after having spent two years pursuing molecular-biological research at Harvard).

When Andrew Lever worked at Harvard, before coming to Peterhouse, he became the first to identify the RNA encapsidation signal of HIV-1. His group has recently identified an RNA secondary structure of this region, and, in association with members of the LMB, he has solved the 3D structure of this critical piece of RNA. Interestingly, the mechanism by which the viral protein attaches to this structure and unwinds it during recognition is akin to the process of recognition by tobacco mosaic virus, solved by Butler and Klug.

Molecular biology as a topic continues to excite other Fellows of Peterhouse. Malcolm Ferguson Smith, now Emeritus Professor of Pathology, introduced, in the mid-1990s, the powerful technique of cross-species chromosome painting in the study of chromosomal evolution. And Suzanne Dickson, a physiologist trained in Edinburgh, is deeply immersed in studying the biochemistry of leptin, a central topic in understanding obesity. Her principal objective is to understand how the central nervous system controls body composition by receiving and initiating endocrine signals.

At another level altogether, and at a location south of the city of Cambridge, Peterhouse has established a Technology Park, the intention being to attract the research and development laboratories of molecular-biological companies to the exciting and fructifying atmosphere of the Colleges and University of Cambridge. It is encouraging that the first major tenant at this site is Peptide Therapeutics, Plc.

References and Notes

1. This is the name (used in Cambridge since the sixteenth century) given to a Cambridge Examination leading to a first degree. The word derives from the Greco-Latin for a "three-legged stool" and refers to the object on which the examiner used to sit while listening to candidates offering oral defenses of their work, before the introduction of written examinations.

2. This is what Bernal's acolytes (like Dorothy Crowfoot, later Dorothy Hodgkin, Isidore Fankuchen, and Max Perutz) called him, because "he knew everything."

3. Bernal, J. D.; Crowfoot, D. *Nature* **1934**, *130*, 794.

4. Hodgkin, D. M. C. *Biog. Mem. Fellows R. Soc.* **1980**, *26*, 17.

5. Fitzbillies is a confectioners shop in the heart of Cambridge, not far from the site of the Old Cavendish Laboratory.

6. Kendrew, J. C. *Dictionary of National Biography*, 1971–80; Oxford University Press: Oxford, 1986; p 53.

7. A manganese-rich silicate mineral of the pyroxenoid family, $(Mn,Ca,Fe)SiO_3$.

8. Bernal, J. D.; Fankuchen, I.; Perutz, M. F. *Nature* **1938**, *134*, 523.

9. See Perutz, M. F. *The New Yorker,* August 1985.

10. On 5 November, 1997. From the time that he became a Fellow of Peterhouse, Kendrew undertook a number of the tasks which, if properly undertaken, make college life so agreeable and charming. Kendrew's skills as a Steward, which entailed liaising regularly with the Kitchen Manager, are still discussed at Peterhouse. One Emeritus Fellow of the College recalls that, every Monday morning, Kendrew spent part of his time discussing the menus for the week with the Kitchen Manager and his staff and also checking that the army of helpers in the Cavendish Laboratory, engaged in reading intensities of diffraction spots using microdensitometers, had correctly recalibrated their instruments. To all these tasks he brought the same military precision and flair that greatly facilitated his progress as a molecular biologist.

11. Names of all the Nobel prizewinners from the LMB; F. Sanger (twice), M. F. Perutz, J. C. Kendrew, A. Klug, C. Milstein, G. Köhler, F. H. C. Crick, J. D. Watson, and J. E. Walker.

12. Writing his review of Hans Krebs' book *Reminiscences and Reflections* [*Nature* **1982**, *296*, 512], Perutz said "I cannot imagine Francis Crick declaring that he demolished my cherished 1949 model of haemoglobin 'honestly, in good faith and in a spirit of helpfulness' (words used by Krebs) rather than admitting to a certain mischievous satisfaction at the dastardly deed."

13. Sanger was a co-recipient of another Nobel Prize for chemistry in 1980 for his work on determining the base sequence of nucleic acids. In 1977, Sanger's team at the LMB published the complete nucleotide sequence of the DNA of the virus ϕX174. (This entailed determining the order of some 5400 nucleotides along a single circular DNA strand.)

14. On completing, with Watson, his work on DNA, Crick returned to his efforts on protein structure and completed his Ph.D. in 1953 at the age of 37.

15. Perutz, M. F. In *Science and Society;* Moskovits, M., Ed.; Ananse Press: Concord, Ontario, 1995; p 45.

16. Rubies are alumina (Al_2O_3) containing chromium ion impurities, which should confer upon the (white) host a green hue. However, Orgel showed in a classic paper in *Nature* (1958) that because of structural mimicry, whereby Cr—O bond lengths in rubies are closer to the lengths of Al—O bonds in the host Al_2O_3 than they are to Cr—O bonds in Cr_2O_3, a red, rather than green, color results.

17. Crick, F. H. C.; Griffith, J. S.; Orgel, L. E. *Proc. Natl. Acad. Sci. U.S.A.* **1957**, *43*, 416.

18. L. E. Orgel to J. M. Thomas, November 4, 1998.

19. Hargittai, I. *The Chemical Intelligencer* **1999**, *5(1)*, 56.

20. This led to their theory of interallelic complementation (see *J. Mol Biol.* **1964**, *8*, 161), a topic close to the heart of a former student at Peterhouse, J. R. S. Fincham [21].

21. Fincham, J. R. S. *Adv. Enzymol.* **1960**, *22*, 1.

22. Davies, P. *The Filth Miracle: The Search for the Origins of Life;* Allen Lane: London. 1998.

23. Crick, F. *Lile Itself: Its Nature and Origin;* Simon & Schuster: New York, 1981; p 88.

24. Klug, A. In *Les Prix Nobel en 1982;* The Nobel Foundation, 1983; p 93.

25. Rhodes, D.; Klug. A. *Sci. Am.* **1993**, *268*, 56.

John Meurig Thomas taught at the University of Wales (Bangor and Aberystwyth) for 20 years before taking up the headship of Physical Chemistry (in succession to J. W. Linnett) at Cambridge in 1978. In 1986, he succeeded Sir George (now Lord) Porter as Director of the Royal Institution of Great Britain, London, where, as part-time Professor of Chemistry, he does most of his experimental work (on the design and synthesis of solid catalysts). Since 1993, he has been Master of Peterhouse. He was knighted in 1991 for his services to chemistry and the popularization of science.

Van 't Hoff and the Scientific Imagination[a]

Keith J. Laidler[b]

Nobel prizes were first awarded in 1901 and, 100 years later, we appropriately pay tribute to van 't Hoff, the first recipient of the prize for chemistry. The choice was a happy one since he has a good claim to be regarded as, in Sir James Walker's words, "the greatest chemical thinker of his generation" [1]. Well before he reached the age of 40, he had completely transformed the fields of chemical thermodynamics and chemical kinetics. Moreover, a few days after his 22nd birthday, van 't Hoff published his famous paper on the "tetrahedral carbon atom," which had a profound effect on the way chemists think about organic molecules and their reactions.

What was the secret of his extraordinary success? To quote James Walker again, he had "no great mathematical or experimental attainment…no striking gift as a teacher." His strength lay in his "native inspiration and unflagging ardour…". From his early years, he was a strong believer in the importance of imagination in science, and this was the title of his inaugural address in 1878 at the University of Amsterdam.

People who do not know much about science, and even a few who do, often think that scientists scorn the use of imagination and intuition in their work and instead stick rigidly to hard facts. Those of us who have done research in science, however, know that this is far from the truth. In the long run, we must always check our ideas carefully against the observational and experimental evidence. In approaching our conclusions, on the other hand, we find it best to let our minds roam over all the possibilities, however wild they may seem at first. The key to success lies in our ability to make a wise selection from all of the explanations that have run through our minds.

There are two essentially different ways of approaching the truth. The method used in the early years of civilization is the intuitive method—the truth we feel. This is the method that has always been predominant in philosophy, sociology, and religion. Scientists, and many others today, make primary use of the empirical method, which places the greatest emphasis on evidence and takes nothing for granted. The essential feature of the method is that it *works* and is the best method we can think of for our particular purposes. But it would be on oversimplification to say that in science empiri-

cal methods have entirely replaced intuitive ones. That may be true for the final conclusions, but van 't Hoff and many other highly creative scientists have been mainly intuitive in their approach to their research. Their success lay in their imaginative originality and in their skill in coming to well-considered conclusions.

Jacobus Henricus van 't Hoff was born in Rotterdam on August 30, 1852, the son of a practicing physician [2]. Since he had a brother also called Jacobus, he was called Henry (rather than by the Latinized form of his second name). His education was a liberal one, and he developed wide interests which embraced music and literature. He became especially interested in chemistry, which at the time was still in a rudimentary state; water, for example, was written as HO, although the students were told that the newfangled formula H_2O was coming into vogue. While at school, he became sufficiently interested in chemistry to make some experiments at home, and he enterprisingly charged spectators a modest fee, which he used to replenish his supply of chemicals. At the age of 17, he became a student at the Polytechnic School at Delft and, at the end of two years, in 1871, enrolled at the University of Leiden, where he studied mainly mathematics.

He later complained that his university studies were too matter-of-fact and uninspiring and that he would have become a dry and shriveled scientist had it not been for his reading of philosophy and literature. He was particularly influenced by the French philosopher and social theorist Auguste Comte (1798–1857) and by the English polymath William Whewell (1794–1866). On the literary side, he was inspired by the German poet and essayist Heinrich Heine (1797–1856) and the Scottish poet Robert Burns (1759–1796). His favorite author of all, however, was the English romantic poet Lord Byron (1788–1824). His letters abound in quotations from Byron and references to him, and he even composed many Byronic stanzas in English as well as verse in his native tongue. He always enjoyed singing and playing the piano and taking long solitary country walks.

On the continent of Europe at the time, it was common for students to attend several universities before finally registering at one of them for the doctoral degree. The University of Leiden offered no special facilities in chemistry, and after a year van 't Hoff decided to go to the University of Bonn,

[a] *Chemical Intelligencer* 2000(4), 37–41.

[b] Department of Chemistry, University of Ottawa, Ottawa, ON K1N 6 N5, Canada (deceased)

attracted by the fame of the chemist Friedrich August Kekulé (1829–1896) and also by the romantic surroundings of the town. He later wrote that "in Leiden all was prose—the town, the country, the people. In Bonn all is poetry." His letters home, however, show that he soon became disillusioned, melancholy, and even somewhat bitter. He found Kekulé cold and indifferent, obviously not recognizing the genius that lurked inside his reserved and dreamy student. Van 't Hoff accomplished little research in Bonn, but his time was not entirely wasted, since he derived from Kekulé some insights into chemical constitution that soon led to his recognition of the nature of asymmetry in organic compounds.

In 1873, van 't Hoff enrolled at the University of Utrecht and after three months passed the examinations required as preliminary to the doctoral degree. Early in the following year, he went to the "Ecole de Médicine" in Paris to work with Charles Adolph Wurtz (1817–1884), a pioneer of organic synthesis. Here he was much happier because Wurtz, in contrast to Kekulé, was a genial and exuberant man who attracted many able and enthusiastic students to his laboratories. Van 't Hoff, however, still did little research but quietly picked up and reflected on many new chemical ideas. In the summer of 1874, he returned to Utrecht, where for a few months he carried out some rather routine research on cyanoacetic and malonic acids. Later in the year, he presented an undistinguished thesis to the University, which awarded him the Ph.D. degree.

Van 't Hoff's formal academic career was thus less than outstanding, but during it he independently made a scientific contribution of great importance which he decided not to include in his thesis. In September 1874, three months before submitting his Ph.D. thesis, he published privately a pamphlet on the "tetrahedral carbon atom" and its relation to optical and geometrical isomerism [3]. Today his ideas are recognized to have been essentially correct, and it is hard to understand why at the time they were an object of ridicule in some quarters. The German organic chemist Hermann Kolbe (1818–1884) made a particularly intemperate attack on them in 1877, going so far as to say that the ideas are "not far removed from witchcraft and spirit rapping." It is thus easy to understand why van 't Hoff thought it advisable not to include this work in his Ph.D. thesis. It is rather a sad irony that he obtained his degree for work that is now completely forgotten and that work for which he is today famous might have denied him a degree. As is well known, Svante Arrhenius (1859–1927) was less fortunate; in 1884 his pioneering dissertation relating to electrolytic dissociation so little impressed his examiners that he only gained his doctorate by a narrow margin.

For a year and a half after obtaining his Ph.D. degree, van 't Hoff was unemployed, but in 1876 he was appointed to a rather unsatisfactory lectureship in physics at the State Veterinary School in Utrecht. His theory of the asymmetric carbon atom then began to make an impression on his fellow scientists. Much help was given to him by Christoph Hendrick Diederik Buys Ballot (1817–1890), professor at the University of Utrecht. Although primarily a mathematician and a meteorologist, he had rather broad interests and exerted a wide influence. In 1845, for example, he had made a delightful test of the Doppler effect in the countryside near Utrecht. An open railway carriage containing an orchestra of trumpeters was pulled past a group of musicians. The musicians confirmed that as the train approached them the pitch was higher and that it became lower after the carriage had passed them. He made another important intervention by lending support to van 't Hoff in a letter to a Dutch journal, and this perhaps helped to bring about the offer in 1877 of a lectureship in theoretical and physical chemistry at the University of Amsterdam. There, for the next 18 years, van 't Hoff carried out his pioneering work on chemical kinetics and thermodynamics. In 1878, he was promoted to be professor of chemistry, mineralogy, and geology. Later in that year, he married Johanna Francina Mees, the daughter of a Rotterdam merchant, whom he had known for many years. They had two daughters and two sons.

His inaugural address at the University of Amsterdam was memorable. It began with a restrained reply to Kolbe, who had so harshly criticized his theory of the asymmetric carbon atom. He then went on to say that he had made a study of the way in which the best scientific work was carried out and was convinced that it was important to use one's imagination to the full; real progress was not made through slavish examination and analysis of experimental data. His peroration was a quotation from the English historian Henry Thomas Buckle (1821–1862): "There is a spiritual, a poetic, and for aught I know a spontaneous and uncaused element in the human mind, which ever and anon, suddenly and without warning, gives us a glimpse and a forecast of the future and urges us to see truth, as it were by anticipation" [4].

The professorship at Amsterdam was far from being a sinecure. With only two junior colleagues, van 't Hoff was responsible for all the instruction in organic and inorganic chemistry, crystallography, mineralogy, geology, and paleontology and had to conduct practical classes for 100 medical and 20 science students. He nevertheless carried out experimental and theoretical research and directed the research of students. The first six years of his professorship, from 1878 to 1884, were astonishingly productive. He and his students carried out crucial, well-conceived, and well-executed experiments which gave him completely new insights into both chemical thermodynamics and chemical kinetics. During that period he published little, saving everything for his pioneering book *Études de dynamique chimique*, which appeared in 1884. The book is typical of van 't Hoff's style; the amount of experimental work that he had carried out in those six years was not large, but the work was so aptly executed that in the book he was able to clarify fundamental questions of great importance.

Van 't Hoff's expression "chemical dynamics" requires a little explanation, since today it usually refers to the important branch of chemical kinetics that is concerned with the details of the individual molecular events that occur during the course of a chemical reaction. Earlier, however, chemical dynamics had a broader significance. Wilhelm Ostwald, for example, in his 1904 Faraday Lecture, defined chemical dynamics as "the theory of the progress of chemical reactions and the theory of chemical equilibrium," and this is the sense in which van 't Hoff used the expression. His great success in thermodynamics and kinetics sprang from the fact that from the beginning he perceived the close relationship between the two fields. His fundamental ideas about the rates of reactions and about the behavior of chemical systems at equilibrium developed in parallel, his understanding of one strengthening his understanding of the other.

In proceeding in this way, he was unusual. Both fields were still in their infancy, and the relationship between the two was scarcely appreciated. When the *Études* appeared in 1884, only a dozen or so significant papers on chemical equilibrium or kinetics had appeared, and among chemists as a whole there was little interest. Van 't Hoff was a pioneer in writing the First book covering either field, and even modern textbooks in the fields follow his lead closely as far as fundamental principles are concerned.

Before about 1860, it was thought that when a chemical system is at equilibrium, all reaction has ceased, but it gradually became clear that this was not the case; instead, chemical reaction is occurring at equal rates in the forward and reverse directions, and this clearly links the two fields of thermodynamics and kinetics. The basic equation for a chemical equilibrium was arrived at in 1862 by the Norwegians Cato Maximillian Guldberg (1836–1902) and Peter Waage (1833–1900), but their reasoning was unsatisfactory. Beginning at about the same time, the English chemist Augustus George Vernon Harcourt (1834–1919) and his mathematician colleague Willian Esson (1838–1916) made some important studies of the kinetics of certain reactions. Before the *Études* appeared, that was about all that was known about chemical dynamics.

The *Études* is packed full with important and pioneering ideas, which can here be mentioned only briefly [5]. Van 't Hoff began his book with kinetics and gave a number of examples of reactions of different kinetic orders, introducing his differential method for analyzing experimental data. He distinguished clearly between the order of a reaction and the number of molecules entering into the reaction. One of his most important contributions to kinetics was to elucidate the relationship between the rate of a reaction and the temperature. The fact that this relationship is now universally called the Arrhenius equation is a tribute to van 't Hoff's generosity.

When Arrhenius discussed the equation in a paper published in 1889, he clearly acknowledged that it had been proposed in the *Études,* but in his later publications van 't Hoff modestly allowed it to be assumed that the credit should go to Arrhenius!

Van 't Hoff's derivation of what is now known as the Arrhenius equation is a good example of success arising from considering two fields—thermodynamics and kinetics—side by side. He first arrived at the equation for how an equilibrium constant varies with the temperature. I say "arrived at" rather than "derived," since he did not give a rigorous proof of the relationship but wrote it down intuitively on the basis of the way in which the dissociation pressure of a solid such as ammonium chloride varies with the temperature. He then noted that under certain conditions an equilibrium constant for a reaction is the ratio of the rate constants in forward and reverse directions. This being so, rate constants must vary with temperature in essentially the same way as equilibrium constants, since otherwise their ratio could not lead to the right relationship. This is a simple and obvious point, and it is hard to understand why it was not immediately grasped by others. But nearly 30 years after van 't Hoff's book appeared, the Englishmen Harcourt and Esson were still urging the adoption of a quite different type of temperature-dependence equation for the rate of a reaction, an equation that was inconsistent with the condition that van 't Hoff had emphasized. They made only a casual reference to van 't Hoff's work, which they obviously had failed to understand.

In this they were by no means alone. Although today the *Études* makes easy reading for any scientist, at the time of its publication it was too difficult for the majority of chemists, who for the most part lacked any facility even with the simplest mathematics. This was partly because there was still a prejudice in the chemical world against the introduction of theory into the subject. We have seen an example of this in Kolbe's reaction to van 't Hoff's structural ideas. When the distinguished German physical chemist Wilhelm Ostwald (1853–1952) suggested to Emil Fischer (1852–1919) that physical methods would be helpful in solving some of the problems of organic chemistry, he was abruptly told "I have no use for your methods." Fischer was the second winner, in 1902, of the Nobel Prize for chemistry.

Thus, in writing the *Études,* van 't Hoff overestimated the ability of his readers. His derivations, while clear and logical, left out some of the mathematical steps, which few chemists of the time would have been able to fill in for themselves. In his Memorial Lecture [1] published in 1913, James Walker commented that when he first read the *Études,* he had a "mingled feeling of revelation and bewilderment.... Here, I thought, was the real thing at last, hard to comprehend, certainly, but something definite. What I understood was excellent. What I did not succeed in understanding seemed, somehow, even better." Walker became a distinguished physical chemist himself and wrote a textbook on the subject [6]; if he found the book "hard to comprehend," most chemists of the day would have found it impossible. The book nevertheless gradually began to exert a strong influence. One who did understand and appreciate it was Svante Arrhenius, who on its publication had just—and only just—gained his Ph.D. degree. Arrhenius wrote a very favorable review of it in Swedish and sent a copy to van 't Hoff, whose wife was able to translate it with the aid of a dictionary. This initial exchange of ideas between the two men had important consequences for the understanding of fundamental principles of physical chemistry. Arrhenius later worked in van 't Hoff's laboratories, and their association was particularly important in the development of Arrhenius's theory of electrolytic dissociation and in the refinement of van 't Hoff's ideas about osmotic pressure. Arrhenius always referred to van 't Hoff in terms of the highest praise and hung his portrait prominently in his house in Stockholm.

Van 't Hoff's treatment of chemical thermodynamics appears in the latter part of the book. Several important relationships appear for the first time, including what came to be called the van 't Hoff isochore and a version of what is now known as the Gibbs-Helmholtz equation. There is also a clear statement of what is now called the Le Chatelier principle, and here we get another example of van 't Hoff's generosity. Henri Louis Le Chatelier (1850–1936) proposed his principle also in 1884, the year of publication of the *Études,* but in his later books van 't Hoff was always happy to give all the credit to Le Chatelier. The *Études* also contains the fundamental ideas about osmotic pressure, which van 't Hoff continued to develop in subsequent years.

Van 't Hoff was not, of course, entirely alone in developing the theory of thermodynamics at about that time. Important work was also being done by the German physicist and physiologist Hermann Ludwig Ferdinand von Helmholtz (1821–1894). After distinguished work in a variety of fields, Helmholtz returned to thermodynamics in about 1880 and covered similar ground to that covered by van 't Hoff. Both these men had a sound understanding of practical problems in chemistry, but the same was not true of the American physicist Josiah Willard Gibbs (1839–1903). It is now recognized that in the 1870s Gibbs had published important contributions to thermodynamics. They had practically no impact on the scientific world, however, for several reasons. Besides publishing in an obscure scientific journal, Gibbs wrote in an opaque and abstract style that made his papers largely incomprehensible. Worse still, Gibbs knew nothing of chemistry, and although he gave an abstract treatment of chemical equilibrium, he never came aware of the existence of an equilibrium constant, a function that both

Helmholtz and van 't Hoff found of great value in their treatments.

The *Études* was, in my opinion, van 't Hoff's most splendid achievement, the jewel in his crown. It is probably the most original book ever written on any aspect of chemistry; in one stroke, at the age of 32, he clearly revealed the fundamental principles of chemical reactions for the first time. Having made in the book so many innovations, van 't Hoff could do little more in the next few years than develop his ideas and obtain additional experimental support for them. In papers published in 1886, he gave a rigorous derivation of the expression for the free energy change (which he called the work of affinity) in a chemical reaction. In the following year, following his discussions with Arrhenius, he extended his osmotic pressure relationship to take account of the dissociation of an electrolyte, in so doing giving convincing support to Arrhenius's ideas, which were still the object of some opposition. He also considered other properties of solutions, the so-called colligative properties, that are related to osmotic pressure. In the next few years, he and his students made some significant experiments on explosions in gases and on reactions at surfaces.

In 1896, when he was still only 44, van 't Hoff accepted a professorship at the University of Berlin, attracted there by the fact that he was completely freed from administrative duties and from a heavy teaching load, which were beginning to overwhelm him in Amsterdam. In Berlin he was required to give only the occasional lecture and was able to devote himself entirely to research. He embarked on a completely new line of research, a study of the marine salt deposits at Stassfurt, Germany. His decision to enter this field was partly due to the influence of Wilhelm Meyerhoffer (1864–1906), who had been his student in Amsterdam. Together with a group of collaborators and students, he greatly clarified the fundamental principles relating to salt deposits and introduced the concept of solid solutions. As a result, petrologists today regard him as one of the founding fathers of their field.

Van 't Hoff was never a robust man, and late in 1906 his health began to fail as a result of the onset of pulmonary tuberculosis. To his disappointment, he found that he had to spare himself, bringing his research gradually to a close and devoting himself to revising some of the books he had written. As late as 1909, he was able to carry out some research on a reaction catalyzed by the enzyme emulsin. On the evening of March 1, 1911, he died peacefully, in his 59th year.

Aside from his scientific attainments, van 't Hoff was a man of unusually fine character and personality. One of his students, the British physical chemist Frederick George Donnan (1870–1956), said of him: "To those who had the privilege of working with him he was endeared by the unaffected friendliness of his nature." Shakespeare's words in *Julius Caesar* apply perfectly to him:

His life was gentle, and the elements
So mixed up in him that Nature might stand up
And say to all the world "This was a man!"

References and Notes

1. Walker, J. "Van 't Hoff Memorial Lecture," *J. Chem. Soc.* **1913**, *103*, 1127.

2. There are many biographical articles about van 't Hoff; the following contain references to most of what has been written about him: Snelders, H. A. M. In *Dictionary of Scientific Biography;* Gillispie, C. C., Ed.; Charles Scribner's Sons: New York, 1976; pp 578–581; Laidler, K. J. "Chemical Kinetics and the Origins of Physical Chemistry," *Archive for History of Exact Sciences* **1985**, *32*, 43; Fleck, G. In *Nobel Laureates in Chemistry;* James. L. K, Ed.; American Chemical Society: Washington, D. C., 1993.

3. The paper was in Dutch, but a French translation of it appeared later in 1874, at the suggestion of Buys Ballot. Kolbe's caustic criticism was published in 1877. For some reason, Kolbe was contemptuous of any structural ideas, including those of Kekulé on the structure of benzene. Kolbe's vitriolic attack made a sneering comment on the fact that van 't Hoff taught at a veterinary school, which hardly seems a valid objection to a scientific theory.

4. Buckle wrote a monumental, and in its time highly influential, *History of Civilization in England* (1857–1861) in which he practiced what he called a scientific method of writing history, taking into account a country's climate and other factors.

5. For details, see my article "Chemical Kinetics and the Origins of Physical Chemistry," *loc, cit.* or my book *The World of Physical Chemistry;* Oxford University Press: Oxford, 1993.

6. Walker, J. *Introduction to Physical Chemistry;* Macmillan: London, 1899.

Index

B. Hargittai and I. Hargittai (eds.), *Culture of Chemistry: The Best Articles on the Human Side of 20th-Century Chemistry from the Archives of the Chemical Intelligencer*, DOI 10.1007/978-1-4899-7565-2, © Springer Science+Business Media New York 2015